# The West Indies

*Before and Since Slave Emancipation*

JOHN DAVY

CAMBRIDGE
UNIVERSITY PRESS

CAMBRIDGE UNIVERSITY PRESS

Cambridge, New York, Melbourne, Madrid, Cape Town, Singapore,
São Paolo, Delhi, Dubai, Tokyo, Mexico City

Published in the United States of America by Cambridge University Press, New York

www.cambridge.org
Information on this title: www.cambridge.org/9781108020732

© in this compilation Cambridge University Press 2010

This edition first published 1854
This digitally printed version 2010

ISBN 978-1-108-02073-2 Paperback

# CAMBRIDGE LIBRARY COLLECTION

*Books of enduring scholarly value*

## History

The books reissued in this series include accounts of historical events and movements by eye-witnesses and contemporaries, as well as landmark studies that assembled significant source materials or developed new historiographical methods. The series includes work in social, political and military history on a wide range of periods and regions, giving modern scholars ready access to influential publications of the past.

## The West Indies

John Davy (1790–1868) was an English doctor and brother of the chemist Sir Humphrey Davy. After graduating from Edinburgh University, in 1814 Davy became Inspector General of Army Hospitals, and he was elected a Fellow of the Royal Society in 1834. In his capacity as Inspector General, he spent 1845–8 living in Barbados and visiting other Caribbean Islands. This volume, first published in 1854 describes the society and culture of Barbados and other islands, including Trinidad, Tobago and St Lucia. Based on Davy's notes and observations made while stationed on the island, the book describes in vivid detail the disparities in education, quality of life and behaviour between the freed slaves, indentured servants and plantation owners of Barbados and other islands. Davy's sympathetic account provides valuable first-hand descriptions of the social conditions and tensions which existed after the Emancipation Act of 1834.

Cambridge University Press has long been a pioneer in the reissuing of out-of-print titles from its own backlist, producing digital reprints of books that are still sought after by scholars and students but could not be reprinted economically using traditional technology. The Cambridge Library Collection extends this activity to a wider range of books which are still of importance to researchers and professionals, either for the source material they contain, or as landmarks in the history of their academic discipline.

Drawing from the world-renowned collections in the Cambridge University Library, and guided by the advice of experts in each subject area, Cambridge University Press is using state-of-the-art scanning machines in its own Printing House to capture the content of each book selected for inclusion. The files are processed to give a consistently clear, crisp image, and the books finished to the high quality standard for which the Press is recognised around the world. The latest print-on-demand technology ensures that the books will remain available indefinitely, and that orders for single or multiple copies can quickly be supplied.

The Cambridge Library Collection will bring back to life books of enduring scholarly value (including out-of-copyright works originally issued by other publishers) across a wide range of disciplines in the humanities and social sciences and in science and technology.

# THE WEST INDIES.

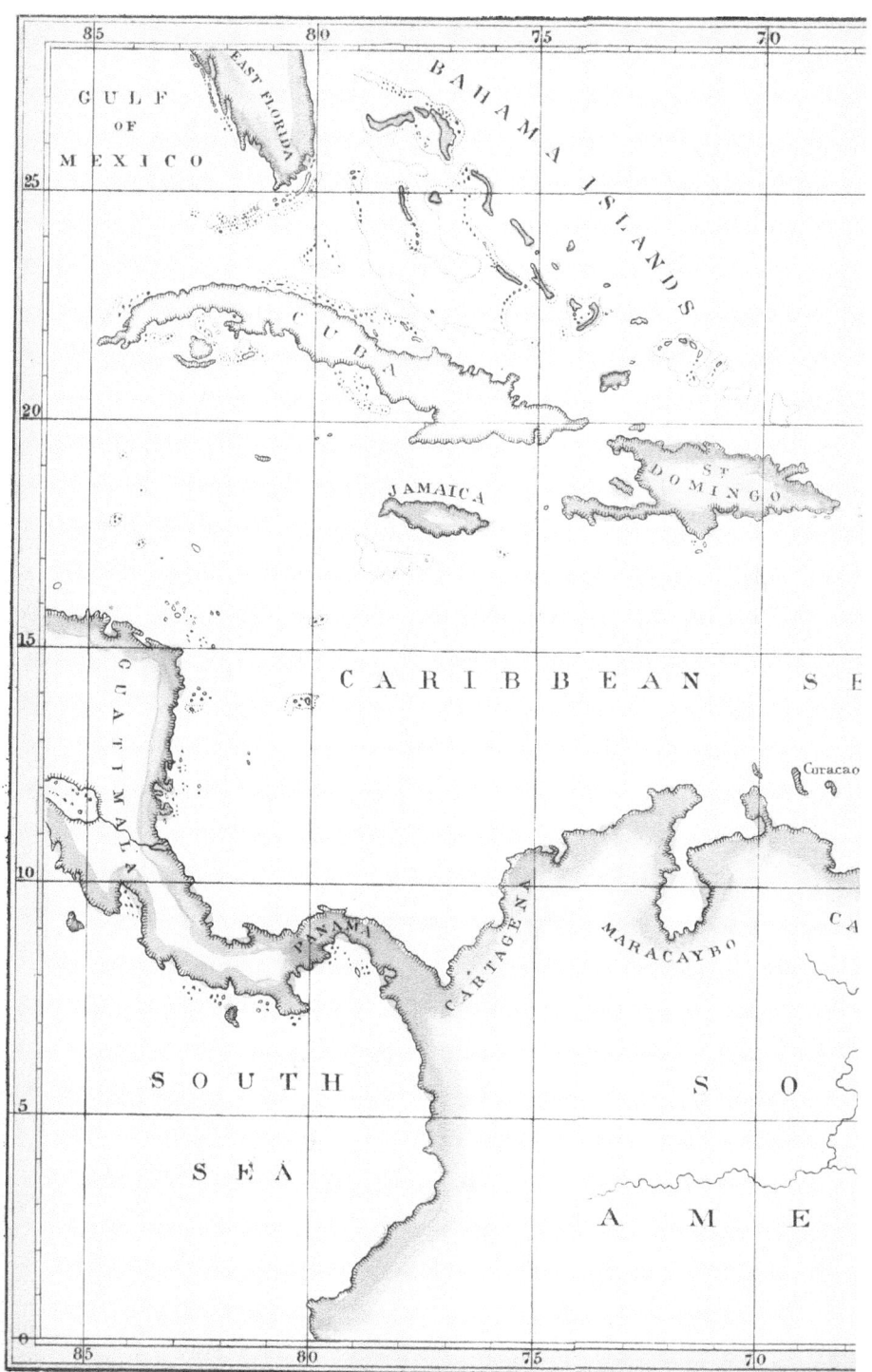

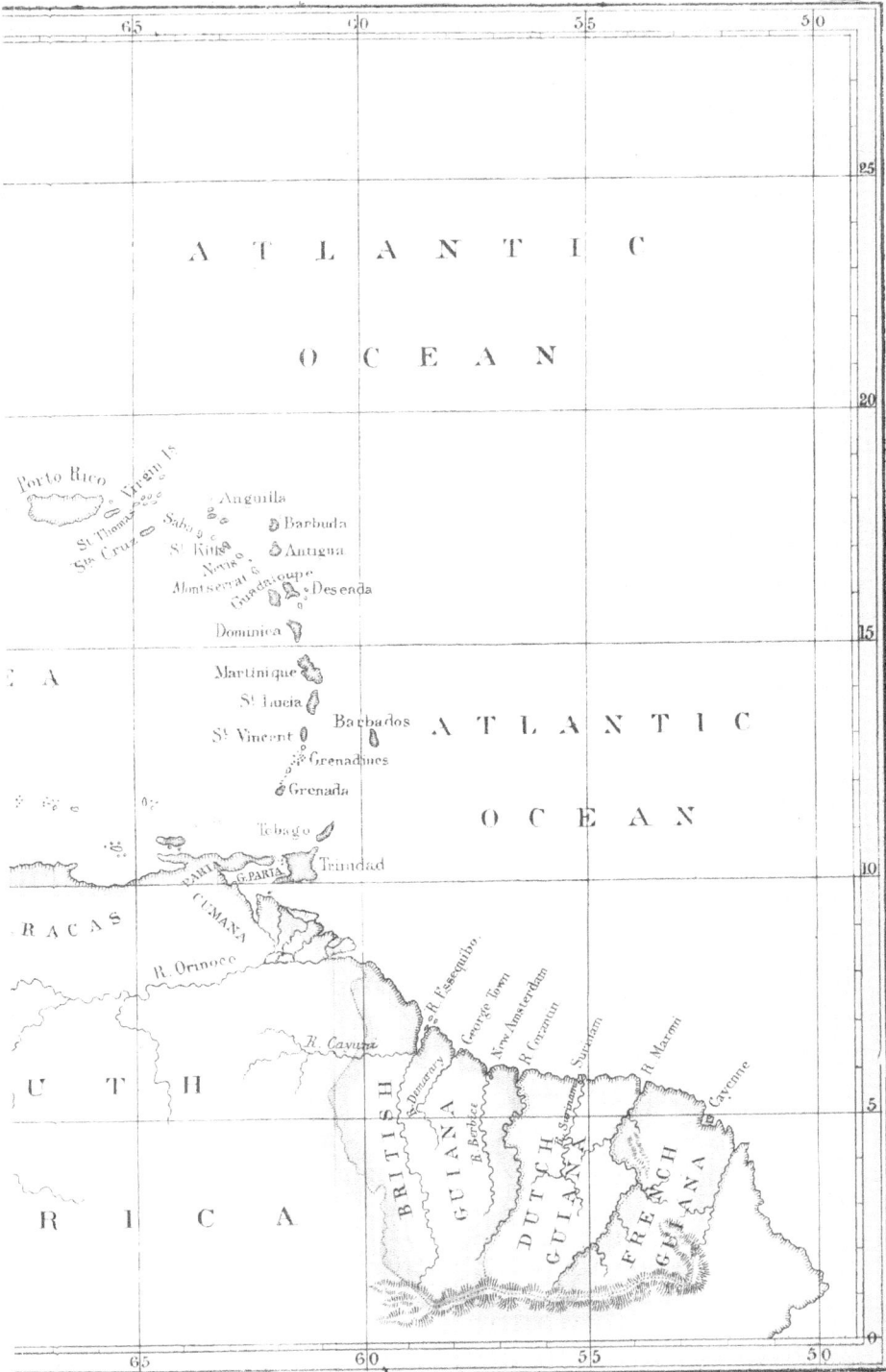

THE

# WEST INDIES,

BEFORE AND SINCE

# SLAVE EMANCIPATION,

COMPRISING

THE WINDWARD AND LEEWARD ISLANDS'
MILITARY COMMAND;

FOUNDED ON 'NOTES AND OBSERVATIONS COLLECTED DURING A THREE YEARS'
RESIDENCE.

---

BY

JOHN DAVY, M.D., F.R.S., &c.

INSPECTOR GENERAL OF ARMY HOSPITALS.

---

LONDON: W. & F. G. CASH, 5, BISHOPSGATE WITHOUT.
DUBLIN · J. M'GLASHAN AND J. B. GILPIN.
BARBADOS: J. BOWEN.

1854.

PRINTED BY T. KNAPP, FARINGDON.

# DEDICATION.

*To*

*The Members of the General Agricultural Society,
and, of the several District-Agricultural Societies
of Barbados.*

GENTLEMEN,

PERMIT me to inscribe this work to you. I
am induced to make the offering from a two-fold
motive;—one, arising out of a grateful recollection
of the many kindnesses I experienced, whilst residing
amongst you, and especially at your meetings, which
I had the privilege of attending as a honorary
member ;—the other, from a feeling of respect,
reflecting on the exertions you have made, the
example you have set, in the advancement of
tropical agriculture, and, especially of late years,
in the midst of difficulties of no common kind, and
which in some of the other colonies have been all
but overpowering.

Societies for the promotion of the sciences and arts,—institutions of modern times, probably, you will agree with me, in considering, as no bad criterion, if not of the condition, at least of the interest taken in the several sciences and arts to which they are devoted, and not least so, in the instance of agriculture. Compare England and Ireland, in this point of view,—not to take too wide a scope;—in the one, we see a most active agriculture, promoted by innumerable agricultural societies, self-supporting; whilst in the other, we witness altogether the reverse,—a backward, and hitherto languishing state of agriculture, and,—it is a remarkable fact, without, or at least as far as I can learn, almost without any agricultural societies, excepting those under government patronage and support.

Taking a view of the history of your island, I know no circumstance relating to it, of which its inhabitants may be more justly proud than its societies of this kind,—societies which you have yourselves instituted from a conviction of their usefulness; which have been under no control but your own, and have received no support, no aid either from the home or from the local government. It is remarkable too, how early you took

the lead in this matter, and were in advance, I
need not say, of the other colonies, but of most
countries, even European, England being the principal exception. The first formed of your societies,
—if I may be allowed to particularise for the information of others,—was that entitled, "The Society
for the improvement of Plantership," instituted nearly
half a century ago, viz., on the 8th of December,
1804. I know of no other that preceded it, similarly constituted, even at home, excepting one,
" The Bath and West of England Agricultural
Society," founded in the autumn of 1777. Existing
in activity for many years, and eminently useful,
this your earliest society was followed by another,
the " St. Phillip's District Agricultural Society,"
which was established in 1843; and, this, emula-
tion being excited, the advantages accruing becoming manifest, by four others in rapid succession,—
the " St. Thomas' District Agricultural Society," in
1845; the " General Agricultural Society," in 1846;
the " Leeward District Society," in the same year;
and the " Cliff District Society," in the present year.

What other colony has shewn the same intelligence
and enterprise, and what other has been so amply
rewarded? Since the first society was established,

the produce of your island, in the instance of its staple, sugar, has more than doubled. In 1804, it did not exceed 10,000 hogsheads.* This last year, with an unfavorable season, it reached 40,000 hogsheads, and the preceding with a more auspicious one, the extraordinary amount of 50,000.

Other circumstances, no doubt have favored you ;—but were it not for your intelligent exertions, of which, I hold your societies to be indices, they would have availed but little. What has been the circumstance in which you have differed most from the sister colonies? Has it not been chiefly in the large proportion of available labour? But, one of the first objects to which the society for the improvement of planter-ship directed attention was to ameliorate the condition of the labourers and to promote their increase; and, the St. Phillip's Society, when slavery was abolished, exerted itself with the higher aim, that of ameliorating the labourers' condition, as men, as thinking, responsible moral beings, forming a part of the community, in contradistinction to what they had previously been, in the degraded state of slavery. These were wise measures as well as humane ones, and you

* See *West India Common Place Book :* it is there stated, p. 18,—that to 1805, the yearly average crops had been no more than 9,554 hogsheads; but, then, a good deal of cotton was grown.

have been rewarded for them. Had the other colonies acted with the same views, how different might have been their present condition. Their emancipated slaves might have become efficient lalourers, sufficient and more than sufficient to have carried on successfully the cultivation of the estates, without recourse to the expensive and hazardous measure of a new traffic in labour,—that of immigration,—open to abuses and pregnant with evil, second only to the old one in slaves. And, in proof, it may be adduced that the one colony which differed least from you in the treatment of its labourers, viz: Antigua, has next to you, been most prosperous, or, has suffered least under its adverse circumstances.

Barbados has been called Little England; and not undeservedly, whether we consider its institutions, its independence and self-government, its agriculture, and may I not add, these your societies. How marked is the contrast between your condition and that of the West Indies generally! Bryan Edwards, dedicating his work on the West Indies, in 1793, to the then reigning monarch George III, designates them, as being "the principal source of the national opulence and maritime power." What a change since then! I will not allude to the causes; but conclude, with the hope, founded on what has been accomplished with you, and, I cannot but think, greatly by your means, that the West Indies may become in reality, what they were supposed to be,—and, more, an instance of a well organized society; so that their future, in the page of history, may afford compensation and some consolation for the past,—giving proof demonstrative,—that even regarding the lowest motives of human action, slavery may be abolished with advantage. In making this reflection and indulging this aspiration, let me remind you of the melancholy words of a man, who, as a poet in early life, and an historian in later, had given the subject his careful attention. "Take it all in all, (writes Mr. Southey,) it is perhaps as disgraceful a portion of history, (that of these colonies) as the whole course of time can afford; for I know not that there is anything generous, anything ennobling, anything honorable or consolatory to human nature to relieve it, except what may relate to the missionaries. Still, it is a useful task to shew, what these islands have been, and what they are."* This was written before emancipation

* From a letter to John May, Esq., relative to a proposed history of the West India Islands, by Captain Southey, the author of the *Chronological History of the West Indies.*

had taken place : may it be for you worthily to follow up that great act, and whilst taking the lead in improving your agriculture, forget not the improvement of the people. The last Sunday but one, that I attended Divine Service in your Cathedral Church, I heard a discourse,—and, it was from your respected Bishop,—such as would have touched the heart of Southey, had he been a listener, and might have mitigated his censure,—a discourse, in which the preacher, not only raised his voice against slavery, (slavery was its special subject, and this sermon the first, I believe, that ever denounced its evils from a West Indian pulpit,) but also in the true missionary spirit, exhorted to the practice of the christian virtues so essential to freedom.

*I have the honor to be,*

*Gentlemen,*

*Your obedient humble Servant,*

J. DAVY.

Lesketh How, Ambleside,
Dec. 22, 1853.

# PREFACE.

THIS work is offered as a contribution towards the History of the West Indies. The materials of which it is formed, have been obtained from several sources. Partly from my own observations, whilst on service in the West Indies, in charge of the medical department of the army; partly from Official Returns, especially Parliamentary Papers, commonly designated "Blue Books," for free access to which I have to express my thanks to Mr. Mayer, Librarian of the Colonial Office; and partly from friends in the Colonies, who, on many points of enquiry have kindly afforded me information, either orally, or by letter.

Residing mostly in Barbados, the head quarters of the military command, and, visiting the other stations of the command occasionally only in the performance of my duties as Inspector General of Hospitals, I had

it in my power to become more minutely acquainted
with it than the others ; and, in consequence I have
entered more into detail respecting it, than them ;—a
difference of treatment I believe that the circumstances of the island, even alone, would have justified.

The period of my personal experience was a critical
one: extending from July, 1845, when I landed in
Barbados, to November, 1848, when I left it to return
to England, it comprised events of no small importance
to the colonies, especially those distressing ones consequent on the alteration in the duties on their staple
produce.

Bryan Edwards, in his excellent History of the
West Indies, thought it right to vindicate the character of the aboriginal races of America from various
degrading aspersions : a similar duty has fallen to
me, and which I have conscientiously undertaken
in regard to the negroes of African descent and the
people of color of mixed races,—having the firm
conviction that the low and degraded state in which
they were sunk and from which they are but slowly
emerging, has been owing not to any inherent inferiority of nature or of mental capacity, but to the
dire circumstances of their former condition in the
state of slavery.

On the evils of slavery, now so well understood, I have not thought it advisable to enter in detail: were there any doubt remaining on the subject,— any question of slavery being justifiable, I could easily have adduced instances of its horrors, under cruel masters,—instances well authenticated, which I heard either from the individuals, the subject of brutal treatment, or from managers and others who had witnessed the acts, without being able to prevent them,—and this, at a time be it remembered, when in consequence of the abolition of the slave trade, the proprietors of estates had every worldly motive to take care of their labouring men and women, as much so as they have now to be careful of their labouring cattle.

Were I to advert specially to the objects I have had in view in composing this work, I might, place this—the vindication of the colored races of man, foremost ; and, next to it, a desire, in giving an account of the present condition of these colonies, to state fairly and free as much as possible from bias, the case of the planters ;—with the hope, even now, that the disadvantages under which they labour in comparison and in competition with the planters in the slave colonies, in consequence of the sliding scale of abolition of the differential duties, so near its completion, will have, before it is too late, re-consideration, and be followed by some regulation of the sugar duties, so amended as to be beneficial to the planters and to the consumers, serviceable to the great cause of slave-emancipation in the competing sugar-growing countries, recently lured by the graduated abolition of these duties to resist emancipation, and, in brief, be secure of the approval of all, but of the veriest advocates of slavery.

What the alteration in these duties might be, I have ventured to hint at in the concluding chapter; in which also, though not without hesitation, I have glanced at other defects relating to these colonies;—believing that inquiry temperately conducted, is always useful, and that suggestions arising out of inquiry and not from a mere love of change, may well be exempt from the charge of presumption.

# TABLE OF CONTENTS.

### CHAPTER I.—INTRODUCTORY.

### CHAPTER II.—BARBADOS.

### CHAPTER III.

### CHAPTER IV.—BARBADOS.

CHAPTER V.—ST. VINCENT.

CHAPTER VI.—THE GRENADINES.

CHAPTER VII.—GRENADA.

CHAPTER VIII.—TOBAGO.

CHAPTER IX.—ST. LUCIA.

CHAPTER X.—TRINIDAD.

# ERRATA.

Page i (Dedication), line 6, *for* a *read* an honorary.

,, v, do. ,, 4, *for* lalourers *read* labourers.

,, 3, line 14, *for* implement *read* implements.

,, 8, ,, 8, *for* westward population *read* population westward.

,, 27, ,, 4, *for* effects *read* affects.

,, 30, ,, 8, *for* agritulture *read* agriculture.

N.B.—The foot-notes above the Table should have been below it.

,, 68 (Note), line 2, *for* desribes *read* describes.

,, 231-2. N.B.—The quotation in these two pages should have been inserted as a foot-note, following the one given.

,, 265, line 22, *for* then *read* than.

,, 283 (Note), line 2, *for* Jereme *read* Jeremie.

,, 313 do. line 3, *for* increased *read* decreased.

,, 370, line 6, *for* to it by *read* by it to.

,, 385 (Note), *for* that island *read* Rat-island.

,, 402, line 21, *for* teaches *read* taiches.

,, 511, ,, 25, *for* formed *read* favored.

,, 512, ,, 19, *after privileges, add,* with the exception of Demerara, which has a Mayor and Town Council.

,, 517, ,, 27, *for* effects *read* effect.

,, 465, ,, 16, *for* . Accidents *read* , accidents.

# CHAPTER I.

THE Windward and Leeward Islands' military command, as the designation implies, is one of great extent; situated in the Caribbean Sea, it stretches through at least thirteen degrees of latitude, or about 780 miles measuring merely from shore to shore, exclusive of the widely-spread interior of British Guiana. Extending from about 5° to nearly 13° North, and from 59° to 63° West, it includes four civil governments; viz., that of the Windward Islands, formed of Barbados, St. Vincent, St. Lucia, Granada and the Granadines, and Tobago; that of the island of Trinidad adjoining the continent of America, separated from it only by the Gulf of Paria and within sight of its mountains; that of British Guiana (part of that continent) a country of vast extent, a considerable portion of it even now a *terra incognita*, and its boundaries hardly yet settled; and lastly that of the Leeward Islands, consisting of Dominica, Antigua, Montserrat, Nevis and St. Christopher's;—each government (Trinidad excepted) a Bishop's See.

B

Lying in the way of western navigation, the out-
skirts of the great continent, the islands enumerated
were well called The Antilles, (Anti-illas) and the
Lesser, to distinguish them from the Greater, Jamaica,
Cuba, Porto Rico, and Hispaniola or St. Domingo.

Whether viewed in relation to their civil or natural
history, their inhabitants or productions, their past or
present state, they abound in matter of interest, and
surely of instruction, and afford ample scope for further
and useful inquiry, especially political and economical,
keeping in mind the great changes that have taken
place in their condition since they first became known
to us, and that most memorable one pertaining to slave
emancipation—unexampled in the world's history—of
which they have been the scene.

A rapid sketch of the more important of these
changes may not be without its use as an introduction
to what is to follow; their past is essential to the com-
prehending their present, and to the forming an
opinion of their future.

When first discovered, little more than three centuries ago, they were found either desert, covered with
dense forests, or thinly inhabited by a rude and warlike race of Indians, truly savage, the Caribs; a people
little acquainted with the useful arts, little addicted to
agriculture, leading a nomadic, predatory life; by
means of their canoes, passing from shore to shore, not
unlike the northern sea kings; finding a precarious
subsistence by fishing and hunting, and in the wild
fruits and roots of the forest; seeking their prey every-

where more like beasts of prey than men, and occasionally under the impulse of hunger or revenge, or of
a taste acquired by the indulgence, feeding on their
fellow men.

Short as is the historical period that has elapsed, the
race of these aborigines has now almost died out; the
only remnant of them in our islands being a few families in the wilder parts of St. Vincent and Trinidad,
and these almost in their original savage state, but
without its former energy, and from passive indolence
threatened soon with utter extinction. In the other
islands they are almost forgotten, and are only occasionally brought to recollection by some primitive wedge-
shaped implement,—remains of them, now and then
turned up, made either of hard stone or of softer conch
shell.*

In even a briefer period—the colonization of these
parts not having been earnestly entered upon till the
commencement of the 17th century—a new population
has sprung up in them, and with a few exceptions has
rapidly increased; a population of a very mixed kind,
daily becoming more and more miscellaneous, and presenting strange contrasts with the habits of those they
have supplanted.

* Those of shell are often found in the lands now under cultivation in
Barbados, close to its shore. They were probably attached to a handle
and used as a hoe, in the cultivation of the little maize that was grown by
the Caribs in their occasional visits to the Island. Those of hard stone
were probably used as hatchets and axes, for felling trees and cutting wood,
and as weapons in war. They bear, both in their form and material, a
close resemblance to the implements of the Celts and of the old Scandina-
vian people.

The first settlers in Barbados were chiefly English, in the time of the great civil war and of the commonwealth,—exiles and banished men, free and bondsmen, roundheads and cavaliers,—the latter, distinctions, which with their party feuds, they found it there inconvenient to continue, and which they soon ceased to use by mutual assent.*

Nevis and St. Christopher's were settled about the same time under like circumstances; the latter in part by English and in part by French, remaining divided for a period of nearly a hundred years—a period of almost uninterrupted hostilities. About the same time also were settled Antigua and Montserrat; the former by English, the latter chiefly by Irish emigrants. Of the colonies which came under British dominion later, either conquered or ceded, the scanty white population of Tobago and Grenada is, or rather was, chiefly of Scottish origin, with an admixture of French in the latter; that of St. Vincent of mixed Scotch and English, as is also that of St. Lucia with an admixture of French; and likewise of British Guiana and Trinidad, in the former with a portion of Dutch, in the latter of Spanish and French.

To their first settlement by Europeans, various causes conduced,—a disturbed state at home, religious and political persecution, the love of adven-

---

* Ligon, who was in Barbados between 1647 and 1650, states in his account of the island, that, " among the better sort of planters, a law was made inflicting a fine on any one calling another Roundhead or Cavalier."

ture, the thirst of gain. These new regions were a land of promise, offering wealth, security and enjoyment, not only to the enterprising and imaginative, but even to the prudent and calculating citizen. In consequence, the most favored of them, as Barbados, St. Christopher's, Nevis, were rapidly peopled, and as rapidly brought into cultivation and made productive.

Comparing the numbers of their white inhabitants, the great majority of them free, within fifty years from the period of their being first taken possession of, with the numbers of the same now, the difference is strongly marked. Then they were three or four times as great as at present. We are surprised now, at the large white force that was then available, how Barbados could contribute above three-thousand volunteers; and Nevis, with St. Christopher's, one-thousand, to join the expedition in the time of Cromwell, which effected the conquest of Jamaica.*

Their early prosperity is not less worthy of notice, especially associated with its causes,—a virgin soil of great fertility, rich productions amply remunerating the cultivator, all of them luxuries fetching high prices in the market of Europe; first tobacco, cotton, and indigo,

---

* Long, in his History of Jamaica, written about 1774, says "that Barbados, in 1676, was reported to maintain 70,000 whites and 80,000 blacks, in all 150,000. The peopling of the island," he adds, " had been encouraged by grants of ten acres of land to poor settlers and white servants who had fulfilled the term of their indentures." From their land they procured the necessaries of life ; many of them, moreover, gained money by making cotton hammocks, which they sold in the island or exported to the French and English colonies. These small properties, he states, were gradually bought up, their owners seeking their fortunes in the adjoining countries.

with the minor articles of profit; afterwards sugar, with greater profit; free trade, free navigation, few or no fiscal duties; little competition; self government and self defence;* with mental qualities fitted to make the most of advantages—those qualities which belong to enterprising settlers, especially of the same race,—zeal, industry, perseverance, frugality, which we still witness, carrying the American westward population without stop or check.

Wealth and the enervating effect of a tropical climate ere long became antagonistic in these favored regions, tending to introduce causes of an opposite kind, the effects of which soon became apparent and are now too manifest.

The high atmospheric temperature rendered field labour irksome, and more than that—other circumstances disposing—destructive of health. The African was found fitted for this labour: slavery in consequence was permitted by the home government, after the example of the Spanish. Slaves in large numbers were imported, and soon the cultivation was almost entirely carried on by them, with little skill, no implement but the hoe, and with a great cost of life.

With the introduction of slaves a change took place, and that of an unfavorable kind in the description of agri-

---

* Till so late as 1780 no permanent military force was sent from home. In that year the 89th regiment was stationed in Barbados, the House of Assembly protesting, as contrary to usage. Barracks (colonial) were then constructed by a vote of the House, no doubt to escape the billetting system before in use when military aid, on occasions, was supplied by the mother country.

culture. The number of small proprietors diminished; the same variety of productions ceased to be cultivated; more capital was invested, plantations were enlarged; one or two articles, sugar or cotton, chiefly the former, engrossed almost entirely the attention. The larger proprietors now began to become absentees, taking up their residence in England, trusting the management of their estates to attornies, with increased expense and often diminishing profit. Coeval with which, were restrictions on trade, increased duties, augmented colonial expenditure from increase of salaried offices, many of them not needed but for ministerial patronage, and to meet it, an augmentation of colonial rates and taxes, and duties both on imports and exports, and last and not least an augmenting competition with the foreign growers, especially the French, of West India produce.

For many years the condition of the colonies was in a fluctuating state, oftener to the proprietors a matter of complaint than of gratulation, the latter especially towards the end of the reign of William and Mary. The cultivation of the sugar cane more and more engaged in, was almost as speculative an undertaking as mining. With favorable seasons and under favorable circumstances, whilst our islands chiefly supplied Europe with sugar, as was the case till near the peace of Ryswick in 1697, the crops were good and the profits large. Such in elated hope seem to have been calculated on for the future as regular and unexceptionable. The mode of living was in accordance, luxurious and profuse. With droughts and bad seasons, disasters from

floods, hurricanes and earthquakes, not of rare occurrence, with war and disturbed trade, and interrupted
supplies, reverses ensued, difficulties arose, debts became contracted, estates mortgaged, cultivation checked,
altogether productive of general distress.

All the details in the history of these colonies bear
out this statement. The early period of success, a period
of limited production and high prices, hardly reached
the end of the reign of Charles II.—certainly not of
his successor, when additional duties were imposed*—

* In 1625, on Muscovado sugars, 2s. 4d.; on refined, 7s. per cwt. The
imposition of these duties was attributed to the influence of the sugar
bakers, an interest that has always more or less interfered with the colonial.
Previously, the duty, the custom, had been five per cent.,—the ordinary duty
on all imports, an *ad valorem* duty, exclusive of the four and half duty in
the plantations. " In the book of rates made *anno* 1660, all commodities
were generally rated at five per cent., as near as could be to their value ;
accordingly white sugar from our plantations being then worth £5 per cwt.
paid a subsidy of 5s. per cwt., and brown muscovado sugar 1s. 6d."

" Some considerations about the intended subsidy, humbly offered with
respect to sugar."—No date, Lib. B. M.

In " The Groans of the Plantations," 4to. Lond. 1689, the principal
causes of the early prosperity, and its decline, are strongly described. They
are well deserving of attention even now.

" In former times we accounted ourselves a part of England, and the
trade and intercourse was open accordingly, so that commodities came
hither as freely from the sugar plantations as from the Isles of Wight
and Anglesey. But upon the king's restoration we were in effect made
foreigners and aliens ; a custom being laid upon our sugars amongst other
foreign commodities."—

" Heretofore we could ship off our goods at any port, or bay, or creek ;
and at any time, either by day or night. But now, since the king's restoration, we must do it at those times and places only at which collectors of the
customs please to attend."

" Heretofore we might send our commodities to any part of the world,
but now we must send them to England, and to no place else."

" Heretofore the things we wanted were brought to us from the places

The more or less disastrous and fluctuating period followed close after, and may be said with little interruption to have continued to the present time. A vastly increased production outstripping the demand so reduced the price of produce as to render it almost profitless. This appears to have been the main cause, others, —such as those above alluded to, conducing in con-

where they might best be had. But now we must have them from England, and from no other place."

" Heretofore we might send to Guinea for negroes, when we wanted them, and they stood us in about ten pound a head. But now we are shut out from this trade, and a company is put upon us, from whom we must have our negroes, and no other way ;" doubling (he states) their cost.—

The occasion of these " groans" was the imposing of additional duties on sugar, on the coming of King James to the throne; viz , seven groats (4s. 4d.) on muscovado per cwt., and 7s. on sugars fit for use ; 1s. 8d. and 5s. were the previous duties. " The act passed," (says the writer,) "and " the plantations are ruined."—" The duties fall so terribly upon our improved sugars, that it doth quite discourage and confound us ; our ingenuity is baffled, and our industry is cut up by the roots."

Bryan Edwards' graphic account of modern, not the earliest colonization, is the following; it should never be forgotten ;—we are too apt to forget when criticising the measures of foreign states, those of our own government ;—the parable of the mote and the beam. He thus writes. " The leading principle of colonization in all the maritime states of Europe (Great Britain among the rest) was commercial monopoly. The word *monopoly* in this case admitted a very extensive interpretation. It comprehended the monopoly of supply, the monopoly of colonial produce, and the monopoly of manufacture. By the first, the Colonists were prohibited from resorting to foreign markets for the supply of their wants ; by the second they were compelled to bring their chief staple commodities to the mother country, and by the third to bring them to her in a raw or unmanufactured state, that her own manufacturers might secure to themselves all the advantages arising from their further improvement.—This latter principle was carried so far in the colonial system of Great Britain, as to induce the late Lord Chatham to declare in parliament that the British Colonists in America had no right to manufacture even a nail or horse shoe."—*History of the West Indies*, vol. ii. p. 565.

C

junction with civil dissentions and bad management. It is lamentable to consider the condition of Barbados, taking it as an example, as recorded in 1732—36, when perhaps at its lowest ebb of fortune, when sales of sugar in that island were said to be made at eleven and twelve shillings the hundred.* This was after two destructive hurricanes, after droughts and short crops, and when the fiscal impediments and burdens were greater than they had ever been before. A desertion even of the island was threatened. We are assured that many insolvent planters to escape their creditors fled to the northern islands; most grievous are the stories of distress related. †

Though there was some little relaxation of restrictions in the way of help, trade with the north American colonies having been allowed, and France excluded from the Irish sugar trade; and from an increased demand from increasing consumption, some improvement

---

    * The low price seems to have been partly owing to the necessities of the seller, and partly to the increase of sugar having outstripped the demand. —In the time of Charles II., the quantity of sugar used in England according to Dr. Campbell was about 1000 hogsheads yearly, and about the same quantity exported.—*Long's History of Jamaica.* In 1734, it was asserted that "all the people upon the earth are not sufficient to consume the sugar that is already made."—*Carribeana.* We must keep in mind, that tea and coffee were at that time little used, the price of the former in London up to 1707, being about sixty shillings a pound, and coffee having been introduced not earlier than 1652.

    † The state of Barbados in 1732—6, as described, reminds one of what it was during the panic of 1846.—See "Carribeana," 2 vols. 4to. London, 1741; and "A Voyage to Guinea, Brazil and the West Indies, by John Atkins," 8vo. London, 1735. In the former miscellany a great variety of curious information is to be found by native writers, relative to the then state of Barbados and of the West Indies.

in the price of the staple products, yet no very favorable and well-marked change occurred till after the American war, and the French Revolution and the ruin of the sugar plantations in St. Domingo, the effect of the internecine struggles about that time commenced. Immediately before the Revolution this island was in its most prosperous state in regard to cane cultivation, the forcing of which by an immense outlay of capital, had in no small degree impoverished the mother country, and in that way in turn conduced to the Revolution that followed. It at that time exported a larger amount of sugar than all the British colonies combined, and had an amount of slaves equal to that in all of them.* After the exterminating war—exterminating to the whites, waged by France with the negroes—following the peace of Amiens, and the success and complete domination of the blacks, the sugar estates became neglected, and sugar finally ceased to be exported or even produced. This circumstance, by diminishing the supply of the article, whilst the demand for it was rapidly increasing, greatly inhanced its price in the general market, and promoted the extension of the culture of the cane in our possessions;—an extension that was effected in a very enterprising manner, and

* In 1790, St. Domingo exported to France 150,685,000 lbs. of sugar, 45,274,000 lbs. of coffee, 3,845,000 lbs. of cotton, 1,948,000 lbs. of indigo, 600,000 lbs. of cocoa; and its slave population was then estimated at about 500,000; in that year it is stated that 30,000 slaves were brought from the coast of Africa. From 1726 to 1742, the sugar produced there is estimated to have increased from 33,000 hhds. of 12 cwt. to 70,666 hhds. —*Mem. of the Sugar Trade of the Brit. Col., Lond.* 1793.

at an immense cost both in the purchase of slaves and
in the erection of machinery and of boiling houses for
the manufacture, to the exclusion of other kinds of
cultivation, especially that of cotton.* The reviving
prosperity that followed was very unequally sustained,
varying greatly in our different colonies during the
protracted war—the last war with France—checked as
it was by the foreign sugar trade carried on clandes-
tinely in neutral bottoms, and by the high price of
freight and insurance, and further by a rapid increase
of cane cultivation in the Spanish colonies of Cuba
and Porto Rico.

This more auspicious time, as it certainly was to our
planters on the whole, was not of long duration. And
it is worthy of note, that whilst it lasted, few improvements of any importance are recorded to have been
made or attempted, either in the cultivation of the cane
or in the manufacture of sugar, or in agriculture in
general. With high prices and a presumed exclusive
market, with favorable seasons, the crops were in many
instances largely remunerative, especially in the newly
conquered colonies. Labour—slave labour—tasking
neither skill nor science—*that* was in greatest demand,
procurable almost without limit, with unlimited ca-

---

* This again was not without its consequences; viz., First the rapid
extension of cotton cultivation in South Carolina, where it had not been
attempted till 1783, the year of independence of the United States; and
Secondly, as this cultivation was carried on entirely by slaves, adding a new
and powerful motive for the maintenance of slavery; a motive that was
even strengthened afterwards by—as a boon to our cotton manufacturers—
the free admission of cotton, without distinction as to the country from
whence exported, or the kind of labour by which it was raised.

pital and extended credit ;* the slave trade in all its inhumanities was extended, and the condition of the slave was unmitigated. We are assured that in St. Domingo, since the negroes, the offspring of slaves have become dominant, the black population has doubled in twenty-five years. In our colonies the same race in slavery was, till the slave trade was abolished, constantly diminishing in numbers, the deaths exceeding the births,—the strongest proof perhaps that can be given of their wretched lot.

When towards the close of the war the evils of slavery, owing to the long-continued exertions of a few distinguished philanthropists, attracted more and more attention, and increased efforts were made, first to prohibit the trade, next to effect slave emancipation,— then the minds of the planters began to be roused in anticipation of, and preparation for the great change that was about to ensue ;—then they began in earnest to consider the condition of the slave population and enter on the study of the causes affecting their health and duration of life, and to suggest means,—viewing them merely as animals,—for the strengthening of the one and the prolongation of the other, and for an increased fertility. Now, implemental husbandry had more attention paid to it—the plough was brought to the aid of the hoe, and improved methods both of culture and manufacture, but most of all the former,

* It is stated by Roscoe that the bills of exchange, with which the planters paid for the slaves, were drawn at a longer date than most others, sometimes payable at the end of three years.—*Life of Roscoe, North Worthies,* vol. iii. p. 48.

began to be introduced,—exertions that were not un-
rewarded. Barbados, which took the lead, affords a
happy example, displayed in a great increase of the
colored population, from which she is still feeling the
benefit, and it is believed in a more productive agriculture. In 1805, three years prior to the abolition of
the slave trade, the white population of the island was
estimated at 15,000, the free colored at 2,130, the negro slaves at 60,000; in 1829, four years before slave
emancipation, the numbers of these classes were the
following—the whites 14,959, shewing no increase;
the free colored 5,146, the negroes 82,902, the latter
denoting an increase of about 20 per cent.; and that
this increase was mainly owing to a favorable change
in the proportion of births to deaths,—the result of the
measures adopted,—seems to be demonstrated by the
fact, that before the abolition of the trade, even with
fresh importations, it was difficult to prevent a diminution, so far back as 1757 their number exceeding what
it was in 1805, in the former year having been returned as amounting to 63,645.*

When the great event of emancipation took place,
it was supposed by the home government, that the
sum voted in compensation coupled with the transition
measure of a four and six years apprenticeship (of four
for house servants and artificers, of six for field
labourers) might satisfy the planters and enable them

---

* In the United States the slave population, since great attention has
been paid to the breed, has doubled in the last twenty-five years; now it is
said to exceed three millions [3,200,000].

with advantage to enter on the system of free labour, protected as colonial produce then was by differential duties so heavy as to be prohibitory of foreign competition in the home market. What occurred was not in accordance with the expectations, except in the excluding effect of the differential duties. The apprenticeship plan was not found to answer, it was considered a sham by the negroes, with the name of freemen, they found they were without freedom; and now the want was the more irksomely felt, when they supposed it belonged to them as a right: irritation and general discontent were the consequence. In Antigua, a wise foresight and a well founded confidence induced the local government to dispense with the apprenticeship. In all the other colonies the term was abridged, viz., by two years in the instance of prædial labourers, bringing it to the same as that of domestic servants and artificers. Nor was there cause for regret; the emancipated, in their peaceable behaviour at least, shewed themselves worthy of the boon,—affording an example ever to be remembered, of the influence of justice and humanity in allaying the angry passions, and in promoting good will and order, those best bonds of society; not a single outrage was committed in the excitement of the moment; not a single act of revenge was perpetrated, then or after, that is recorded; there appeared to be a complete oblivion or forgiveness of all past wrongs and hard usage; all bad feelings seemed to be overpowered by one of gratitude for the benefit conferred.

Free labour, however, was not found to answer, at least in the degree expected. Various difficulties arose in connexion with it, one was the fixing of the rate of wages, the labourers being often high in their demands,—the planters low in their offers. Another, the paying money wages, and the substitution in part of an allowance of land and the occupancy of a dwelling. Another, the neglect of, or feeble legislation relative to vagrancy and the engagements of masters and servants. The consequences were in many instances, the desertion of the estates by the labourers, or a scarcity of labour, or insecurity of obtaining it when wanted. The evil, as might have been anticipated, was most felt where land was of least value and was most given in lieu of wages, and where the negro found it more for his advantage to labour on his own account than for the planter on his estate; and this was a growing evil, tending to render the labourer more and more independent, and the proprietor, as I have often heard it expressed, the slave, in his turn, dependent on the labourers. It was least felt in the best populated islands, such as Barbados and Antigua, where during slavery, the planters found their labourers in provisions, not allowing them provision ground, except on a very limited scale, and after emancipation paying money wages.

Other circumstances of a prejudicial kind, of a different order, co-operated. At the time of emancipation, few West Indian properties were unincumbered with debt; a large proportion of the compensation money was absorbed in liquidating these debts, leaving the planters without the capital necessary to secure labour and carry on the cultivation of their estates successfully,—unless, indeed, they raised money for the purpose from the merchant at a high rate of interest, and in a manner became his dependants. Further, the large sum of money that was paid in compensation, most of which was transferred to the merchant, tended to make money transactions easy, (if borrowing at six per cent., the lowest rate of interest, could be so called,) and to promote a spirit of speculation, especially in the purchase of land, not unlike that which prevailed at home towards the termination of the war, purchasers calculating on the war prices of produce during peace, and, not unlike that at home, often ruinous in its consequences; for instance, when the land obtained was bought with borrowed money at a high rate of interest, and moreover at a high price, (occasionally even as high in Barbados as before emancipation, the slaves on the property included) in sanguine expectation of high profits.

To supply the labour required, when failing, great exertions were made to obtain foreign labourers. A new species of traffic commenced, expressed by a new word, that of immigration, carried on mainly with the intent of affording aid to the planter, with little or no regard to the welfare of society and the forming of a well organized community. In British Guiana and Trinidad, where most needed, labourers were obtained from various quarters; from the western coast of Africa, chiefly captured liberated slaves; from Madeira; from India; and are about to be from China; hitherto at a great expense, and except in the instance of the liberated Africans, under terms onerous to the planters. In St. Christopher's, Antigua, St. Vincent, and Grenada, where the deficiency was less felt,—but in these in various degrees,—the smaller number of labourers introduced were chiefly from the mother country, from Malta, and Madeira. The results have been of a mixed kind, not so profitable as was expected, especially at first; and let us hope not so injurious as might have been feared. It is a subject to recur to hereafter.

Whilst in our colonies the quantity of sugar produced was diminishing, owing to land thrown out of cultivation, or indifferently cultivated from want of ample and well regulated labour, it was otherwise in most of the foreign colonies in which slavery was continued, and the slave trade though abolished by treaties, treaties "constantly, feloniously violated,"* was secretly tolerated or encouraged, especially in the Brazils and in Cuba. In Cuba, for example, the extension of cane cultivation was enormous, and hardly less so in Porto Rico. It began after the slave insurrection in St. Domingo, and has rapidly advanced—partly owing to the judicious measures of the Spanish government removing restrictions and encouraging colonization, commerce, and agriculture, as had been done at an early period in St. Domingo, and partly to the increasing

---

* The words of Lord Clarendon in the House of Lords.—Debate on the Importation of Slaves into Cuba, May 30, 1853.

demand in the markets of the world for colonial produce, and especially later, after the continental ports were opened and trade active in 1815, at the end of the war. Most of the sugar grown in Cuba is shipped at Havannah. In 1760 the amount was about 5,000,000 lbs.; in 1800 it had increased to above forty millions.; in 1820 to above one hundred millions; and now it is estimated at from 110 to 120 millions. The total now exported from Cuba, Porto Rico, and Brazil, is calculated to be no less than the enormous amount of 347,000 tons.

Situated, then, as our colonies were, already struggling with difficulties, they were no-wise prepared to encounter greater, and least of all to enter into competition with the Brazilians and Spanish, with ample labour at command, and slave labour. This competition they were forced into by the ministerial measure of 1846, followed by that of 1848—admitting slave grown sugars into our market with a declining differential duty to cease in 1854, when one rate of duty is to be levied, if the act be adhered to, without distinction as to the country whence the sugar is imported,* or the

---

* The war duty on sugar of £1 7s. the cwt. was not abolished till 1826. In March, 1845, the duty on colonial muscovado sugar was reduced from £1 4s. to 14s. the cwt.; and sugar from foreign countries, in which slavery did not exist, was admitted at a duty of £1 3s. 4d. instead of the former prohibitory duty of 63s. In August, 1846, the distinction between foreign sugar the produce of slave labour, and foreign sugar the produce of free labour, was done away with, and all foreign sugars were admitted at a duty for the ensuing ten months of £1 1s., which was gradually to decrease until 1851, when both foreign and colonial were to be admitted at the same rate of duty; muscovado sugar is implied. With this change of duties the use of sugar in British distilleries, previously excluded, was allowed, and

labour, whether free or slave by which it is raised, or the vegetable from which it is obtained.\*

Great was the boon to the highly taxed home population, diminishing as it did the price of sugar nearly one half; but as regards our colonies the measure unquestionably produced much immediate distress, and was in too many instances ruinous in its consequences.

the duties on rum were diminished to put the colonial distiller, it was said, on a par with the home distiller. The act fixing these duties was declared to be a final one. It was superseded, however, shortly, by a new one—that, of 1848, introduced by the same administration, extending the equalization of the duties to 1854, and otherwise modifying the differential duties, especially as regards the qualities of sugars, more favourable to the home and Dutch refiners, and the planters in the slave colonies than in our own colonies, checking skill there, and promoting waste and loss on the voyage.

\* Beet-root sugar, the manufacture of which was encouraged in the first instance by bounties, is now so skilfully and economically made, that it has become a problem whether it cannot compete without protection, even with the sugar of the cane, at least that obtained by free labour. This is certain, that the opening of the British market to slave grown sugar has been in favour of the manufacturers of the beet-root sugar; and its increase by supplying so considerably the markets of Europe, has become a new element in the difficulties of the colonial planters and must be an increasing one, if its production commenced in Ireland and England meet with encouragement and be extended under excise supervision. Those who may feel tempted by the flattering prospectuses of the one or two companies now forming, or attempting to be formed, would do well to refer to the views taken of the subject by disinterested persons and competent judges, such as Baron Liebig and Dr. Knappe. These are to be found in the work of the latter, "*Chemistry applied to the Arts*," vol. iii. He justly, I think, considers it "as by no means a natural growth," but one "resting entirely on an artificial basis," and purchased at a considerable sacrifice of the whole community;" if so in Germany, without sugar colonies, how much more so in this country with our ample means of obtaining cane sugar and in doing so, benefiting our fellow subjects and countrymen. During the French Revolution, nitre by means of chemical skill was made in France; it might be made in England also, but would it be wise to attempt the costly manufacture of it at home in preference to importing it from Bengal? The case of Beet-root sugar compared with that of the cane, is not, I apprehend, unanalogous.

As introduced, it was a hurried measure, and certainly not a prudent one, discouraging as it did production in our own colonies with free labour, and encouraging it in foreign colonies with slave labour; and so actually and effectually encouraging the maintenance in these countries of slavery itself.

How different might have been the result had the old differential duties been sustained, excluding the sugars subject to them from our market, and reducing greatly the duties and various burdens on our own colonial sugar, and all sugars grown by free labour! Cheap sugar would have been secured to the British people;—the cost of production it should be remembered is about a penny the pound, a fraction more or less; consumption would have increased to counterbalance the reduction of duty as affecting the revenue; colonial industry and enterprize would have been stimulated; and a strong motive—the strongest, that of self-interest—would have been given to the Brazilian and Spanish governments to abolish slavery with all its threatening risks. And that with the encouragement needed, an adequate supply of sugar might be obtained by free labour can hardly be questioned, since even now, of the vast quantity of sugar imported into this country and consumed, viz. 301,000 tons (this was the estimate in 1850 even, double that of 1823—then 150,000 tons, when the duty was £1 7s. the cwt.) 253,000 tons were the produce of free labour, 48,000 only of slave labour.

That some good, more or less, will be derived from

the difficulties that the planters in our colonies have
to struggle with, must, I think be admitted; indeed
some good amidst a vast deal of misery is already to
be witnessed resulting from them,—as will be more par-
ticulary pointed out hereafter,—in a stricter economy,
in improved methods of culture and manufacture, and
a more earnest and vigilant superintendence. Where
the difficulties are fewest and most easy to be dealt
with, as in Barbados, Antigua, St. Kitt's, and St. Vincent, there it may be anticipated the exertions making
will not be without some success; and it is believed by
the sanguine—and I cannot help believing—they will
afford proof that free labour can successfully compete
with slave labour; indeed, it is a question now, even
now, whether the condition of Barbados—the condition
of the others is more doubtful,—is not as prosperous as
that of Cuba or Porto Rico, merely materially considered, labour being cheap and abundant, provisions for
the supply of the inhabitants largely grown, an increased extent of land in cane cultivation, and the produce of sugar greater during the last two or three
years than ever before obtained.

The market price of sugar and the cost of production
must regulate the profits of the planter; the former
(duties apart) must depend on the supply and the demand. When the supply was extremely limited, when
the trade was entirely in the hands of the Portugese,
before the cane was introduced into the West Indies,
we are informed that sugar in England was sold at £8
the cwt. As the supply increased the price fell, and as

the one fluctuated so did the other vary. In 1689, it is stated that the price of Muscovado sugar in Barbados, was commonly twenty shillings per cwt., and of white sugar, fifty shillings.* From 1744 to 1778, the prices of sugars in the London market varied from thirty-three shillings to sixty-six shillings the cwt., in the last mentioned year the price was forty-seven shillings, and shortly after, following and owing to the disturbance of the trade and diminished supply by the insurrection in St. Domingo, it rose to eighty-one shillings, and this of raw sugar of the lowest quality from the West Indies.† Now, had the slave trade and slavery not been abolished in our colonies, who can say, reflecting on these fluctuations, that the condition of the planters generally would have been less disastrous than at present? This we are sure of, that the cultivation of the cane would have been vastly augmented in the virgin soils of British Guiana and Trinidad, and the slave population of these countries enormously increased, with proportional increase of danger of outbreak, insurrection and ruin. Further, overproduction,—a supply exceeding the demand,—would probably have resulted, with its constant attendant, distress, to the producers—a distress that would have been most felt in the smaller islands, the soils of which were called exhausted, and the culture of which requiring manure was expensive—these very islands,

* " The Groans of the Plantations, 4to, London, 1689.
† A Report from the Committee of Warehouses of the United East India Company relative to the culture of sugar.—London, 1792.

which are now struggling vigorously, one of them at least, with a good chance of success with free labour, with security from insurrection and the horrors of servile war,—a security, they all enjoy in common, with the conscious satisfaction of being with their families in a more healthy moral, and social condition. Every political mistake has its penalties; these are the inexorable fates; politics cannot be separated from morals with impunity. It was a mistake and a crime the substitution in the West Indies of slave labour for free labour, and Nemesis, the avenger, *filia justitiæ*, the never failing, however dilatory, has exacted the fine and inflicted and is still inflicting the punishment; and it is curious to see, witnessing the condition of the several colonies, how nicely that is graduated. Las Casas might well in his latter years, after his experience of the treatment of Africans, brought as slaves to the West Indies, with the humane intention in the first instance of relieving the native Indians, have had his fears that "he might after all fail to stand excused for it before the Divine Justice."*

Of the form of government of these colonies a very brief preliminary notice may suffice. In all of them excepting Trinidad, St. Lucia, and British Guiana, there is one identical form, following the analogy of the government of the mother country, being tripartite, composed of a Governor or Lieutenant Governor, the representative of the Sovereign, owing his appoint-

* Note to Mr. Ticknor's Hist of Sp. Lit. He quotes Quintana, who wrote from original authorities.—Vol. i. p. 519.

ment to the colonial minister, and commissioned by the Crown; of a deliberative and executive Council similarly appointed, bearing a faint analogy to the House of Lords; and of a House of Assembly, strictly resembling the House of Commons, both in the manner of the election of its members and in its functions; with the marked exception that the executive does not take a part and a lead in its proceedings, the Governor having only the power of exercising his veto.

In Trinidad the popular representative part is wanting, as it is also in St. Lucia; the government in each is conducted by a council presided over by the Governor. In British Guiana there is the Court of Policy with the Combined Court; the former, the executive, composed of an equal number of official and non-official members; the latter formed of the preceding with the addition of a certain number of financial representatives elected by the people; both presided over by the Governor. By this latter court supplies are voted.

These modes of government last noticed give little satisfaction; they may be considered as provisional; in time probably, when the people are better fitted for self government, they will be superseded by the first mentioned, which in theory is so excellent, and when well conducted excellent, but like all things of human institution, liable to failure, especially in the smaller islands, in which the elements of discord and petty party strife are commonly more active than is compatible with the common good; and the proceedings of those concerned in the administration of affairs not unfrequently are

such, as if designed to create ridicule, and show how
ill proportioned such an elaborate organization is to so
small a community.  The judicial affairs of these colonies are conducted much in the same manner as at
home.  Each has its own judicial establishment.  In
each is a Chief Justice, an Attorney and Solicitor General, and Stipendiary Magistrates, who in the older
colonies have relieved the Justices of the Peace in most
of their duties.  In the same colonies the laws enforced
are after the common law of England, with sta-
tutes of their own enacted to meet local emergencies.
In Trinidad and British Guiana, in Grenada, St. Lucia,
and Dominica, the laws partake more or less of the
institutions of the original colonists, Dutch, Spanish,
and French.  Without exception however, the inhabitants have the privilege of the Habeas Corpus act, of
trial by jury, and of appeal, and of petitioning, with all
the rights of British subjects, and most of the advantages, with the exception—a marked exception—of
want of federative union binding together and giving
strength to their disjuncta membra.  In each at least
of the larger islands there is an organized police; and
in all the latter a body of troops, under the command
of one General Officer, the only example of a union of
force; and in each government a Clergy presided over
by a Bishop.

The expenses of the establishments are defrayed
partly by the imperial treasury and in part from the
colonial; the governors, and troops, from the former;
the law officers, the police, and the clergy from the

latter, by funds raised by duties chiefly on imports, a
land and export duty in some instances, and a few assessed taxes,—altogether, especially when best selected,
no-wise oppressive, and light in comparison with home
taxation.*

Of the scenery, soil, and climate, a few slight preliminary notices may also suffice.

Of the first it may be truly said that in no part of our
globe of equal extent, is there greater variety, or more
varied beauty of landscape, which may easily be imagined when its parts, its elements are kept in mind ;—
a sea rivalling the Mediterranean in purity and colour,
like that sea almost tideless, and with like shores ;
islands most of them more picturesque, and bolder in
their forms than those of the Grecian Archipelago, not
barren like them and naked, but commonly covered
with evergreen foliage, either wild or cultivated, in
which, be it the one or the other, whether round the
planter's house as in India, or in the primeval forest,
differing from what is witnessed in India, palms of several kinds being conspicuous and marked ornaments—
indeed in many instances in which the charms of scenery are greatest, wild palms are in profusion ; Tri-

* The tendency of late years has been to diminish the assessed taxes
and export duties and to increase the import duties ; making thereby the
labouring class contribute to the revenue. The advocates of this system of
taxation maintain that the amount of duty effects but little the retail price
of the imported articles ; that depending more on the supply and demand
coupled, as is commonly the case, with little competition. Those opposed
to it are of opinion that the absentee proprietor is unduly relieved from contributing his share to the public revenue ; which would be a cogent argument, were the proprietors commonly prosperous.

nidad is an example; there these beautiful plants
having no natural enemies, such as they have in the
tropical forests of the old world in the elephant, spread,
self-sown, not needing the protection of man for their
growth and preservation.

As to climate; for a tropical one it is mild and equable; in its ordinary qualities partaking more of the
oceanic than of a continental character. A cloudless
sky is rare, as is also a sky completely overcast, and
fogs are almost unknown, excepting amongst the mountains, and the higher vallies. On the coasts where the
principal towns are situated, the thermometer in the
hottest season rarely exceeds 88° by day, and in the
coolest rarely falls below 70° by night; about 80° may
be considered as near the yearly mean. On the higher
grounds the temperature varies with the elevation, diminishing rapidly in ascending, so that even at a height
of from six to eight hundred feet above the sea level,
an agreeable coolness may be found without interruption, conjoined with great equality, the thermometer
at these elevations seldom exceeding 75° and 80°, and
at night never falling so low as to render fires necessary. Hot and parching winds such as are experienced
in India and the Mediterranean are entirely unknown;
calms beyond a few hours' duration are uncommon;
the atmosphere is almost constantly agitated by the
gentle trade winds. Droughts of long duration are
very rare, as is also protracted and continuous rain;
seldom a week passes without a shower; and it is uncommon for field labour to be interrupted by rain be-

yond a few hours. Such is the favorable aspect of the
climate most generally considered. Its unfavorable
traits are the destructive hurricanes, to which it is
more or less liable; the occasional though rare devas-
tating torrents of rain; the blights brought by currents of air, not unfrequently ruinous to vegetation;
and malaria, that mysterious something in the atmosphere, or exhalation from the earth,the cause of fevers,
which, of uncertain recurrence are so fatal to Europeans
and the unacclimatised. Even with these drawbacks
nature in point of climate has been too kind and indul-
gent; its influence here disposing to rest and the passive enjoyment of life rather than to exercise and a life
of vigorous exertion, especially in the instance of the
natives of cooler regions.

A like beneficence may be traced in the soils of these
countries. Almost without exception they are of a
good quality; not easily exhausted; commonly easily
worked; and well repaying all judicious outlay made
for their improvement. These advantages of soil have
not been, I may assert, justly appreciated. For more
than a century it has been the usage to speak of their
exhaustion in the islands that have been longest colonized,—a period little exceeding two hundred years;—
forgetful of the experience of the world, how good
soils with moderate care are inexhaustible even in
thousands of years, as we witness in Sicily and the
best parts of Italy; forgetful now how, that very land,
as in Barbados and St. Christopher's, supposed a century ago to have been worn out, is still productive;

and perhaps, most of all overlooking what experience has clearly proved, that agriculture, carried on in virgin soils, needing little care or skill for profitable production, is least likely to create or advance either, or to conduce to an elevating and enduring prosperity.

These colonies are well called sugar colonies, from the great staple, now perhaps too exclusively so, of their agritulture, the sugar cane, the culture of which has been so pregnant with good and evil, and has had so marked an effect on the destinies of a large portion of our fellow men, affording a striking example of great effects from an apparently small cause, and these produced in a very limited portion of time. Little was it imagined four centuries ago, when sugar was used as a medicine, and was to be purchased only in the shop of the apothecary, that it would become as it now is, almost a necessary; that the plant yielding it, indigenous in China, and probably in India, and then sparingly grown as an exotic in Egypt, Sicily, and the south of Spain, would spread round the globe and become the principal production of some of its finest regions, giving employment to millions of labourers, freight to thousands of ships, and in the discussions connected with the interests involved in the trade disturbing the councils and threatening the peace of nations.

*The following detail may convey some idea of what is generally
expressed in the text, and also of the probability of an augmented cane cultivation with an increasing demand for sugar
in the world at large.—*

## PRODUCTION OF SUGAR IN 1828.

### 1. British Possessions.

|  | Tons. | Tons. |
|---|---|---|
| West Indies | 210,500 | |
| Mauritius | 18,000 | |
| British India | 7,800 | |
| | | 236,300 |

### 2. Foreign Labour.

|  | | |
|---|---|---|
| Java | 5,000 | |
| Siam, &c. | 10,000 | |
| Beetroot, Europe | 7,000 | |
| | | 22,000 |

### 3. Foreign Slave Labour.

|  | | |
|---|---|---|
| Cuba and Porto Rico | 65,000 | |
| Brazil | 28,000 | |
| French Colonies | 50,000 | |
| Danish and Swedish | 10,000 | |
| Dutch Guiana | 10,000 | |
| Louisiana | 20,000 | |
| | | 183,000 |
| | | 441,300 |

## PRODUCTION OF SUGAR IN 1850.

### 1. Free Labour.

| | | |
|---|---|---|
| British Possessions | 260,000 | 260,000 |

## 2. Foreign Free Labour.

|  | Tons. | Tons. |
|---|---|---|
| Java | 90,000 | |
| Manilla, Siam and China | 30,000 | |
| United States—Maple Sugar | 70,000 | |
| French, West Indies, and Bourbon | 60,000 | |
| Europe, Beetroot | 190,000 | |
| | | 440,000 |
| Total Free Labour Sugar | | 700,000 |

### 3. Slave Labour.

|  | | |
|---|---|---|
| Cuba | 250,000 | |
| Porto Rico | 46,000 | |
| Brazil | 110,000 | |
| Dutch, West Indies | 13,000 | |
| Louisiana | 124,000 | |
| | | 543,000 |
| Grand Total | | 1,243.000 |

Knapp's Chemistry applied to the Arts, vol. iii. p. 420. from the *Economist*.

### SUGAR CONSUMED PER HEAD PER ANNUM.

| | |
|---|---|
| In the countries of the league (German) | 4.08 *lbs.* |
| France | 6.5 |
| Spain | 3.5 |
| Holland | 14.5 |
| Belgium | 7.0 |
| Russia | 0.5 |
| Ireland | 4.5 |
| England and Scotland | 21.0 |
| The United States | 14.5 |
| Cuba | 56.0 |
| Inhabitants of Venezuela (according to Codazzi) | 100.0 |

—*Op. Cit.*

# CHAPTER II.

BARBADOS.

Its relative importance.—Sketch of its geological structure.—Features of its scenery.—Notice of its soils, minerals and springs, its climate and seasons.—Meteorological observations and incidents.

THOUGH small in its dimensions, little exceeding the Isle of Wight,* Barbados has hitherto been considered, and justly, one of the most important of the West Indian Islands. Its position to the windward of all the others; its being the seat of government and head quarters of the troops; the wealth of its inhabitants and their numbers; its comparatively good state of cultivation and productiveness, are some of the principal circumstances which have conduced to its being so highly estimated. Other circumstances might be mentioned as aiding: such as its having been one of the earliest colonized; its never having been conquered by a foreign enemy; and not the least remarkable and worthy of note, its having struggled through all the difficulties it has had to encounter—many of them formidable—displaying a progressive improvement, and promising moreover to afford, if not already affording an example of the triumph of free labour over slave labour, and a vindication of the cause of humanity,

* Its form is an irregular triangle; the extent of its sea coast, inclusive of sinuosities, about 55 miles; its greatest length about 21 miles; its greatest width about 14, comprising in its area about 106,470 acres, equal to about 166 square miles.

F

even materially viewed. It is not vastness of space
be it remembered that gives importance to a country ;
witness those spots in which we have been, and are so
deeply interested, such as Palestine, Attica, and so
many others, narrow in their limits, wide in renown—
the people inhabiting them, and their deeds having
earned them their renown.

Barbados physically is happily constituted, and
hence a main cause of its prosperity. Its rocky shores,
guarded by coral reefs affording few landing places
have been its main protection from invasion in time of
war. Without lofty mountains, the highest of its hills
hardly exceeding 1100 feet, with little rocky surface,
and that little calcareous, it is fitted for cultivation
throughout; and further, being situated in the wide
ocean fully exposed to the trade winds, its surface
generally moderately elevated, with an almost total
absence of marshy ground, it is, for a tropical island,
nearly as favorable to animal as to vegetable life ;
and from the equability of its climate, peculiarly favorable to the health of the colored races, especially those
of African origin.

Viewed as a whole, its scenery is of a very mixed
kind, mainly depending on its geological structure.
Of this a brief sketch may suffice.

Geologically considered then, it may be described as
consisting of two well-marked regions very different
from each other in their features, in manner of formation as it must be inferred, and in the nature of the
materials of which they are constituted.

The larger portion rises from the sea by successive terraces, broken more or less by transverse rents or gullies, separated by parallel vallies and table lands, in which are innumerable cavities and basin-like hollows or depressions, the whole giving the idea of having been modelled under water by currents, and of having been raised from the bed of the sea, in which undoubtedly it was formed, and so raised by a force acting continuously and regularly. The smaller portion on the contrary descends to the sea from a central height divided by vallies and hills, a miniature alpine country, each valley with a continuous descent uninterrupted by gullies, and without depressions or basin-like hollows. This smaller region seems as if produced in a different manner from the preceding, as if raised by a force acting violently beneath and suddenly, by which it has been raised up in its present form and all its existing ruggedness.

Further, as to their component parts: whilst the larger region consists of beds nearly horizontal, of calcareous marl, and of calcareous sandstone and freestone, abounding in many places in sea shells, coral &c. resting on clay; the smaller region consists of strata, or beds of a chalk-like matter, of different kinds of clay and of different kinds of sandstone, mostly siliceous, with strata of volcanic ashes intermixed, and occasionally beds or masses of bitumenous matter and coal, and these more or less inclined, in some places indeed almost vertical, and with the exception of certain of them, (the beds of chalk-like matter, that

abound in the skeletons, mostly siliceous, of microscopic animalcules,*) containing very few organic
remains.

In regard to age, the proofs seem to be satisfactory
that the larger district is of recent formation in the
history of our globe, and that the smaller is more
ancient, though it may have been thrown up later.
Accordingly, the majority of the shells in the former are
of existing species, and probably all of them are.  It
is worthy of remark that where they occur in the
higher terraces, they are in the form of casts, where in
the lower they are little altered, retaining often a
portion of their animal matter, and even their original
colour but little changed ;—a difference favoring the
conclusion adverted to, of the gradual elevation of this
district.

* When I arrived at Barbados in 1846, I was assured that no organic
remains had been discovered in the " Scotland district" (the name given
to the smaller geological region, from its mountainous character).  In one
of the first excursions I made into it, I detected the cast of a shell in an
almost perpendicular stratum of conglomerate, forming a part of that bold
and picturesque hill, consisting of quartz, gravel and sand, of a chalky
whiteness, improperly called Chalky Mount.  Soon after I ascertained by
microscopical examination of a portion of chalk-like matter, the presence
in it of many forms of infusoria.  The specimen was from the Pike of
Teneriffe, a hill so called from its remarkable form composed of this chalk-
like matter, capped by a mass of coral rock.  Extending the examination
I found these infusorial remains more or less in the chalk-like matter
wherever it occurs.  Specimens were afterwards sent by Sir Robert Schomburgk to M. Ehrenberg.  This distinguished microscopist detected in them
three hundred and sixty one species of animalcules, so he considers them,
which he has divided into seven families, and forty-four genera.—(See Hist.
of Barbados, by Sir Robert Schomburgk, p. 558.)  Later I detected in the
calcareous freestone, forming the summit of Bissex hill (a hill almost
isolated), besides shells, the spines of echini, and the teeth of squali.

The organic remains in the latter, both the infusoria and shells* point to an earlier period of formation; whilst the manner in which some of its heights are capped with coral rock, and the sides of some of its hills have scattered over them detached masses of this rock, seem to denote a later projection, or up-heaving; and, that this was sudden and violent, the forms of the hills and vallies may be held to indicate, of which further proof is afforded in accumulations of matter little cohering, having the character of volcanic ashes† or tufa. But whether this part of the island was ever the site of an active volcano, it may be difficult to decide. There are basin-like hollows on its verge, in their form of very crater-like character, ‡ from whence it may be imagined that eruptions have proceeded, coeval with the up-lifting process.

Of the eleven parishes into which the island is divided, the smaller geological district comprises two: viz. St. Andrew's and St. Joseph's, and parts of five others, viz. St. Philip's, St. John's St. Thomas', St. James', and St. Lucy's. It occupies chiefly the windward, or north east coast, extending coast-ways, rapidly becoming narrower, from Waite's bay in the

* See the list of fossil shells in Sir R. Schomburgk's *Hist. of Barbados*, p. 562.

† Good examples of the kind occur in the neighbourhood of Bissex hill, and on the shore of Skeetes bay in the form of greyish or blackish tufa, coloured by carbonaceous matter.

‡ One of the most remarkable of these is on the estate called Castle Grant, bordering on the Scotland district.

latter named parish, to Skeete's bay in the first mentioned.

The larger district contains the greater portion of these five, and the whole of the parishes of Christchurch, St. Michael, St. George, and St. Peter's. The shores of both these districts are not without their peculiarities;—that of the former is with few exceptions low and shelving, its beach formed of a mixture of siliceous and shell sand; its inland boundary, a line of steep cliffs formed principally of coral rock, often broken and rent in a very remarkable manner. The shore of the latter offers a contrast to the other, marked as it is to a great extent by rocky beaches, and precipitous cliffs; and, where sand occurs, consisting almost entirely of comminuted shells.

This latter line of coast is not without interest as illustrating change, the effects of the formative and destructive powers, which are at present in action. Thus in some places where the waves break with the greatest violence, favouring the escape of carbonic acid from the sea water holding it in solution, sandstone may be seen in the act of formation—the loose sand becoming cemented by carbonate of lime, separated from the water previously held in solution by the acid gas; whilst in others, the perpendicular or overhanging cliffs, formed chiefly of calcareous matter, rising out of comparatively deep and cool water, are acted on by this water as a solvent, from the presence in it of the same acid gas, and are becoming excavated, often exhibiting a deep grooved line corresponding to the high water mark.* Further in some situations the sea appears to be invading the land,† in others the land encroaching on the sea.‡ The first is a rare occurrence, the second less so, and where it is met with, is more striking and remarkable especially connected with other appearances seeming to tell the same story, such as rocks now distant from the shore similarly scooped on their seaward side, as if they had been washed by it, and partial collections of pebbles and sand, more or less distant from the sea, as if they had once been a portion of its shore. ‖

To revert to the scenery of the island: in these two

* Good examples of the formative and destructive agency alluded to may be seen at Long Bay, and in its boundary cliffs, and at Maycock Bay. The destructive excavating process is apparent in all the rocks of a calcareous kind that, scattered here and there, rise out of the sea along the coast. From its action they are almost invariably of a mushroom form, and when in groups have a very singular appearance as seen from the shore.

† At Oistins is an example of the sea invading the land. There an old battery has been destroyed by its inroad, and one or two of the cannon are now almost constantly under water.

‡ At the Crane, almost within the memory of man, the sea not only washed the cliffs, but rose to such a height as to allow boats to float close beneath it, so as to admit of their being loaded from above; which was effected by a crane placed on the margin, whence the name of the spot. Now the water is more than one hundred yards distant, and where it before flowed a kind of garden has been made and various fruit trees planted, including a grove of cocoa-nut trees already of a respectable size. Moreover it is worthy of note that the hollow of the cliff, the result of the wearing, excavating action of the sea, is now converted into a store room, merely by the erection of a confining wall.

‖ Examples of old sea beaches considerably above, and distinct from those at present washed by the sea, are to be met with on the southern coast, between Bridge Town and Hole Town; and in St. Lucy's towards its north-west extremity, a striking instance occurs of a rock several hundred yards inland, bearing marks of having been worn by the sea.

regions it is remarkably different, conforming in a great measure, as already observed, to their geological structure. The smaller region may briefly be said to be distinguished by picturesque wildness arising out of its deep vallies, bold and often precipitous hills, and in one spot mural precipices. The larger, on the other hand, may be described as approaching to tameness in all its features, its gently descending and expanding vallies, its terraced and high table lands, marked by smoothness and a certain regularity, and yet pleasing in no small degree to the eye, especially in conjunction with a widely spread culture and verdure, and the various circumstances of art, marking a well conditioned and prosperous country. The gullies, the vast rents and chasms, which occur in this district, intersecting as already described some of the ridges and table lands, are in their aspect quite an exception to the above, being distinguished for their wildness, often picturesque in a high degree, many of them indeed perfect wildernesses of rocks, trees, and shrubs, strangely intermixed, presenting in their beautiful rudeness, altogether without culture, a marked contrast to the adjoining grounds, all cultivated with care.

Commonly there is a certain connection observable between the geological structure of a country and its soils, as well as its scenery. In a large number of instances this connection is most intimate, so that the rocks constituting the hills and mountains being known, the soils of the plains and vallies may be predicated with considerable accuracy; and for this obvious reason, that the latter by a process of decomposition and disintegration are derived from the former. This is remarkably the case in primitive countries, where the existing soils appear to be formed almost entirely by the process alluded to. Occasionally, however, it is otherwise; there are countries in which there is no immediate relation, or at least of dependency, between the soils at the surface and the strata on which they are incumbent, the soils not being derived from the rocky strata on which they rest. This is strongly marked in the instance of tertiary geological formations. Barbados affords examples of both. In the Scotland district there is observable more relation between the soils and the rocky substrata than in the larger district and that of more recent formation. Nor is this surprising or inexplicable: owing to the steepness of the hills there, their declivities are powerfully acted on by heavy rains, and there is a constant tendency in consequence to the denudation of their sides. Accordingly in this part of the island, where chalk is the substratum, a white calcareous soil, very little different from chalk, is found at the surface; where a bed of clay lies beneath, the surface soil is found to be stiff and argillaceous; where sandstone is the basis, there the soil resting on it is little more than sand. These remarks apply chiefly to the steep declivities of the hills; less so to the vallies, especially where they open out into little plains. There, there is found a certain uniformity of soil, which perhaps may

be considered as a mixture of all the several ingredients of the higher grounds washed down by torrents
and commingled. It is a well marked alluvial soil,
abounding in siliceous sand, containing more or less
clay with a small proportion generally of magnesia
and lime.

In the other and larger portion of the island, where
there is less, and often no relation between the quality
of the soils and the rocky beds on which they rest, in
many places, even on the same estate, there may be
found a considerable variety of soil and sub-soil; in
one spot a calcareous marly soil; in an adjoining one
a stiff clay; and near at hand to this a loose light soil,
containing a good deal of siliceous sand, and no small
proportion of calcareous earth. The principal varieties of these soils may be conveniently classed under
a few heads according to their composition, on which
their nature and qualities depend, and in which, notwithstanding the remark above made, as to admixture,
some order of distribution is observable.

On the higher grounds a reddish brown soil is predominant, containing a large proportion of siliceous
matter in a very finely divided state, with a certain
portion of clay, and an admixture in small quantities
of lime and magnesia, the whole coloured by peroxide
of iron. This soil presents itself even at the edge of
the cliff, bordering on, and bounding the smaller hilly
region,—a situation where certainly it could not have
been brought after the ground on which it rests—coral
and shell limestone—was raised from the depths of the

sea, there being no higher ground near from which it could have been conveyed by the action of water.*

Another quality of soil is that which prevails between the terrace elevations.† It contains more clay than the first mentioned, a large proportion of silica, and commonly but a very small proportion of lime and magnesia. Its colour is variable, red and brown are its predominant tints on the higher grounds. In some places, at a certain depth, it is of a greyish or bluish hue, from the presence of protoxide of iron; a hue which it retains only so long as it is excluded from the air; on exposure, as when the soil is turned up, the protoxide readily passing into the peroxide, the colour changing to yellow or red.

A third variety of soil is that which differs but little from calcareous marl, is incumbent on a substratum or subsoil of marl, and consists chiefly of carbonate of lime, fragments of sea shells with a small proportion of clay and silica, and a smaller one of magnesia. It is generally of a light colour. It occurs in some parts of St. Philip's, especially its north eastern part, and in some parts of St. Michael's and of St. Lucy's.

A fourth variety is a dark soil, in some situations almost black, a colour which it owes to vegetable matter in a peculiar state of decomposition, approaching the state of peat. This soil commonly contains a good

* A striking example of this is to be seen at the top of Horse hill, immediately above the road leading down to St. Joseph's Church : good examples of it also offer on the upland estates of Bloomsbury, Welchman's hall, and Blackman's.

† It is well marked in the elevated valley of the Sweet Bottom.

deal of clay, with a sufficiency of calcareous matter, and of silica and magnesia. It occurs most commonly in low situations towards the sea coast, where the drainage in consequence is imperfect, and there is a tendency when there is an excess of rain, to stagnation. One more variety of soil may be noticed, a calcareous, argillaceous marl, of a grey colour, consisting of alumine, carbonate of lime, and silica in well adjusted proportions, with some carbonate of magnesia. The most remarkable example of it is the Codrington College estate, below the limitary cliff, where thrown up into steep hills and ridges, and depressed into narrow vallies and ravines, this marl forms a little district almost apart—allied however to the Scotland district, of which it may be considered as an offset or extension, and is equally remarkable for the barrenness of its aspect, its real fertility, and its abundance of water, all depending on the same cause, the nature of the soil and subsoil.

Other varieties of soil might be pointed out. They occur mixed with one or other of the principal varieties above enumerated, for none even of these, the principal varieties, are free from admixture to any extent. Few estates, for instance whether situated high or low, are without marl; few are without deposits of stiff clay. In one estate in St. Lucy's there is a substratum of carbonate of lime in the form of minute ovoid granules; in one in St. Michael's the same substance is met with in the form of minute rhomboidal crystals uncohering, after the manner of sand ; in one in

St. Thomas' there is a deposit of siliceous earth in a finely divided state, essentially different from, but in appearance perfectly resembling chalk. These may be considered as curiosities, but they are significant in their indications, and instructive. In connection with the general geological history of the island they afford proof that the soils of this larger division were deposited as already observed when the rocky substrata on which they rest were lying at a considerable depth in the ocean where they were formed; and that the great variety met with must have been owing to the causes then in operation, rivers bringing down the detritus of the mountains of the continent, currents in the sea distributing them, and probably submarine volcanic eruptions, of which there are indications here, disturbing them, and adding new materials to complicate them the more.

As regards productiveness, quality, and fitness for culture, these soils—the soils of Barbados generally—may be considered above par, and with proper culture inexhaustible.*

After what has just been stated I need hardly remark that an opinion too commonly entertained that the planters of Barbados have to contend to disadvantage with a poor soil is entirely unfounded, the reverse happily for the agriculture of the island being remark-

---

* I may refer those interested in the subject, for details in proof, to a discourse delivered before the General Agricultural Society of Barbados, which has been published both in the *Agricultural Reporter*, and in a small volume entitled, *Lectures on the Study of Chemistry*.

ably the case ; indeed I know no country of the same
extent so highly favored by nature in regard to its
soils, whether one considers their variety, their suita-
bility for admixture, or their individual goodness.

Barbados thus rich in its soils, according to a common law, is poor in its minerals. The list of those
hitherto found is very limited. Bituminous soil, asphaltum, mineral pitch or tar, anthracite, iron pyrites,
two or three varieties of iron ore, black oxide of manganese, alum, nitrate of lime, gypsum, calcspar, hyalite,
sulphur, constitute the majority of them. With the
exception of coal and mineral pitch they are all of little
importance, occurring either rarely or in scattered minute quantities, so that those which admit of useful
application could hardly be collected with profit, or
even defray the expense of collecting. Most of them
have been found only in the Scotland district. There
bituminous coal has been discovered in several places,
and worked in two or three, and probably will be found
in many more. Mineral tar exuding at the surface is
of common occurrence, and where it appears, there probably beneath, coal and asphaltum will be found on
excavating. Where the coal has been worked it has
been found included in siliceous sandstone, or in limestone. The form in which the coal presents itself is
remarkable—not in regular seams, or beds or strata,
but in irregular masses or bunches. Its quality also
is remarkable; of the same mass, one portion may
have the character and properties of bituminous coal,
swelling up and caking in burning after the manner

of the best coal of the kind, whilst another may exhibit all the properties of asphaltum, liquefying in burning. It is further remarkable that in one spot anthracite or native coke has been discovered at the surface, in the soil, and adjoining one of the most productive of the coal pits. This discovery was made on the Codrington College estate. The form in which the anthracite appears is precisely that of the bituminous coal, viz. small-columnar; and that it is derived from this coal can, I think, hardly be doubted. How the change, however, was effected is open to conjecture. It may have been from a dike of igneous rock coming in contact with the soil,—a dike that has disappeared in consequence of decomposition or disintegration. When examining the spot, this was the conclusion I was disposed to adopt, from the quality of the clay in which the anthracite was included. Or, it may have been in consequence of a large fire kindled there, the heat of which might have reached the coal, and dissipated its volatile part, in brief, converted it into coke. In favor of this idea it may be mentioned, that in the sea cliff below—the burnt cliff as it is called—it is known that there was once a smouldering fire, the bituminous matter there exuding having been kindled either by chance or design. Hitherto in every instance the coal has been worked only superficially, and yet in more than one the bunch has been worked out. Whether accumulations of it exist of such a magnitude as to yield a profit to mining operations on a larger and more expensive scale, remains to be determined.

Nitrate of lime and alum are mentioned in the list of minerals. The latter from the situation in which it was found, a hill in St. Joseph's adjoining Chalky Mount, was evidently derived from the decomposition of iron pyrites mixed with clay; the former was detected mixed with an incrustation of common salt, in the cavernous side of a limestone cliff, about a mile from the sea in the parish of St. Michael; the nitric acid, the acid of the salt, it may be inferred was derived from the atmosphere of which it is always probably an ingredient. Nitre—nitrate of potash, I have sought for in Barbados, and also in the other islands, but without success; nor have I heard of its having been discovered anywhere in the West Indies, owing no doubt to the nature of the soil, and the composition of the rocks not favoring its production.*

The climate of Barbados it has been already noticed is mild for a tropical climate and equable, rarely subject to any great fluctuations—such as are marked by the barometer—of atmospherical pressure, or to any great vicissitudes of temperature as denoted by the thermometer. During my residence on the island, the greatest variation witnessed in the barometer (and it was during a gale approaching to a hurricane) did not much exceed one quarter of an inch.† There, as else-

---

* The circumstance most favorable to the production of nitre, and which occurs wherever this salt is abundantly obtained, is the presence of lime in the soil or decomposing rock associated with a mineral containing potash, which is liberated in the process of its decomposition.

† In the great hurricane of the 12th of August, 1831, the barometer fell from 30° to 29 25°; according to Dr. Baxter, Inspector General of Hos-

where, except on the ocean, remote from any land, the hottest part of the day is not when the sun is at its meridian, but between 2 and 3 p.m. After sunset—immediately after—the temperature rapidly falls, and continues to fall, but slowly through the night; at 10 o'clock it is commonly at 80°; just before sunrise it is often below 76°. The season of greatest heat is from about the middle of May to the middle of November, when the wind is often south of east and variable. The coolest season is in the months of December, January, and February, when the trade wind, the N. E. is most constant. The latter part of November, and the earlier

pitals, it was so low at 2. a. m.; this was at the commencement of the hurricane.

Horary variations of the barometer are distinct in the West Indies, in common with other tropical regions. The following is the result of Sir R. Schomburgk's half-hourly observations.

MAXIMUM.

|  | h. | m. |  |  |  |  | h. | m. |  |
|---|---|---|---|---|---|---|---|---|---|
| Winter Solstice | 9 | 30 | a.m. | ... | ... | ... | 9 | 0 | p.m. |
| Vernal Equinox | 9 | 45 |  | ... | ... | ... | 9 | 15 |  |
| Summer Solstice | 9 | 0 |  | ... | ... | ... | 9 | 45 |  |

MINIMUM.

|  | h. | m. |  |  |  |  | h. | m. |  |
|---|---|---|---|---|---|---|---|---|---|
| Winter Solstice | 3 | 45 | p.m. | ... | ... | ... | 4 | 0 | a.m. |
| Vernal Equinox | 4 | 15 |  | ... | ... | ... | 3 | 30 |  |
| Summer Solstice | 4 | 15 |  | ... | ... | ... | 4 | 15 |  |

This regularity and very small normal variation of the barometer coupled with the abnormal fall of the mercurial column on the approach of a hurricane, has been taken advantage of to give warning of, and make preparation for the conflict. Whilst Sir William Reid was governor, according to directions from him, any extraordinary fall of the barometer, and its degree in relation to threatening danger was signalized.

H

portion of May, and the whole of March and April belong rather to the cool season than the hot; being intermediate, however, they are somewhat uncertain. The mean annual temperature may be stated as about 80°. These remarks apply to the south west and leeward coast. The windward coast fully exposed to the N. E. trade wind,—the coolest wind that blows within the northern tropic—is even more temperate, it is be-believed,—for few, if any, thermometrical observations have been made there by which the difference and its degree can be determined. And more temperate still, of course, are the high grounds, their coolness increasing with their height; indeed the hill climate of Barbados is truly delightful: there we have warmth without heat; coolness without cold. There at elevations from 800 to 1000 feet above the level of the sea, the thermometer by day is commonly below 75°; there even in the hottest season, and the hottest time of the day, its ascent above 80° is of rare occurrence in the shade and within doors; and there in addition, the comfort of coolness is enhanced by the almost total absence of those troublesome insects which prevail more or less in the lower grounds. It is only those who have had a personal knowledge of night heat, and the insect night pests of the tropics who can duly appreciate the enjoyment of a bed not requiring mosquito curtains (always unpleasant from their closeness) for protection, and yet requiring a blanket (a single one sufficing) for warmth.

In relation to rain, Barbados on the whole may be considered favored. In favorable years the quantity that falls may be estimated at between 60 and 70 inches; and in unfavorable under 50 inches. The following table shews the results of observations with the rain gauge for two short periods, a century nearly intervening.*

* The earlier are from Dr. Hillary's work on Barbados, published in 1759. He notices the year 1754 as an unusually wet one; the rain that fell in June, 19.78 inches, he says occasioned destructive floods and was greater (the monthly fall,) than ever before known.

The later are from the quarterly reports made by the Inspector General of Hospitals, on the health concerns of the stations. At my suggestion, on my going to the West Indies, the Secretary of War, then the Hon. Sydney Herbert, authorised the supply of rain guages for the Command, at the public cost. Since then, they have been used at every station where there has been a medical officer, and it is to be hoped will be continued in use.

| | 1752 | 1753 | 1754 | 1755 | 1756 | 1846 | 1847 | 1848 | 1849 | 1850 | 1851 | 1852 |
|---|---|---|---|---|---|---|---|---|---|---|---|---|
| January .. | | 0.37 | 5.63 | 1.20 | 0.45 | 2.84 | 2.46 | 4.37 | 1.60 | 0.22 | 1.51 | 2.75 |
| February.. | | 1.03 | 0.89 | 1.41 | 1.27 | 0.90 | 0.65 | 1.35 | 1.42 | 1.50 | 1.67 | 1.18 |
| March .... | | 2.21 | 3.53 | 0.66 | 1.52 | 0.94 | 1.61 | 3.44 | 2.25 | 3.55 | 1.32 | 0.64 |
| April .... | | 0.24 | 4.22 | 2.17 | 0.37 | 5.85 | 2.22 | 2.66 | 3.05 | 0.55 | 1.60 | 5.48 |
| May ..... | | 1.31 | 14.65 | 6.62 | 1.12 | 2.75 | 0.98 | 4.24 | 1.82 | 4.15 | 3.35 | 8.26 |
| June .... | 10.03 | 2.44 | 19.78 | 5.84 | 3.37 | 6.89 | 1.51 | 2.13 | 7.34 | 12.97 | 5.42 | 1.90 |
| July .... | 8.46 | 6.67 | 7.52 | 5.70 | 6.75 | 4.96 | 2.92 | 6.38 | 5.18 | 7.54 | 5.95 | 2.88 |
| August ... | 8.72 | 3.47 | 4.69 | 6.28 | 3.89 | 8.72 | 5.84 | 3.36 | 8.16 | 9.23 | 5.77 | 6.92 |
| September | 7.89 | 8.77 | 6.10 | 4.56 | 7.69 | 4.68 | 11.66 | 6.49 | 4.18 | 6.00 | 5.89 | 3.65 |
| October .. | 12.14 | 8.17 | 4.07 | 9.54 | 5.44 | 4.66 | 5.16 | 7.21 | 3.96 | 13.51 | 7.36 | 2.67 |
| November. | 12.96 | 1.33 | 4.66 | 4.40 | 3.75 | 5.60 | 5.79 | 6.00 | 1.91 | 8.65 | 3.90 | 6.22 |
| December . | 2.25 | 2.11 | 11.27 | 8.91 | 5.44 | 16.34 | 2.56 | 10.17 | 4.35 | 5.75 | 3.60 | 4.37 |
| | | 38.12 | 87.01 | 57.29 | 41.06 | 65.13 | 43.36 | 57.80 | 45.22 | 73.62 | 47.34 | 46.92 |

Ordinarily, there can hardly be said to be any dry season; a month without a shower is rare indeed. The wettest months are commonly those of variable winds, July, August, September, October and November; the driest are almost invariably January, February, March and April; December, May, and June are more uncertain, oftener however what may be called rainy than dry months. Continuous rain for many hours is unusual; it commonly falls in showers of short duration, followed by sunshine in a clear, or only partially clouded sky. The days in which in some portion of the 24 hours there is more or less of rain, exceed in number those in which there is no rain, and that largely.* Generally the showers are moderate, but occasionally very heavy, as much as two, three, and even four inches of rain falling in as many hours. Most frequently also

---

* During three years, with some unavoidable interruptions, that my attention was given to the rain guage, in the neighbourhood of Bridgetown, the days i. e. twenty-four hours in which more or less rain fell were 590, whilst those in which there was none, were only 285. Year by year the results were the following :—

|  | 1846. | 1847. | 1848. |
|---|---|---|---|
| Some rain ... ... | 199 | 165 | 229 |
| None ... ... ... | 70 | 139 | 76 |
| Total days observed | 269 | 304 | 305 |

Notwithstanding the large proportion of days not without rain, it is worthy of remark that during the whole of this period, the sun was noticed as entirely hid for a whole day only once; it was spoken of by the natives and described as a very unusual occurrence. A cloudless serenity of sky, supposed to be a common property of the tropical sky, I need hardly remark, is rather the exception than the rule as regards Barbados, and I may add, the West Indies generally; such a state of atmosphere belongs rather to the summer season of the Mediterranean than to any tropical region, of which I have any knowledge from actual observation.

they are partial; sometimes, and that when heaviest, extending over a very limited portion of country. Their cooling effect is remarkable, a light shower often occasioning a reduction of 4 and 5°, so that the air before hot, almost suddenly becomes cool. It is a question hardly yet solved, whether most rain falls by day or by night,* and also, which are the parts of the island that have the largest proportion and fall of rain, and which the smallest, which are most favored, which least,—rain and fertility being in close relation even as cause and effect. The common opinion is that showers

---

* During the year 1848, from January until October, I noticed the quantity of rain that fell by day and at night, i. e. from 6 p.m. to 6 a.m, and from 6 a. m. to 6 p. m. The daily and monthly differences were in some instances considerable, whilst the totals differed but little. The following were the results :—

Day.—January, 2·634. February, ·741. March, 2·305. April, 1·450. May, 1·127. June, 2·155. July, 2·910. August, 1·310. September 3·671. October, 4·170. Total 21·573 inches.

Night.—January, 1·742. February, ·610. March, 1·136. April, 1·210. May, 3·120. June, ·975. July, 3·475. August, 2·050. September, 2·320. October, 3·045. Total 20·163 inches.

During the same ten months the amount of water lost by evaporation measured by day and at night, I found to be as follows :—

Day.—January, 3·40. February, 4·85. March, 5·69. April, 6·65. May, 6· 10. June, 5,00. July, 4·96. August, 4·72. September, 4·32. October, 4·18. Total 49·87 inches.

Night.—January, ·86. February, 1·44. March, 1·43. April, 1·68· May, 2·70. June, 1·62. July, 1·42 August, 1·06 September, 1·02. October, ·81. Total 14·04 inches.

According to which, compared with the preceding, the loss by evaporation was greater than the amount of rain in the same time. It is to be kept in mind that the evaporation was from a surface of water and must have been very much greater in proportion than from the surface of the Island generally. The observations were made in the neighbourhood of Bridgetown ; the evaporator (a tin vessel with perpendicular sides, of the same superficial area as that of the funnel of the rain guage) was placed in an open gallery in the shade, exposed to the wind.

are more frequent in the hilly district, and in the higher grounds, and that the total quantity of rain is greater there than in the lower, and especially the S. E. portion and extreme north of the island; an inference probably correct.*

Another question, and one more difficult of solution is, whether the tendency to drought is greater at the present time than formerly, now that the island is almost entirely cleared of forest, than when it abounded in wood. It would be out of place to enter here on the discussion involved in this question; I may briefly remark, that notwithstanding the high authorities that

* The following table shews the quantity of rain, measured by the pluviometer which fell in each month of 1848, (considered about an average year in relation to rain, the produce of sugar being 28,000 hhds.,) and in four different parts of the island,—three of inconsiderable elevation, the fourth, that marked St. John's, about 700 feet above the level of the sea, and in a central situation.

|  | St. Michael's. | St. Philip's. | St. John's. | St. Lucy's. |
|---|---|---|---|---|
| January... | 4·37 | 3·89 | 6·70 | 4·41 |
| February | 6·29 | 3·38 | 2·61 | 1·72 |
| March ... | 3·44 | 3·21 | 3·81 | 1.47 |
| April | 2·66 | 1·82 | 2·02 | ·91 |
| May | 4·24 | 8·05 | 8·13 | 4·03 |
| June | 2.13 | 2·00 | 1·04 | 2·59 |
| July | 6·88 | 6·83 | 6·23 | 5·60 |
| August ... | 3·36 | 4·66 | 8·72 | 9·20 |
| September | 6·49 | 5·32 | 3·42 | 7·30 |
| October... | 7.21 | 10·10 | 9·01 | 16·23 |
| November | 5·56 | 5·92 | 6·43 | 3·27 |
| December | 4·10 | 7·67 | 10·42 | 3·75 |
|  | 56·73 | 62·85 | 65·84 | 60·48 |

In 1850, which was a very favorable year, the produce of sugar having been about 38,725 hhds.) the amount exported independent of what was reserved for island consumption, about 2000,) the following were the quan-

support the doctrine that trees promote rain, it seems to me more than doubtful, and this whether we consider the theory of its formation—the mixing together of currents of air loaded with moisture of different temperatures—or whether we have regard to the results of experience.

Hills and mountains appear to be the great refrigatories of nature, and the generators of rain. Compare the hilly and mountainous districts of our own country with its plains and lowlands, how great is the difference in the quantity of rain that falls in each; in the former often three, four, or even six times as great as in the latter, though in the latter trees more abound. That forests with a dense canopy of foliage may conduce to coolness of climate, and render it of a more equable moisture may readily be admitted, and in this respect exercise a beneficial influence; but it does not

tities of rain which fell in the several months and in the same places, with one addition, also at no considerable elevation, viz :—Whitehall, St. Peter's.

| | St. Michael's. | St. Philip's. | St. Lucy's. | St. John's. | St. Peter's. |
|---|---|---|---|---|---|
| January... | ·22 | 0·50 | 1·12 | 1·80 | 3·355 |
| February | 1·50 | 1·75 | 3·50 | 2·31 | 5·093 |
| March ... | 3·55 | 0·70 | 0·80 | 0·83 | 1·829 |
| April ... | ·55 | 1·70 | 4·36 | 2·81 | 4·226 |
| May ... | 4·15 | 5·27 | 2·72 | 6·10 | 3·329 |
| June ... | 12·97 | 12·40 | 7·80 | 11·24 | 8·516 |
| July ... | 7·54 | 8·34 | 8·28 | 10·42 | 11·270 |
| August... | 9·23 | 5·45 | 9·28 | 5·73 | 11·770 |
| September | 6·00 | 4·12 | 2·36 | 3·54 | 2·940 |
| October | 13·51 | 8·88 | 16·17 | 5·72 | 10·696 |
| November | 8·65 | 7·71 | 10·97 | 10·14 | 12·142 |
| December | 5·75 | 6·95 | 6·39 | 5·73 | 8·651 |
| | 73·62 | 63·77 | 73·75 | 66·37 | 83·817 |

equally follow, that acting thus they will promote
rain.* The road is kept moist by the shade of overhanging trees, and the air there cool and damp; yet
under the shade of trees we never see dew. An instance in point is recorded of the destruction of woods
followed by a diminution in the flow of running water,
but not by a decrease in the fall of rain.† Further, we
have no proof that Barbados is more subject now to
drought than when it was first colonized. The chroni-
cles of the island shew that this calamity is of irregular
occurrence, and that some of the severest examples of
it took place at an early period.‡ Moreover it should
be kept in mind, that low and succulent plants, such
as the cultivated crops of this island are, afford no inconsiderable protection to the soil from the heating ef-

* It is even maintained that forest trees in growing, appropriating a
large quantity of water, have a drying effect on the atmosphere, and must
tend consequently to diminish the quantity of rain. *M. A. De la Rive on
Glaciers. Ed. Phil. Journal, January* 1852.

† Marmato in the province of Popayan is situate in the midst of enormous forests, and in the vicinage of valuable mines. The amount of the
discharge of the streams here accurately measured by the work performed
by the stamping machines which they drive, was observed to decrease stea-
dily as the wood was cut down; within two years from the commencement
of the clearing, the decrease of the flow of water had occasioned alarm.
The clearing was now suspended, and the diminution ceased. "From the
rain gauge," it is added, "it appeared that the fall of rain had not diminished concomitantly with the flow of the streams." *See the Edinburgh
Philosophical Journal for January* 1851, *p.* 163.

‡ In 1629—30, four years after the arrival of its first colonists, when little
cleared of wood, it suffered severely from a long continued drought, (called
the starving time) which by occasioning a scarcity of provisions reduced
the planters to the utmost extremity.—*Poyers' Hist. of Barbados* 1807.—
Mem. of the island of Barbados 1743. Pere Lebat writing in 1693 states
that there are no streams of water in Barbados, and that water is sometimes so scarce there, as to be even dearer than beer or wine.

fect of the sun's rays, and that owing to their lowness, their cooling effect by night from radiation is even greater than that of forest trees.

The proportion of aqueous vapour in the atmosphere, *cœteris paribus*, being according to the temperature, it is necessarily large in Barbados. From the observations I made when there, I was led to infer that in 1000 parts of atmospheric air, the quantity of this vapour is often as much as 30, and seldom less than 23, estimated by volume, which by weight will amount in one instance to nearly ten grains of water in the cubic foot of air, and in the other to little less than seven grains, denoting an average at least double that of the climate of England.*

Such an abundance of atmospheric moisture has a great and varied influence, and may be considered as an important element of the climate. Owing to its presence, evaporation is moderated, dews and showers and clouds are of easy production; the face of the sky is ever changing; owing to it also, the tendency to decay in animal and vegetable substances is great, and their preservation is rendered difficult; indeed here the productive powers of nature may be said to be the chief conservative powers; no where are the influences of life and death more strongly marked.

---

* Comparing the moistened with the dry bulb thermometer, (one of the best of hygrometrical instruments) I have rarely even in the driest weather seen a difference exceeding 7° by day; in ordinary states of the atmosphere the difference was about 5°; after sunset with wind it seldom exceeded 2°, and without wind or cloud then all difference ceased and dew appeared— the effect of radiation.

Radiation of heat—that property by which heat from the surface of the earth escapes into free space, exercising a cooling influence,—generally is less strongly marked in tropical than in temperate, and in these than in the arctic and antarctic regions; commonly diminishing with increase of atmospheric heat and moisture, and *vice versa;* thus whilst in England there is often a difference of 12° between two thermometers, one placed on the grass, fully exposed to the sky at night, the other suspended a few feet in the air, in Barbados near the level of the sea, I have never observed the difference to exceed 6° and that very rarely, in serene weather and a cloudless sky, without wind, (not a very frequent occurrence there); a difference of 2° or 3° may be about the average difference in the early part of the night, as at ten o'clock, and 4° or 5° in the early morning, at, or shortly before sunrise.*

This low degree of radiation so conducive to an equable temperature no doubt is wisely ordered, and is favorable both to a healthy state of animals and vegetables natives of the tropics, all of which seem to suffer when exposed to great vicissitudes, being peculiarly sensitive to changes of temperature, and to the chilling effect of any sudden reduction. The umbrella, and its diminutive the parasol are by the colored race almost as much used by night as by day, especially by moonlight, nominally as a defence from the moon's rays, but truly I believe as a protec-

* In Sir Robert Schomburgk's History of Barbados, in the appendix some observations of mine on the subject are to be found.

tion from the cooling effect from exposure to the clear sky, the result of radiation.

In an island on which commonly the clouds drop plenty in the form of rain, it might perhaps be expected that springs and rivers, or at least rivulets, would be abundant; but it is not so excepting partially, and this in consequence of geological causes. Over the larger portion of Barbados, that consisting chiefly of coral and shell limestone, and freestone, the ordinary drainage is chiefly subterraneous.* The inhabitants are chiefly dependent for their supply of water on ponds, wells, and cisterns;—the ponds natural hollows, their bottoms impervious to water from a lining of clay ; the wells, pits artificially sunk, varying in depth from 270 feet to $2\frac{1}{4}$ feet;† the cisterns made either of wood or masonry, attached to dwelling houses and collecting the rain that falls on their shingled roofs. The few streams that there are and springs occur chiefly in the lower grounds, little removed from the sea shore, little above high water mark. In the other part of the

---

* By means of fissures, often with funnel-like mouths called " sucks," from the office they perform. They are of common occurrence ; occasionally they become obstructed in their subterraneous channels, and are converted into ponds.

The gullies of this portion of the island are also concerned in its drainage; and after heavy rains, more than the ordinary channels can convey away, they become water courses, and sometimes the beds of torrents.

† In five of the parishes from which returns were obtained in 1847 it would appear that there were 427 wells, the majority of them exceeding 100 feet in depth, and the deepest generally situated in the higher parts of the island. *See Agricultural Reporter* 1847, *p.* 89.

island, its smaller portion—the Scotland district, and
that designated "below cliff," springs and rivulets are
less rare, indeed each valley has its stream, which after
heavy rains is swollen into an impassable torrent.
From their ordinary small volume of water these
streams have not yet been used as a mechanical power,
nor to the extent they might be for the purposes of irrigation. In the whole island there is not a single water
mill, and excepting in two instances, irrigation has not
been attempted. As to quality of water there is a
marked difference according to its source. Cistern rain
water collected from the roofs kept clean, is, I believe
the purest, though even this is not absolutely pure,
not even after having been filtered; it commonly on
examination affords traces of vegetable matter, and
also of saline, viz. common salt, chloride of calcium,
and carbonate of ammonia. The water of springs may
be considered next best, and the purer in proportion to
their copiousness; the matter impregnating them is
chiefly carbonate of lime. The next in point of purity
is the water of the higher wells, those viz. which are
inland, and on more elevated grounds, which differs
but little in composition from the best spring water.
Lowest in the scale is the water of the wells near the
shore, and of the ponds; the former is often impregnated with a considerable proportion of saline matter,
which it may be inferred is derived from the sea by
percolation; whilst the latter commonly contains an
undue proportion of organized matter, either dead or

living—either vegetable or animal, or both, rendering it more or less unwholesome, and sometimes poisonous.*

Mineral waters can hardly be said to occur in Barbados, at least of any strength or well marked efficacy when used medicinally. One well to which my attention was directed I found slightly impregnated with sulphuretted hydrogen, derived I believe from the decomposition of sulphate of lime and magnesia, by the action on it of vegetable matter; this was from Golden Ridge Estate in the parish of St. George. A feeble spring in St. Andrew's in the Scotland district, at the Estate called Vaughan's has more the character of a mineral spring, and may perhaps have some medicinal power, though less than is commonly attributed to it. A specimen of it which I examined I found of specific gravity 10,108; it differed chiefly from brackish sea water in containing a larger proportion of sulphate of lime, and of sulphate of magnesia; a trace of iodine was detected in it.

Of the accidents, if I may use the expression, to which the climate is subject—the awful phenomena of hurricanes and thunder storms—a very brief mention may suffice, they, especially the former, having been so often described. The period of variable winds, and of

---

* Pond water is reputed to be poisonous when a greenish scum appears at its surface, said to occur in severe drought. I have been assured that poultry, horses and cattle, have been taken ill, after drinking of such water, and that death in many instances has rapidly followed. A specimen of greenish scum which I examined, I found to consist chiefly of microscopic animalcules.

greatest heat, that extending from July to the middle of November is commonly called the hurricane season, all the great hurricanes recorded having occurred in one or other of these months. Sir Robert Schomburgk in his history of Barbados has given a list of those which have taken place in the West Indies since their first discovery, amounting altogether to 130,* of which 21 have more or less desolated Barbados. Of these the more remarkable and severe are the following, viz. that of the 31st August 1675, which was ruinous to the island for a while, when almost at the height of its prosperity, and from the effects of which, and of the disasters that followed, drought, short crops, scarcity, epidemic disease, insurrection of slaves, losses at sea from the enemy's privateers, it was long in recovering. That of the 10th October 1780, said to have lasted 48 hours, which was even more destructive than the preceding; the loss of property it occasioned was estimated at £320,564 sterling, and of life at 4326 souls; those of the 13th October 1819, and of the 11th August 1831, the last only second to that of 1780 in the widespread ruin it effected. For minute details I may refer to works expressly on this subject, especially to Sir Wm. Reid's treatise on the laws of storms and variable

---

* 130 is the sum total of his list. In a note he states "From the year 1494 to 1846, or a period of 352 years, I have found recorded 127 hurricanes and severe gales, which committed more or less injury in the West Indies. Of this number occurred in the month of March 1, June 4, July 11, August 4, Sept. 28, Oct. 28, Dec. 2; and of 13 I have not succeeded in finding the months recorded."

winds, in which he developes the vortical theory founded on a wide induction of facts.

Though so destructive in their immediate operations, hurricanes are not purely evil in their consequences; there is reason to believe that they have been often beneficial. After some of them, especially the last which occurred in Barbados, the seasons were more favorable, vegetation more active; there was improvement in the health of the people, certain diseases even disappeared,* benefits, in the opinion of many, more than compensating for the instant losses sustained.

Invariably, hurricanes are accompanied by a disturbed state of the atmosphere in relation to electricity, and it must be inferred, by an unusual accumulation of aqueous vapor;† and hence the terrific thunder and lightning, and the deluging rains, their common accompaniments. During the hurricane season, and only during it, thunder storms are of common occurrence, and then almost invariably they are attended with rain and that often very heavy and damaging to the lands, producing what is commonly called "a wash."

* Particulars in accordance with the above, I heard related by many who were exposed to the last severe hurricane, that of 1831; delicate ladies even were not exceptions; I was assured by one that she never had so keen an appetite and never felt better than immediately and for some time after,—an experience she said, that accorded with that of others. It may be asked, is this state of vegetable and animal vigor subsequent on a great natural catastrophe so destructive in its instant effects, one of compensation, of which so many instances are witnessed in the economy of nature?

† The sun before a hurricane has been seen of a blue colour:—may not this have been owing to the surcharged state of the atmosphere with aqueous vapor?

# CHAPTER III.

THE native population of Barbados is less mixed than that of any other of our West Indian colonies. It consists of two principal classes, the white and the colored; the one chiefly of English descent, the other of African, including half castes. All born to the soil, all speaking the same language, and having the same habits and usages, they may be considered, as they truly are, a united people with common interests binding them together; and since slavery has been abolished, and prejudice relative to color overcome—as it is in a great degree—happily free from any great element of disruption or dissension likely to set class against class and endanger the peace of the community. An advantage this certainly not inconsiderable, especially compared with the state of the island before emancipation, when a servile war from slave insurrection was always more or less apprehended; it had more than once taken place; and when in consequence every white man was an enrolled soldier, and liable at a moment's warning to be called out on military duty.

According to the last census, that of 1851, the total population of the island was then 135,939, which gives the large proportion of 817 to the square mile—a proportion surpassing that of any European country, Malta excepted, and equal even to any Asiatic, not excepting China. Compared with the number afforded by the preceding census, that of 1844, amounting to 122,198, the increase has been not quite one and a half per cent. per annum. In this census the distinctions of race and color have not been avoided as in the former, shewing, I would hope, that there is no risk of giving offence, as was then imagined, in making them. Referring to the return of the grand total, it would appear, first, that 62,272 are males, 73,667 females; secondly, that 134,820 are Creoles or natives, 589 Europeans, 530 other foreigners; thirdly, that 15,824 are whites, 30,059 colored or half castes, 90,056 negroes.

Relative to the whites, they may be divided into two classes—the poor labouring portion of them, constituting the majority, and the smaller portion consisting of those in easy or in affluent circumstances.

The former are in many respects remarkable and not less in appearance than in character. Their hue and complexion are not such as might be expected; their color resembles more that of the Albino than of the Englishman when exposed a good deal to the sun in a tropical climate; it is commonly of sickly white, or light red, not often of a healthy brown; and they have generally light eyes and light colored sparse hair. In make they bear marks of feebleness; slender and

K

rather tall, loosely jointed, with little muscular developement. In brief their general appearance denotes
degeneracy of corporeal frame, and reminds one of
exotic plants vegetating in an uncongenial soil and
climate. In character, morally and perhaps intellectually
considered, they show marks also of degeneracy, not
less than physically. They are generally indolent and
idle, ignorant and improvident, and often intemperate.
Is it surprising then that they are poor, and objects of
pity, or of contempt? Beggars are few, and these almost
invariably persons of this class. What they are they
have been made undoubtedly by circumstances, and
this in the course of a few generations, the majority
of them being the descendants of indentured servants
introduced as labourers at an early period of the settlement of the island. Previous to emancipation, they
were of far more importance than subsequently; then
the militia which for a long period was the sole defence
of Barbados, was principally composed of them;*
then those who were not small proprietors had a certain
allowance of land granted to them by the larger proprietors on the condition of performing military service

---

* The planters for every sixty acres of land, had to provide a man for the
militia, who was called out once, at least, each month, and as much oftener
as was necessary. He was supplied by his principal with a gun and ammunition, and had a house and two acres of land free of rent, on which he
raised some vegetables and kept a cow, or two or three goats. Very idle
himself, his wife worked with the needle, and got money by making clothes
for the negroes. Thus they continued to live in comfort and plenty.
During the last war the militia force of the island amounted to 12,000, of
whom, the greater number were whites, the lesser, freemen of color.

according to a law to that effect. So situated, easily supporting themselves and families with little exertion, it is not surprising that they acquired the habits which now unfortunately distinguish them. After emancipation, the law alluded to ceased to be in force; and the militiamen ceasing to serve, they were permitted no longer to retain rent free the land before allowed them; and hence, with their acquired habits, in a great measure their present miserable condition. The share that the influence of climate has had in their degeneracy is more obscure. That it has not been inoperative, I think cannot be doubted, especially as their women who are industrious, and who are exempt from most of their vices, exhibit in their countenances the same sad unhealthy color, though in a less degree, and much the same peculiarities of bodily frame, which are also to be seen, I think, or traces of them, in the white Creoles of the superior class. In the West Indies generally, and most of all perhaps in Barbados, there is a tendency to an unhealthy state of the skin, and of the mucous membranes; the lips, from the exposure to the sun and wind, are very apt to become chapped and ulcerated, and the face inflamed; in consequence of which the planters, and even many of the poor whites, feel under the necessity of wearing what is called a face cloth, a mask of white linen covering the whole of the countenance, with the exception of the eyes and nose. Further, apthous affections of the mouth and throat and primæ viæ are common amongst the whites of all classes; and the fever almost peculiar to the

West Indies and the warmer portions of the American
continent, viz., that fatal scourge, the yellow fever, is
characterised by a concentration of diseased action in
the same parts. What the noxious cause is, whether of
the fever, or of the peculiar color and other peculiarities of the Creole white, remains to be ascertained;
the latter—the—color is the most remarkable, as the
skin of the Aborigines of America, though commonly
called red, being painted of this color, * is really brown
of different degrees of intensity, differing but little
from that of the Hindoo, or Malay.

The poor whites or "red legs," as they are contemptuously called, from the red hue of their naked
legs, are located most in the more distant parts of
the island—distant in relation to Bridgetown, its
capital and chief sea port—viz. in the Scotland district, and in the poorer portion of St. Philip's and St.
Lucy's. Now that they are obliged to support themselves as they can, they are variously employed.
Those who possess a little land, or who rent a few acres,
cultivate chiefly those crops which require least labour,
and the smallest means, such as ground provisions, arrowroot, aloes, and perhaps a little cotton. I have seen
one of them at work on his ground in a manner not a
little characteristic; a hoe in one hand, an umbrella in
the other, which he held over his head, and a face

---

* According to Pere Labat, the Caribs used a vegetable substance called
rocou, the same probably as the urucu of Southey, as a paint; he desribes it
as red as blood.—Curious particulars are to be found in this author's work,
of their manner of using it, and of adorning their persons.

cloth over his face. Some who have been taught to read and write, are engaged as book-keepers by the proprietors of the larger estates, with a pay of about six dollars a month, and board and lodging. Some are chiefly occupied in fishing, and that of a simple kind, by means of the casting net, and are to be seen exercising their skill on the shore, almost among the breakers, apparently at the risk of their lives. Some gain a livelihood as carters and grooms, and some as field labourers, a kind of occupation which, when slaves only were employed in field labour, would have been resisted by them as an insupportable degradation, and even now is only engaged in from necessity, and with good reason, for they are ill fitted for such work. In the wilder parts of the country here and there, a school is to be found, chiefly for poor children of their own class, kept in aid of subsistence by a white man or woman. When by the former, it is for boys, and reading and writing, and the first rules of arithmetic are taught; when by the latter, for girls, and the teaching is limited to reading and plain needlework. As is too often the case in our village schools, and in some indeed of higher pretensions, the children in reading are more exercised in the pronouncing of words, parrot-like, than in the understanding of them; and to use the expression of an old dame, who courteously allowed me to question some of her scholars, they are "nonplused" when asked the meaning of all but the words in most familiar use.

Some attempts, not I believe very vigorous or long

continued, have been made to raise these poor whites from their degraded state, but hitherto without success. During the time of slavery an effort of the kind was made by a man of no common character, a man distinguished alike for his philanthropy and energy, a Mr. Steel,* a relative and contemporary of the cele-

* The Hon. Josiah Steel (Honorable as a member of council) gave no ordinary proof of active humanity in coming to reside on his property, (a large one consisting of three estates, of between seven and eight hundred acres) at the advanced age of eighty, and effecting during his few surviving years, viz. from 1873 to 1790 many reforms equally as regarded his own interests, and the comfort and well being of his slaves whom he wished to prepare for emancipation. An interesting account of his labours is to be found in a pamphlet by Mr. Clarkson, published in 1823, entitled, "Thoughts on the necessity of improving the condition of the slaves in the British colonies, with a view to their ultimate emancipation, &c. &c."

The memory of this good man, I may add, is still enduring, with illustrative anecdotes of his doings, some of which were related to me by a gentleman of the island when riding by the pond, still bearing the name of the hapless Indian woman, the subject of the pathetic tale of "Inkle and Yarico," given in the *Spectator*, the outline of which some have supposed had been communicated by him;—it is to be found in Ligon. "Yarico's" pond, which is on the Kendal plantation in the parish of St. John, according to tradition was so called from the unfortunate and ill-requited woman, having washed her baby there, on giving birth to it unexpectedly when by the water side! The anecdotes my friend mentioned were chiefly relative to some abuses then common from which the absentee proprietors suffered. The repeating of one may suffice.—Mr. Steel, on his taking possession of the plantation house, seeing a fine sheep, and desiring that it might be killed for the table, was told, it was the manager's, and the same in succession of all the good things, even to a turkey which met his view, making him doubtful that anything was his; to solve which doubt, he requested a general clearance of all laid claim to by his factor. Before going to Barbados he had filled the office of vice-president of the London Society of Arts, Manufactures and Commerce; shortly after his arrival there, he instituted a society of arts, &c , with a like object, but which was not long continued, owing, it is said, to unworthy jealousies. Two publications, however, proceeded from it, one entitled, "Institution and first proceedings of the Society for the encouragement of Arts, Manufactures and Commerce, established in Barbados in 1781;" the

brated Sir Richard Steel of "the Spectator." He tried
to form them to industrious habits by in-door ccupa-
tions, such as knitting and weaving, but in vain. The
late Governor, Sir William Reid, from his experience
in Bermuda of the efficiency of white Creoles in the
capacity of sailors, has suggested to the local government the making of an effort to ameliorate their condition, and at the same time benefit the commerce of
the island and its resources, by engaging them, especially their sons, as apprentices in trading vessels.*
Whether the suggestion, so deserving of being carried
into effect, has in any way been acted on I am ignorant.
Debased as the majority of these men are, it would be
injustice not to remark that they are not all alike
degraded, and that instances occur from time to time
of individuals, who by their intelligence, activity, and
good conduct, better their condition, and raise themselves into the aristocracy of the land. All of them
indeed have the aristocratic feeling in its worse sense,
the class pride acquired in time of slavery, when they
were an important portion of the privileged order; and
it is marked in their manners and bearing, which, blunt
indeed and coarse, are those of free men. It is to be

second, " The abridged Minutes of the Society continued," Barbados, 12mo.
1785. I quote from the Catalogue of the Library of the Royal Society of
London. Though printed in Barbados, I did not find either of them in the
public libraries there.

* From a return made in 1847, it appears that there were then registered
40 vessels belonging to the port of Bridgetown, the smallest of 6 tons
with a crew of two men, the largest of 146 tons with a crew of 8 men and
boys, making a total tonnage of 1613 tons, and total of crews, 245 men and
boys, a shipping employed chiefly in coast and intercolonial trade.

hoped, and I believe with hope, that notwithstanding all their failings there is some innate and inbred worth, such as can hardly be separated from freedom; and that in consequence they are not to be considered altogether irreclaimable. Could they be removed to New Zealand or Canada, or to any of the British colonies more suited to the European—their English constitution, I have little doubt they would turn out useful and successful colonists.

Between the preceding and the superior class of planters, there is a striking contrast. The latter are commonly well educated, and well informed, courteous and hospitable, reminding one of the best description of country gentlemen at home. Nor is this surprising, considering that many of them have been educated in England, and that the majority of them have resided more or less in England. Though the climate is unfavorable to bodily exercise, it is not so to moderate mental exertion, but, I think, conduces to it; especially its equable temperature, and the equable circulation of the blood connected with that temperature. The Creoles of this class may be adduced in proof. Rarely in society anywhere does one meet with more good sense and intelligence; and both are to be witnessed in the debates of their House of Assembly, the majority of the members of which are white planters; and in the manner in which the public meetings, of frequent occurrence, are conducted, and the agricultural societies of which no less than four are now in existence. Proof too is afforded in the past and present literature of the island; in its general good government (chiefly local) with few exceptions, and though last not least, in its progressive improvement and present comparatively flourishing condition. No colony in the West Indies can boast of so many men who have acquired honored distinction; no other has produced so many publications relating to its history; in no other are there so many collections of books; and I may assert without hesitation that it surpasses every other in the number and well ordered state of its useful institutions.

This class, it is believed, and I write on the authority of some of the best informed belonging to it, has benefitted much by the great measure of slave emancipation, and by circumstances and events, some of them disastrous at the moment, which have since occurred or were little anterior to it. Their difficulties have roused them to exertion, and that exertion has told in various ways, all of a beneficial kind. Since the last great hurricane there has been a decided improvement in the public health, and especially in that of the planters and their families, attributable it may be in the first instance to an immediate purifying effect, not an unusual result of such an elemental war, but chiefly, afterwards, to an altered manner of living, more simple and less luxurious; to having better dwellings—the successors of the old ones swept away by the storm,—to extended cultivation, and to the increased use of carriages, with an extension and improvement of the roads, now in most parts of the island (excepting in the Scotland district)

good carriage roads, many of which before were little
more than bridle paths or cart tracks.*

And what the hurricane did for the physical atmosphere of Barbados, emancipation effected for its moral
and domestic atmosphere, it purified that in a remarkable manner, and to the matron ladies and their
daughters, always exemplarily correct, was an incalculable comfort. Licentiousness, whatever it might have
been before, was almost entirely banished from society:
young men no longer exposed to the same temptations
as before, acquired new ideas of correctness and purer
tastes and habits, all of an elevating kind and favoring
the developement of the higher energies.

The poor whites I have compared to exotic plants,
withering under, or barely existing in, an uncongenial
soil and climate. The superior class bear comparison

* According to an official return, the number of carriages in 1848 was
810, of which 436 were four-wheeled, 374 two-wheeled.

According to another return, the number of horses imported in ten years,
viz., from 1833 to 1842 was 8318, and of mules 409. The number of the
former in the island in 1841, was 4052; and of the latter 301. The mortality amongst the horses has been estimated at about 25 per cent., certainly
a high one, but owing I apprehend not so much to the climate, as to the
description of animals imported. They are chiefly from the United States,
and seldom free from blemishes, or young.

In the Scotland district, the roads now are probably much in the same
state, as they were in other parts of Barbados thirty or forty years ago.
Speaking of their then condition, I remember hearing a gentleman say, it
was formerly near a day's journey coming from town to the Spring estate,
(where the remark was made) a distance of about seven miles.

Now there are about 200 miles of public roads intersecting the country,
the greater portion good, kept in repair by means of a road rate, and a tax
on carriages and horses; land contributing 6d. an acre, houses, 1½d in the
pound rent; 20s. each carriage, 6s. each horse.

with the same plants carefully cultivated and protected in their conservatory life, flourishing tolerably, and not unproductive. To drop the metaphor, experience seems to show, that with the comforts required, and using the precautions necessary, the white race can enjoy a fair proportion of health in this island and in the West Indies generally, and be equal to all the exertions necessary both for the business of superintending their own estates, and for all ordinary business in which the brain, the intellectual organ, is more concerned than the muscles and sinews. Many of their habits, and some of the circumstances in which they are placed are favorable to health and its enjoyment; such as the early hours that are commonly kept, residence out of town and mostly in elevated situations, and on pleasant spots; the perfect ventilation of their houses, an abundant and wholesome diet, a large proportion the produce of their estates; no overstraining of any of the faculties, with a moderate exercise of them, and a sufficiency of bodily exercise, either carriage or on horseback.

It may perhaps be asked, how is it, if they can bear the climate so well, that their numbers have not increased? In reply it may be said, that were a selection made of those who have married young, their families would shew a great increase, especially of late years; many a large family I was acquainted with, exceeding in numbers the European average, with eight or ten offspring all in health, and who had never been saddened by any loss by death. That which has prevented any great increase—for that there has been some increase since the time of emancipation can hardly be doubted,—is, in part the little disposition shewn to contract early marriages from prudential considerations, or to enter that state at all; and in part in a less degree, under the same motives, the disposition to emigrate and attempt the bettering of their circumstances in the other colonies in which their skill, with cheap land is expected (often delusively) to be better rewarded.

I may be thought by some to have taken too favorable a view of this class, but I trust I have not, although writing with the pleasant and grateful recollection of many kindnesses conferred by them and never failing courtesies, exempt from a single jar, during more than a three years residence and intercourse with them. Were I to find any fault, I would say not that they are naturally indolent or without enquiring minds, but that mentally they are not sufficiently communicative, or in their enquiries sufficiently persevering. The periodical literature that has been attempted, has commonly languished after a while and come to an end for want of support. Few of their institutions have shewn a healthy growth however promising at commencement, or a continued improvement, and that owing to the same want. As is often' observable in other small societies, there appears to be here a deficiency of faith in the powers of science and in the energies of mind directed to useful and improving purposes, and consequently a lukewarmness in all scientific matters,—and in high thoughts and aspirations, without which no people, no society has made distinguished progress.

The climate bears the blame, but not I think altogether justly considering what has already been accomplished, and how in the East Indies under stronger motives so much more has been effected. Other causes are probably more concerned, such as mainly the little encouragement given by the Home government to merit, the little or no aid afforded by government to any liberal institutions in our colonies;— its total neglect of science there, and more than that, its absolute discouragement by fiscal exactions of the introduction of practical science,—for example, as applied to the improvement of the quality of sugar in the manufacturing process. Under these circumstances, is it surprising that the libraries should consist chiefly—not of scientific works that strengthen the mind, but rather of works of light and elegant literature, not indeed uninstructive, but more entertaining? Is it surprising that an attempt by private individuals to found a school of practical chemistry should have failed, or that the only periodical having any pretentions to practical science, " the Agricultural Reporter," is continued with difficulty, though the subscription to it is only one dollar a year, and it is the only publication of the kind in the West Indies?

Having as I have expressed, a high opinion of the capacities of this class, I cannot but regret that they are not just to themselves. It would be well for them to consider that they have a high calling,—that they

are the influential minds of the country, and that
their own interests are intimately connected with its
prosperity: perhaps, I would hope it, a better time
is dawning, that is, a time of greater and more profitable exertion. The hope I have is chiefly founded
in what I know they are capable of; in the endur-
ance, with I believe some improvement, of their agricultural societies, and not least in their self-reliance,
on which they are now thrown by the measures of
free-trade, though to them only partially free; and by
the conviction, that it is only by greater exertion that
they can better or even sustain their present condition.

The colored population, numerically at least six
times the proportion of the white, is principally composed of emancipated slaves and their children, the
oldest of whom do not exceed the age of nineteen.
Moreover, as the slave trade was prohibited in 1808,
the great majority are Creoles or natives, indeed I do
not recollect meeting with a single person of color, excepting soldiers belonging to the West India regiments
(which are recruited in Africa) who had any knowledge of their father land. Fortunately no nice distinctions in relation to color and its shades are made
here as in the slave states of America, and no opprobrium or disgrace is attached to the intensity of hue of
skin, or other marks distinctive of purity of African
blood. They consequently freely intermarry, and in consequence of that, in all probability in another century,
or even in a shorter time, there will be such an amalga-

mation of the colored races, that distinctions will be no more appreciable from color than in any European nation. At present we may conveniently speak of the whole as divided into two classes, the half castes in whom there is a considerable admixture of European blood, and of the larger portion in whom there is little or no such admixture, and who are nearly if not entirely of pure African descent.

The half castes, in animal developement are commonly well made and muscular, well grown, lithe and slender, active both in body and mind, with capacity of mind probably not inferior to that of the European, and in conduct, good or bad, very much according to circumstances of education and example. They are charged with being volatile and fickle, prone to pleasure and amusement, and less to be depended on than either African or European. Whether the charge be just it may be difficult to say, it probably has some foundation in truth, and, depending on mental qualities not well trained and directed.

They have eminently been benefited by emancipation; many of them were previously slaves. During the time of slavery, their habits, even of those who were free, were far less correct than they are now. Then a colored woman considered it an honor to become the mistress of a white man, and courted the connexion; now she avoids it, as a disgrace.

The occupations of this class are various. Comparatively few of them are field labourers, they rather select callings in which skill and intelligence are most required. Many of them are engaged in trade, retail dealers and hucksters; many in handicraft work, as cabinet makers, carpenters, smiths, masons, as shoemakers, tailors, &c.; in conducting which they display respectable skill, * and give proof of capacity not inferior to that of ordinary workmen in this country. The better educated often fill the situation of clerks; they commonly write excellent hands, and make themselves very useful to their employers; a few are landed proprietors, respectable men, and, I believe, generally esteemed. Those whom I had the pleasure of knowing, were it not for their color, could have been in no way distinguished from other proprietors, except perhaps, bearing in mind two, who belonged to an agricultural society that I had the privilege of joining, by more than common modesty and gentleness of deportment. A few also, a still smaller number, belong to the house of assembly and the council, one to the latter, two or three are, or were, members of the former. In each ca-

* It is mere skill which they possess, art without science; what they know, they have learnt by imitation; the best of the artificers are ignorant of the merest elements of mechanics : I have heard an intelligent master workman express regret on this account. I may add, that unless exertions are made to correct the evil, probably even handicraft skill will decrease, as the apprenticeship system is now falling into disrepute, many young men starting as artificers with a very imperfect knowledge even of the manual part of their art. Their charges are commonly high, often exceeding the London prices; boots, for instance, 8 dollars a pair : a pair of slippers, 2 dollars, inviting thereby much to the curtailment of business, the importation of ready-made articles, slop articles of bad quality, but by their cheapness insuring a ready sale. The following were ordinary charges and prices when I was in Barbados.—Putting a mainspring to a watch, one dollar; half binding in the plainest manner, an 8vo. vol., one dollar ; the making of twelve shirts, 6 dollars, the materials costing 9 dollars, 35 cents.

pacity they have not discredited their class, either as regards power of expressing their ideas, or in information, or in independence. Improvisation has been called a southern talent; facility of speaking seems to belong to the natives of warm climates; commonly the colored man is more than ordinarily gifted with an easy elocution, as is also the African; so that when educated and instructed they make no contemptible orators,— indeed some of the best speeches from public speakers that I heard whilst I was in the West Indies, were delivered by men of this class, and the best not only in point of manner but matter.

Of the half caste women I can say little, those of the better sort being so little in society. From the few opportunities I had of judging, they appeared inferior in manners, and greatly inferior in information to the men, the natural consequence of a more secluded life, and a more limited and imperfect education.

It was once an ethnological question whether the marriages of half castes are fertile,—one that has been satisfactorily answered in the affirmative, and so far tending to prove, that white and black may have been derived from one original stock, and that mere hue is adventitious, the effect of climate, and providently designed to render each race the more fit for that of which it is the native; the white man for the colder and more clouded regions of the north, the brown for the warmer, and the black for the warmest. Fitted as is the half caste for the West Indies and healthy as he commonly is, and exempt from most of the diseases to

which Europeans newly arrived and unseasoned are
subject, and also from many of those to which the
white Creoles are liable, yet I doubt that his strength
of constitution is improved by the infusion of European
blood; he seems to be more delicate and less enduring,
and to feel the advantage and pleasure of a life of ease
and the avoiding of hard labour almost as much as the
European, preferring riding to walking, and the carriage to the saddle; even tradesmen are little disposed
to take walking exercise. If you send for one, whether a
hair-dresser, tailor, shoe-maker, or cabinet-maker, he
more frequently comes in his gig or on horseback than
on foot, and would be surprised if by so doing he excited attention. On this account they are better adapted
for cavalry than infantry service. The mounted police
of the island, a very useful body of men, are principally composed of this class.

Of the larger division of the colored race, in whom
the African features and character remain least altered,
the opinion of those best informed and most competent
to judge is favorable, and not only as regards their bodily powers but also their mental.

The time is past that the negro was held to be hardly
human, rather a connecting link between the monkey
and man. It is interesting to see how truth ultimately
prevails, and how science—exact knowledge,—aids the
cause of humanity. One after another most of the traits
which were adduced as distinctive and as separating
the races, have been made light of and put aside, and
so I apprehend they will be all. Some stress has

been laid on the woolly hair, the black skin with its
rete mucosum, the thumb nail without its white cres-
cent, the deficiency of nasal cartilage. Accurate research has shewn that the hair of the African does not
differ more from that of the European, than his differs
from that of the Asiatic or of the American Indian under
microscopical examination; that the crescentic mark
is almost as often present as absent in the nail, and that
when not perceived, it is owing to the cuticle having
been allowed to spread over and conceal it, as we often
witness in the nails of some of the fairest hands of our
countrywomen who have not acquired a false taste for
large and conspicuous nails;* that the cartilage of the

* On these points the results of the observations of my friend Dr. Blair,
in British Guiana accord with my own. In a letter with which he has favored me, in reply to one calling his attention to the subject, he states as
follows. " The day before yesterday I made an examination of all the
thumb nails of all the patients in the upper male wards of the colonial hospital. The patients there are in good general health, and suffering chiefly
from ulcers of the lower extremities. They consisted that day of a few Europeans (Irish and German) Madeira-Portuguese, negroes (both African and
Creole) and Coolies chiefly from Madras. The total number of observations
made by me and the resident surgeon simultaneously and with due care, as
to the state of the cuticle, and with deliberation, were seventy nine, with the
following results."

|  | Crescent well marked. | Absent. | Doubtful. |
|---|---|---|---|
| Negro | 4 | 17 | 7 |
| European | 4 | 0 | 1 |
| Coolie | 14 | 13 | 2 |
| Portuguese | 8 | 2 | 2 |

He adds, "The day before yesterday we had the cartilage (the nasal)
demonstrated at the colonial hospital on a dead negro. While making the
thumb examination I also made several observations on the nose of the
negroes, and although in some I found them as distinctly cartilaginous at
the tip as I should think in my own, I must say that in a few coarse faced,
snub nosed Kroomen, the tip of the nose felt on handling it, more like a

nose conforming to the feature is small but not altogether absent; and lastly that the so called rete mucosum, the supposed seat of the coloring matter of the skin, is a nonentity, the seat of that matter not being a peculiar tissue, but merely the inner layer of the cuticle or scurf skin.*

It is well that all this should be made out, and such trivial distinctions abolished. Did we not know how great is the tendency to error in the human mind, and to delusion, we might feel surprise that they had ever been seriously entertained. Had the African races preceded the European in civilization, and had they made slaves of the white men, most likely they would have thought as meanly and have spoken as contemptuously of them, and have adduced proof not less convincing of the white being distinct from the black and a lower species. An enterprising traveller in Africa mentions an anecdote in point. In a discussion with some natives relative to the power of the white man and negro, it is said, "Another free black took upon himself to ridicule the constitution of the white man;" "Ah," he cried, "what

bit of adipose substance than any thing else. I am disposed to think that the negro, and chiefly the pure African has not his nasal cartilage so well developed as the white, but that this is no more distinctive than the alleged absence of the crescentic thumb mark.

* Though there is no rete mucosum, a distinct membrane, yet I think it must be admitted that there is a peculiarity,—some difference of structure in the outer surface of the cutis, that from which by a secreting act the cuticle is formed in the several races, owing to which, under the influence of the sun's rays, and other circumstances, coloring matter is deposited beneath the cuticle; in some copiously as in the negro, in some in a less degree as in the Mulatto and Brunette, and in some not at all, as in the very fair and the Albino.

is a white man? a poor weak creature; he can't bear
Soudan heat, he gets the fever and dies. No, it is the
black man that is strong, strong always; he never
droops or sinks, look at the strength of my limbs."*
And how truthful are these exclamations in the manner in which they are applied. The white man is indeed
a feeble creature in the climate for which he was not
designed, as is the African in that fitted for the white
man. The specialities of each it should be remembered
are not defects, but, as before observed, provisions
wisely bestowed to meet the exigencies to which they
are exposed. The negro in Barbados does not require
a face cloth to protect his face; the black color serves
the purpose; it, like black paint on a white face, pre-
vents the sun's rays from having a burning, inflaming
effect.† The thick crisp hair of the negro, and his
thicker cranial bones, it can hardly be doubted, are also
wisely intended as a protection from the same agency,
as is also the dark color of his eyes. A like remark
applies to his soft skin favoring perspiration by which
it is kept cool, as well as by its black color favoring
radiation. Moreover there is unquestionably in his
constitution or organic frame a something, we know
not what,—an influence or power of resistance which
makes him proof against malaria, and the destructive
fevers resulting from what we call malaria when acting

* Travels in the Great Desert of Sahara, by James Richardson, vol. 1,
p. 342.

† Some experiments on this subject are given in the Author's Researches
Anatomical and Physiological, vol. 1, in continuation of those by Sir E.
Home, published in the *Philosophical Transactions* for 1821.

on white men, which has made the western coast of Africa "the white man's grave," and has rendered active military service so fatal to European armies when employed in tropical climates.*

Another asserted distinction may deserve a passing notice, which, were it true, might be adduced as most telling and lowering in regard to the African character: I allude to the intellectual organ, the brain. By many it has been asserted that this organ in the African is below par in volume and weight compared with that of other varieties of the human race. This assertion is founded on inaccurate observation, as has been shewn in a clear and conclusive manner by Professor Tiedemann, and Sir William Hamilton, whose researches on the subject so carefully and laboriously made, appear free from all objection. The former states that he could detect "no well marked and essential difference between the brain of the negro and European," whence he concluded "that no innate difference in the intellectual faculties can be admitted to exist between them." Farther on he states, "The principal result of my researches on the brain of the negro is, that neither anatomy nor physiology can justify our placing them beneath the Europeans in a moral or intellectual point of view," adding, "how is it possible to deny that the

* From the enquiries instituted in the United States of America, it would appear that *there* the natives of the African race are healthier than those of the European, are subject to fewer diseases, especially the women,—are more fertile, less liable to insanity or idiotcy, to blindness and deafness or bodily deformities, denoting altogether a sounder average organism.

Ethiopian race is capable of civilization? This is just as false as it would have been in the time of Julius Cæsar to have considered the Germans, Britons, Helvetians, and Batavians incapable of civilization."* Sir William Hamilton from his extended enquiries, has come to the conclusion, not only that the negro brain is not below the average in weight and size, comparing it with the brain of other races, but more than that, viz., to use his own precise words, "That the negro Encephalos (brain proper and after brain) is not less than the European, and greatly larger than the Hindoo, the Ceylonese, and sundry other Asiatic brains." † The former of these writers—professor Tiedemann—moreover corroborates his induction by reference to the "many recorded instances of negroes who have made a certain progress in the liberal arts and sciences, and distinguished themselves as clergymen, philosophers, mathematicians, philologians, historians, advocates, medical men, poets, and musicians," adding, "many negroes have distinguished themselves by their talents in military tactics and in politics." ‡

All the information I could collect in Barbados, and in other parts of the West Indies—and my attention was particularly directed to the subject, was quite in accordance with these favorable and hopeful views of

* *Philosophical Transactions*, 1836 p. 525.

† Remarks on Dr. Morton's tables on the size of the brain, in *Edinburgh New Philosophical Journal* for April 1850, p. 331.

‡ Op. Cit. p. 525.

the African character; what I have further to relate will tend to confirm them.*

The greater number of this class in Barbados during slavery had been employed in field labour; a smaller number as artificers, chiefly on the estates of the proprietors. Their occupations are much the same now. In the census of 1844, no less than thirty thousand, (30,005) of the whole population, and these above 18 years of age, are returned as employed in agriculture, of whom more than one half, more by 853, were women.†

All experience proves how well adapted they are for field labour in a tropical climate, and that the humane Las Casas did not at least err in the appreciation of their powers for labour. And as labourers both women and men are allowed to be efficient, and with ordinary motives to exertion, such as fair wages, justly and regularly paid, and liberal treatment, not wanting in

* Since the above was written, I have read with pleasure, " *Hope for Africa ;*" " A sermon on behalf of the Ladies' Negro Education Society, by the Rev. Alexander Crummell, B. A. Cantab. ; an African Clergyman." A discourse of much ability, eloquent without being rhetorical ; in which, in a very convincing manner, he refutes much that has been said of the inferiority of the African race, considered as a race, and of its incapacity for civilization and intellectual improvement, he himself, a highly educated man, though the son of a slave, being a striking instance in confirmation. His sermon is well worth the perusal of all interested in that most important subject, the civilization of Africa, and the success of Liberia, in which republic he has now an appointment as the head of a college for the instruction of the native youth. The sermon bears the date of London, 1853, printed by L. Seeley, Fleet Street and Hanover Street.

† In the last census, that of 1851, the total number employed in agriculture is given as 36,653, of whom 4,541 were under 15 years, 36,653, over that age ; the distinction of sexes is not made.

industry. It is a mistake often committed to suppose that the African is by nature idle and indolent, less inclined to work than the European. It is a mistake I perceive even fallen into by some of the friends of the race. Thus a son of their distinguished advocate Mr. Wilberforce, the present Bishop of Oxford, speaks of them* as "a people who naturally hated labour, and

---

* At a Meeting of the Ladies' Society for negro education in the West Indies held on the 27th of April, 1849.

On the same occasion, the Bishop of Oxford made a statement respecting Hayti, (St. Domingo,) which, if correct, would tell forcibly against the African race, to the effect, that "all commerce and social intercourse had vanished, and if the island were sunk, it would not, he believed, produce such an impression as the loss of one ship or one flag! She no longer lived as a member of the civilized world." This statement with similar ones made by persons not well informed respecting the actual state of the country, the author of the sermon already referred to, shows to be erroneous, and that in a remarkable manner, in a note appended,—he, adducing Hayti, with its black government, as an example of the progress of the race during the last half century. The following are some of the facts which he brings forward in proof of his argument, and to which he begs the particular attention of the reader, that he "may judge for himself whether, should that island sink, it would make no such impression as the loss of a ship, and whether her condition, (as asserted by another) is a vegetable condition."

1. Hayti came into existence as a nation, about the commencement of this century and contemporaneously with the South American republics; but while they have been rent asunder by repeated revolutions and are going to ruin, (e. g. Mexico, Guatimala, and Buenos Ayres;) Hayti has had but *one* revolution, and still remains strong and vigorous.

2. The population of the whole island in 1800, immediately on the assumption of independence, was 500,000. Its present population exceeds 1,000,000, (in 1824, the census gave 935,000). I have not seen any census of the island since that of 1824.

3. In regard of agriculture, trade and commerce, the following facts speak for themselves.—First, with respect to Great Britain, M'Culloch in his Dictionary of Commerce, &c. London, edit. 1834; gives the following statement. " In 1786, the exportation of coffee was about 35,000 tons. In

N

who would sink into absolute indolence from the want
of the proper stimulants to mental exertion." He
makes this remark, comparing them with " our peasants at home who love labour for the sake of labour."
This I have no hesitation in remarking is a mistake
founded on ignorance. What I have witnessed con-

consequence of the subsequent devastation of the island, the exportation for
some years almost totally ceased, but it has now risen to about 20,000 tons."

The amount of the following articles exported in 1832, was estimated as
follows : —

| | |
|---|---|
| Coffee | 500,000,000 lbs. |
| Cotton | 1,000,000 |
| Cocoa | 500,000 |
| Dye-wood | 5,000,000 |
| Tortoise-shell | 12,000 |
| Mahogany | 6,000,000 |
| Hides | 80,000 |

The quantity of sugar exported in 1826 amounted to 32,864 lbs. and it
should be recollected, that about twenty years before, not an ounce of that
article was manufactured on the island. This was the state and condition
of Hayti up to 1834.

Of the present condition of Hayti, Mr. Crummell restricts himself in considering its commerce and agriculture, to its relation with the United
States. From returns of shipping, and of exports and imports between
the two countries, which he quotes, it would appear that the Haytian trade
is of more importance to the Americans than even the Cuban, and that of
the eighty-one different countries with which the United States are engaged
in commerce, Hayti ranks the eighth as to the shipping employed.

The following is part of the Journal of a French Gentleman who visited
Hayti in 1850. " In conclusion I must say, that I found the elements of
civilization in a country which has been supposed to be completely plunged
in barbarism. In all social relations I have only had to congratulate myself on the character of the inhabitants. The highways afforded a security
which appears fabulous. In the towns I met all the charms of civilized life.
The graces of the ladies of Port-au-Prince will never be effaced from my
recollection." Surely these are statements well deserving attention, especially of those who think that the African race are incompetent to self government, or to taking a part in such government,—an opinion widely prevalent,
and even amongst the friends of the race, owing, I cannot but think to want
of accurate information.

vinces me of it. The vigorous, quick walk of the negro
going to his work; the untiring zest and exertions
made by negro lads on a holiday at cricket, not in the
shade, but fully exposed to the sun; the extra labour
of the negro when cultivating his own plot of ground
in propitious, showery weather, often commening before
dawn by moonlight, and recurring to it after the day's
work;—the amount of work they willingly undertake;—
in India or Ceylon each riding or carriage horse is attended by at least two persons, a groom, called in the
latter a horse keeper, and a grass cutter;—in Barbados
one man will, with the aid of a stable boy, or sometimes without any aid, take charge of three horses, act
also as coachman, and make himself otherwise useful:
these are circumstances which have fully convinced me
that he neither hates labour, nor is naturally indolent
when he has a motive to exertion. Other circumstances
might be adduced in corroboration, such as,—to mention one or two,—the willingness with which he undertakes task work, and the satisfaction that, when so
engaged, he commonly gives; the industry and perseverance he displays in reclaiming ground, an acre or
two, or less, which he may have purchased in fee, and
from a waste, bit by bit changing its character to that
of fertility, very much after the manner of the Maltese
peasant, breaking up rocks, collecting soil, forming, in
brief, little "campi artificiali," and out-doing even the
Maltese peasant in one respect, viz. in turning to account
each small portion as soon as reclaimed by cropping it at
once. He who has witnessed, as I have, this indefa-

tigable and provident industry, will be disposed probably
to over-rate rather than under-rate the activity of the
negro, and his love of, or rather I would say his non-
aversion to labour, for I believe, comparatively few
even of our English peasants truly "love labour for the
sake of labour." In the best of them labour is an acquired habit, and habit, according to the old adage, is
second nature, and so too with the negro.

The fact that so large a proportion of women are
employed in agriculture, must to the English farmer
impart an unfavorable idea of the average work accomplished, and justly. For the quality of labour is generally low, partly derived from the time of slavery, but
I apprehend mainly, from women being so much employed; working with men, and paid at the same rate,—
at precisely the same rate in Barbados,—their exertions
may be said to regulate those of the men, and to constitute a standard.

The time it is to be hoped will come, when men will
take the place of the women in the fields, and the latter
restrict themselves chiefly to in-door work and domestic
duties. Engaged as they now are, they cannot attend to,
nor are they competent to undertake those duties. It
is field labour they have been trained to and little else.
During slavery, sex was little considered; the breeding
woman as well as the breeding mare was required to work.
Few, if any of them are acquainted with the use of the
needle; they can neither make nor mend their own
clothes; and those of their children and of their hus-
bands, if they are married, are commonly made by other

hands, and worn out in rags, for they are seldom mended. The care of their children too, is commonly of necessity delegated; whilst the mothers are in the field they are given in charge to some old woman. Even infants before they are weaned are only partially cared for. This condition of the women is every way injurious—to labour, to morals, and even to population. The rate of her wages renders her independent; her habits render her unfit to be the helpmate of man; the marriage tie is commonly a loose one, and often never made; it is considered indeed right and respectable, but the want of it is hardly considered disgraceful. In our own country we see the bad effects of such a system of labour. One witness, speaking of the evils of female labour, especially field labour, says, " The morality of women thus engaged is mostly low, and their language is often filthy and disgusting; there is no great disposition amongst them to attend to religious instruction, and when an exception happens it is frequently frustrated by a want of proper clothing."* Another witness describes them as ignorant of cookery and needlework. If such are the evil consequences in England, it is only surprising they are not worse in Barbados, amongst a people not of christian and free ancestry, but of slaves, and the majority of them only recently taken from bondage, and who previously had little or no religious instruction.† According to a

* See Poor Law Commission Report, 1843, containing "An exposition of the evils of infant and female labour; and "An enquiry into the extent and causes of juvenile depravity," by Thomas Beggs, London, 1849.

† They never went to church; no religious rites were observed by them,

French statist, M. Mathieu, there were 19,349 legitimate children born in the department of the Seine in 1850, and no fewer than 10,035 illegitimate.* The number of the latter in Barbados is greater; it exceeds the former. We may be surprised by the laxity which the one denotes in a country priding itself on its civilization, and in a part of that country including its capital,—taking, it is boasted, the lead in refinement; but surely we have no reason for any feeling of surprise regarding the laxity which is denoted by the other. It would be a satisfaction to see the proportion of illegitimate births decreasing, but this does not appear to be the case, nor the proportion of marriages increasing; † nor perhaps can either result be reasonably calculated on till there is a radical alteration of circumstances in relation to labour, and the women are placed in their true position. The effecting of this no doubt will be difficult, and must, if accomplished, be gradual, by the elevating influence of moral and religious instruction, and instruction likewise in womanly occupations, such as will conduce to make them good wives and good mothers. The present generation it is to be feared are hardly corrigible; it is the rising and the following generations on whom the reformatory process can be

neither of marriage, of christening or burial. Their dead were not even interred in consecrated ground, but in some spot apart on the estate of the proprietor; and their funerals,—a term of respect hardly applicable,—were commonly followed by a merry making at night.

* Annuaire du Bureau des Longitudes.

† Returns on this subject are to be found in Sir Robert Schomburgk's History of Barbados, p.p. 89—90.

tried with hope. If well conducted, and circumstances favor, it can hardly fail of success, judging from the opinion I have formed of the African character, and the instances I have known of well conducted domestic life in negro families of humble condition, but who had the advantage of instruction, and by example as well as precept, in the households of respectable planters.

The present low state of morals in relation to the sexes, no doubt will check a rapid increase of population, and so prevent an excess, and an undue lowering of wages, which always conduces to the degradation of a peasantry. This, and the great facility of living should promote the gradual withdrawal of women from field labour, especially on the introduction of more skilled labour and implemental husbandry, to both of which the attention of the planters is now seriously directed; a subject to which I shall have occasion to revert hereafter.

I have spoken of field and agricultural labour without alluding to its kinds, and that in which women are most employed, and that in which the men only are engaged. In cane cultivation, and in the cultivation of ground-provisions with the hoe,—in time of slavery almost the only implement in use,—hardly any preference is shown. Some planters who can give strict superintendance, would rather engage men, as being stronger, and capable of more work; others prefer women as being more regular, and requiring less watching. In carrying manure into the fields, and distributing it there, and in certain minor operations of field husbandry, and in reaping and carrying canes, they are almost indiscriminately employed. Men are employed chiefly as carters, ploughmen, herdsmen, and grooms, and in the boiling house in all the processes for making sugar; and also as drivers or overseers, and watchmen on the estates. Thus occupied, they have a higher rate of wages than the field labourers. As skilled labour becomes more in request, and implemental husbandry more in use, this class must increase in number; a circumstance that should not be without weight in the elevation of the whole. They will be able better to afford to allow their wives to remain at home, and to have their children educated.

That they are not averse to the education of their children we have satisfactory proof in the increasing number of the schools and of the children frequenting them, and further in the parents' being averse to let their children engage in plantation work at an early age, interfering with their schooling according to usage in times of slavery. Of the better attendance at the schools when there is an improved system of teaching, one example may suffice, that of Codrington College estate school.

When I visited that school in October, 1848, the number of attending children was 250. I have rarely witnessed a busier or more pleasing scene of intelligent activity. Every room, even that intended for a kitchen was crowded: where the smaller children were congregated, —their black heads, and lustrous eyes most conspicuous, I was reminded of a swarm of bees, they were so close

together. Young as these children were, from four to
five years of age, they were learning the letters of the
alphabet. The elder children were instructed in reading
and writing: The second class of boys, whilst I was
present, read a chapter in the New Testament and were
questioned during the lesson as to the meaning of
words, and regarding the names of places and persons;
their attention was kept on the alert, and an interest
evidently excited. The first class of girls were employed at the time in writing, and their writing was
generally good. All these children were of the colored
race, and labourers. Those whose parents were employed
on the college estate, were taught gratuitously; others,
and they were the majority, were not so favored, their
parents had to make a small payment for them. Large
as this school then was, only a few months before, viz.
the preceding Christmas, I was assured, the average
attendance did not exceed forty, and these the estate
children. The vast increase was owing to one individual,
the Principal of the college; to the improved method
of teaching he introduced; to his vigilant superintendence to secure its being carried into effect, and to the
cheering and exciting influence on young minds of a
mind such as his, taking an interest in their welfare
and progress. Though he had been little more than
twelve months in the island, the building of a new
school-house, spacious and of good construction, was,
at the time of my visit, nearly completed, the old one
being much too small, and otherwise unfit for the purpose required. Whilst there was mental training,

industrial training was not neglected. Labour was inculcated as a duty. Between lessons, the boys were employed in road making, and they had completed a good piece of road leading to the new school-house and chapel adjoining. They had made also for themselves a skittle ground, that they might have play as well as work, cut out from the steep side of the hill. The Principal, who was well acquainted with village children and village schools at home, appeared, from the remarks he made, to have formed a very favorable opinion of the intellect and capacity of the colored children. He remarked that he found greater facility in fixing and interesting their attention than that of children of the same age and situation in life in England. Their minds indeed seem to be precocious, and to resemble, in an early aptitude for receiving instruction, more those of southern than of northern Europe, more those of Greek than of English children. Whether the same aptitude will endure, insuring progress with cultivation as they advance in years, remains perhaps to be determined. The little experience hitherto obtained seems to be affirmative.

The improved system of teaching, which I have thus briefly and imperfectly noticed, is, I believe, extending to the other schools of the island, by means of masters, who have had instructions from the Principal himself, they having attended him at the college by his invitation, expressly for the purpose. How incalculable will be the benefit, if it be persevered in, and made general! Then in a few years, or at farthest in one or two generations, a sound education will be widely diffused amongst the people with all its good effects.

Owing to the want of such a system, or rather the want of any good instruction, the great majority of these people, since emancipation, have improved less in moral than in physical condition, to the disappointment of many of their well wishers of sanguine dispositions who did not make sufficient allowance for inbred vices. Regarding these vices, is it surprising that they have no high principle of action,—that they have no great regard for truth or principle, or that they are incapable of exercising much control over their passions; and in consequence that many of them are given to pilfering and addicted to lying,—are quarrelsome and abusive, and in anger apt to exchange blows as readily as words, and to inflict them with harsh severity on all in their power, whether it be a wife, a child, or the dumb beast? This want of control over their feelings, with little or no moral or religious check, makes them often cruel in the treatment of their children and brutal in their conduct towards each other. It is no unusual thing to see women fighting one with another in stand up fight, pugilistically like men; and even less so to see husband and wife exchanging blows, and to hear the horrid screams of the weaker, when overpowered and severely punished; or to see a father or mother flogging a screaming child without mercy, and desisting only from weariness. Those who have any regard for quietude and comfort, should fix their abode in Barbados at a good distance from a negro village; I say so from painful experience: their scolding, their "talk" as they call it, which they consider their privilege, is disturbing at the distance even of more than a hundred yards; and in these outbreaks it is useless to attempt to check them; they defy even the police, taking the precaution, when vociferating their abuse, to stand at their own doors and not in the highway.

With all their failings however they are not without good qualities. They are little addicted to drunkenness, less at least than the European;* they are naturally sociable and cheerful and friendly; self-supporting, doing well without parochial relief or poor laws;† kind to those connected with them, the sick and aged, rarely forsaking or neglecting them. Though passionate, they are not commonly revengeful; and though assaults are common, murders are of extremely rare occurrence, not excluding child murder.

Physically considered, their condition since emancipation has been peculiarly good, and they fully

---

* In 1847 the number of licenses granted for retailing spirituous liquors in Barbados, was 656, yielding a revenue of about £4000 a year, the payment for each license being thirty-two dollars. This might seem to indicate habits of drunkenness, amongst the population; it is to be remembered that there is a large garrison in the island, and a sea port, both enhancing the demand for such drinks; but even with all due allowances, I fear it must be admitted that the colored inhabitants are losing their character for sobriety. Too often, the planters make an allowance of rum to their labourers, so promoting a taste for it.

† Friendly societies, the members of which are of the labouring class, are numerous. The payment is commonly a quarter of a dollar monthly, the same from both sexes. In sickness, a quarter of a dollar per week is allowed; eight dollars are allowed on the death of a member for funeral expenses.

appreciate and enjoy the change. Having hitherto had good wages, the industrious can obtain not only the necessaries of life, but many of its comforts and luxuries. Their houses commonly of wood, consisting of two rooms, are wholesome and tolerably furnished. A good bed, tables and chairs are considered essential. Few of them are without a cow, a pig or goat, or without poultry, and the former whether they have land or not. They have abundance of cheap food, living chiefly on vegetables. No fire is required, except for cooking, and fuel for this purpose is almost always obtained from the estate on which they work. Little clothing is required; they have commonly more than they need,—two or three suits,—one for working days of the coarsest kind, others for Sundays and holidays and occasions of ceremony, such as marriages, christenings and funerals. Vain and fond of finery, when dressed for display, as on Sundays, they are then as much over-dressed, as the contrary when they are in their ordinary every day attire; labouring men may be seen going to church or to chapel with white kid gloves on their hands, their garments in keeping, commonly ready-made imported clothes of fashionable make, accompanied by their wives and daughters wearing smart bonnets, dresses of silk or muslin, and using parasols. They appear to have no sense of the incongruous or ridiculous. Thus dressed, and when they are on their good behaviour, they are not only civil but somewhat ceremonious, and fond of using and receiving the title of *mister* and *missis*, and of making enquiry after each other's health; to them indeed a matter of no small importance. A light hearted people, given to pleasure and amusement, they are fond of dancing and singing; and their merry meetings,—most merry and noisy they are,—have always these accompaniments. Their dances, especially on Saturday night are said to be, not of the most decorous kind. Their favorite instruments are the fiddle, great drum and triangle. When they begin, they hardly know when to stop, exhibiting an activity and energy increasing with exertion, and almost inexhaustible.

During the period of slavery, many were the victims of Obeism;—many of dirt eating, which are now never heard of. The superstition of the one˙ has passed away, with the temptations to practise it; and the disposition to the other has ceased under a better diet, better treatment, and the cheerful solaces of life. Some other diseases also which were formerly rife and destructive, have either greatly diminished or have almost entirely ceased. Tetanus may be mentioned as one of these, but whether owing to some change of climate or to changes in the habits of the people, is doubtful.

Common˙ to all classes, Barbados possesses many advantages; a good local government, for so it generally has been, regulating in a great measure its own affairs; courts of law, and a magistracy above corruption, in which justice, it is believed, is fairly administered; an efficient police; adequate prisons and places of correction kept in good order; a civil hospital under excellent management, supported chiefly by subscription,—very creditable to all concerned,—which might become the basis of a useful medical school ; a lunatic asylum recently established, and as yet a very imperfect institution, hardly adequate to the wants for which required ; good roads,—great facility of communication ; an inland post for the daily delivery of letters throughout the island, established in 1851 ; a college, and schools* of various kinds, all improving in efficiency, affording means of education beyond the average of country districts at home ; a public library ; †

* According to the returns of schools in 1846, the number under the superintendence of the regular clergy was 43, the number of scholars 1842, of whom 1080 were boys, 762 girls. Connected with the Moravian mission 3, scholars 315, of whom 166 were boys, 149 girls. In connection with the Wesleyan mission, 6 ; scholars 444, of whom 259 were boys, 185 girls, making altogether a total of 52 schools, and 2601 scholars, exclusive of Harrison's Free School, and the School of Codrington College, an account of which may be found in Sir Robert Schomburgk's History of Barbados, or in the statistics of the island collected by the colonial secretary. There are besides many private schools, dames' schools and others in which an increasing number of the colored class are receiving some instruction. In these the charge for teaching the alphabet and to spell is ¼ dollar a month, to read ½ dollar, and in addition to write, 1 dollar.

Since 1846 much progress has been made ; it would appear from a report of the Lieutenant Governor, that in 1851, 5000 children were returned, as being taught in the schools connected with the church of England, and 1,200 in the Moravian and Wesleyan schools, exclusive in each instance of Sunday scholars.

† This library, the first attempted in the West Indies, is open to the public without distinction. It was founded in 1848, by Sir William Reid. Still in its infancy it contains only about 700 vols., but these have been well selected, and it is gratifying to learn, as reported by the Lieutenant Governor, that " almost all of them have been read, and read and thumbed so often, that many of them are literally worn out ;" and further, " that the demand for books of light reading is as nothing compared with those of history and the physical sciences." He says, " of the strong desire to learn

churches,—one to each parish; and other places of
public worship, in which the services of religion are
decorously conducted by ministers who have, I believe,
generally the respect of their congregations, and not
least so, those of the Wesleyan methodists and Moravians,—who were, to their honour be it said, the first to
exert themselves in the instruction of the negroes in
time of slavery;—and as regards the people themselves, an intelligent body of proprietors, a rising
middle class, and an industrious peasantry; the whole
though exhibiting extremes,—equally in races, condition, civilization and manners,—forming a tolerably
united body.

Few countries are without their peculiarities; those
of Barbados and of the West Indies generally, are
many and strongly marked. I shall here allude only
to two,—others can hardly escape occurring to the
mind of the intelligent reader.

One relating to the colored inhabitants is the want
of all that is traditional amongst the mass of the
people, whether affecting the inner or the outer man,
the mind or the body, lore or art. Here we do not
see, as even in the wildest parts of Asia, any traces
of trades, of crafts, that have come down from remote
ages, descending from father to son, any specimens of
delicate handiwork, or even of coarse, peculiar to the
people;—witness the poverty of the West Indian de-

among the young colored people of the island, it is pleasing to be able to
speak with the highest satisfaction." Is not the use of this library a proof
of it? Reports (Parliamentary Papers) for 1851, p. 69.

partment of the Great Exhibition in all but the
productions of the soil, unaltered or little more, as in
the instance of sugar and some amylaceous products,
as tapioca and arrow root, and these prepared by simple
processes introduced by Europeans.

In relation to the remote past, their minds are in a
childlike state, almost a blank, bearing hardly a trace
even of any superstitions. Disadvantageous as this at
first may appear, it is not without its advantages, especially as regards the future, and their improvement, if
zealous efforts be made to instruct them, there being no
deep-rooted prejudices to be rooted out, no wide demar-
cations by castes, * no false doctrines to be confuted, no
pride of opinion to be subdued; offering thus almost a
naked field for culture, which, if the soil be good, and
the innate vigor of the African denotes it to be so, can
hardly fail to reward well the pains of those who may
gird themselves to the labour.

The second, relating to the white inhabitants as well
as the colored, is the marked absence of all indications
of regard for the fine arts. I do not recollect having
seen or heard mention made of more than one statue
in marble, and one in bronze, or of a single painting

* Though there are no such distinctions as castes, yet there is a tendency to
them, as there is indeed, in most societies. It is best seen in their marriages;
domestic servants seldom contract marriages with field labourers, or these
with those, or either with tradespeople and artificers. The injurious effect
of this, especially on the labouring class can be easily appreciated, by those
who in this country, familiar with the cottages of the peasantry, know how
much of the comforts and good manners of their inmates, have been introduced by the wives, learnt in service before marriage, in the families of the
higher class of farmers and gentry.

of value in any of our West Indian colonies.* Indeed it is rare to see a picture of any kind, or even an engraving in the house of a planter or merchant, however wealthy either may be. Nor in the construction of their dwellings, either in town or country, or in the grounds about their country houses, or in their gardens, with a few exceptions, is there anything to denote such a regard. Convenience, ease, comfort, neatness, seem more to have been thought of than what is positively pleasing, elegant, and beautiful. The causes which conduced to this neglect of the fine arts are probably many; some connected with the climate, some with the migratory and absentee habits of the more opulent planters, whose fixed homes are in England, rather than in the colonies, especially since communication by steam packets has been established. As to climate, owing to high temperature, and the moisture of the atmosphere, it is peculiarly unfavorable to the preservation of all perishable things. Besides which there are other destructive agencies, some constantly in operation, to be dreaded, others from time to time at uncertain intervals. Of the former, such as insects and mildew; of the latter, the hurricane, earthquake and volcano. Where the doubt of preservation is so great, there is little inducement to obtain works of art worth preserving, or to expend money in ornamental works.

---

* That of marble is a copy of a bust of General Codrington, presented by All Souls' College, Oxford, to Codrington College; that in bronze is a statue of Lord Nelson, in Bridgetown, erected to his memory by subscription.

And the taste for such works, not being indulged, languishes and all but expires, according to what seems a law of our nature. Like causes, in a less degree interfere with ornamental planting and gardening, the more to be regretted, considering how admirably adapted the country is for both, and the charm they impart to it, truly converting into paradises the spots, —the few exceptions alluded to,—where attempts of the kind have been made.

# CHAPTER IV.

## BARBADOS.

Our West India Colonies are without exception essentially agricultural, not a single manufacture existing in them unconnected with agriculture, nor any important branch of commerce, the export trade, being in a great measure limited to agricultural produce, and the import trade limited to the supplying the ordinary and daily wants of the inhabitants deriving their subsistence chiefly from that produce.

Compared with the others, Barbados stands high, both as regards the proportional extent of land under culture, and the skill and energy of its planters.

The superficial area of the island has been estimated at 106,470 acres; of which about 100,000 acres are said to be under some kind of cultivation; about 40,000 acres in sugar canes, the remainder in pasture and provision grounds. This estimate must be considered, I believe, only an approximate one, especially as regards the acres in canes, and the land lying waste or unproductive ; the former varying from year to year, (it has been gradually increasing;) the latter from its situation, being chiefly in gullies, not easily measured. Some of the circumstances favorable to a successful agriculture have been already pointed out. And there are others, not the least marked of which are, first, the large amount of available labour, and of efficient labourers, and next the size of the estates or plantations, few of them being so large as to be beyond the means of their proprietors to cultivate them well, and a large proportion of them being sufficiently large to require both skill and capital to conduct them profitably. In 1840, the number of landed properties was returned as amounting to 1874, of which 1367 ranged from one acre to nine,* and 508, from ten acres to 879, averaging 174 acres.† Since that time, even in the short interval

* 239 of 1 acre       47 of 6 acres
  236 ... 2           51 ... 7
  116 ... 4           34 ... 8
   73 ... 5           22 ... 9

† The following is a return of the estates on which there are windmills, in the several parishes. The first column shows the number; the second the maximum of acres of any one estate ; the third the minimum ; and the fourth the average size derived from taking the mean of the whole. With the exception of five, all these estates exceed in size nine acres, and with the exception of 45 they do not exceed 400.

|  | No. | Largest acres. | Smallest acres. | Average. |
|---|---|---|---|---|
| St. Michael | 45 | 549 | 10 | 161 |
| James | 33 | 620 | 23 | 227 |
| Peter | 44 | 635 | 10 | 184 |
| Lucy | 43 | 384 | 15 | 156 |
| Andrew | 24 | 559 | 19 | 287 |
| John | 39 | 774 | 11 | 229 |
| Philip | 55 | 724 | 13 | 221 |
| Christchurch | 63 | 522 | 9 | 177 |
| Joseph | 43 | 554 | 5 | 114 |
| George | 62 | 879 | 10 | 150 |
| Thomas | 57 | 575 | 4 | 144 |

508

of seven years, the number has increased. According to a return made in 1847 the number was 2998.

First let us direct our attention to the larger estates. They all have with slight differences one common character. On each is the dwelling of the planter, generally a very comfortable abode, and pleasing object, whether built of stone or wood; somewhat oriental in its aspect from the galleries or verandahs by which it is surrounded, and its light decorations, if any, and bright colors; where possible its site is elevated, both for the sake of greater coolness, and for commanding a view of the property.

Though not so often as could be wished, the air of coolness about it is occasionally increased by the pleasing shade and verdure of a grove, or shrubbery, with an approach still rarer, by an avenue of stately cabbage palms, or of a narrow strip of dense wood.* Well de-

---

* When Pere Labas wrote, (about 150 years ago,) more attention appears to have been paid to ornamental planting at the country residences of the proprietors of estates in Barbados than at present, and not less, even more, to luxury of furniture than now. After praising the quality of the houses in Bridgetown, he states that those in the country are even better built, and are large, well ventilated, provided with glass windows, and the apartments commodious, most of them having avenues of tamarisk trees, or of shadock, or other fruit trees, giving to the houses a cheerful, pleasant look. He adds that the wealth and good taste of the planters appear in their furniture which is magnificent, and in the plate, of which all of them have a good deal, so that (a circumstance that still should be kept in mind) if the island were taken, it would be a richer prize than the galleons,—an enterprise, he insists, less difficult than is imagined: he points out how it may be effected, and what a narrow escape it had from his countrymen in 1702,—the expedition fitted out to invade it, having turned aside in search of the galleons,—concluding with the comforting remark,—" patience ! what is deferred need not be lost."

tached from the house in most instances, and yet at a convenient distance, are the other buildings, such as stables, cattle sheds, and sugar works; the latter comprising a windmill of most solid masonry, in form not unlike a Martello tower, and the largest hardly inferior in size; the boiling and curing house, to which is commonly attached a cottage for the manager and his family. In the olden time low huts were provided for the slaves in the neighbourhood; these are commonly still preserved with a portion of garden ground to each, and are occupied by the labourers employed on the estate.

The plantation itself is commonly a continuity of surface, nowise distinguishable from the neighbouring one, having no conspicuous boundaries,* no hedge rows or enclosing walls, the fields or cane-pieces, provision grounds, and pasture lands of which it consists, being separated only by estate roads. Nor as regards the objects on it, the buildings just now described excepted, are there any sufficiently conspicuous to vary materially its uniformity of aspect. A careful observer indeed may notice varieties of colors and tints, according to the quality of the crops or their progress,—green, purple, yellow, often beautifully intermixed. In one spot or more he may see cattle tethered on a raised platform of mould and vegetable matter, for the purpose of making what is called pen manure; in another, flocks ranging

---

* The estate boundaries are very irregular: in making a survey, if there be a rock at an angle, a cross is cut in it; if not, a stone is sunk there with a cross on it; but there is no law or rule on the subject. This information I had from a land surveyor.

at large over neglected pastures. Imagine a succession of plantations, each with its plantation-house, windmill, and farm offices and fields like the above, all having a common accord of look, enlivened, as they often are, by working parties engaged in the never ceasing and varied operations of husbandry, and the reader may form some idea of the country generally in its agricultural aspect,—an aspect always pleasing, and often charming, when all is verdant, bright, and luxuriant, as it is when the canes have nearly reached their maturity, or are in flower, and the yam crop is in its young purple leaf.

The system of agriculture pursued, somewhat varies in different parts of the island, owing in part to the nature of the ground, whether hilly or champaign, and the qualities of the soil, and in part, and probably more, to the views of the planters, and their supposed interests.

Three objects may be considered common and directing ones, on which the system more or less depends, viz., first, the production of as much sugar as possible, for profit; secondly, the preserving a certain extent of land in pasture, for the support of the live stock; and thirdly, the growing of " ground-provisions" consisting of roots and grain, the yam and sweet potato, the maize or indian corn and others; some selected as rotation crops, some, especially the grain, for marketable produce, and in aid of forage.

It is held by judicious and prudent planters, that in the apportioning the land of an estate, not more than one half of its acres should be in canes, the other half being reserved for pasturage and provisions. Each head of cattle, it is estimated, requires for its ordinary support not less than from one and a half to two acres, laid down in grass,—sour grass (*paspalum conjugatum,*) the most common grass of the country. In favorable times, when the price of sugar has been remunerative, the tendency has been to assign more than this proportion to the cane, on the idea, that more profit will result thereby to the proprietor, though it may be necessary to make a greater outlay in the purchase of foreign manures to supply the place of the estate deficiencies ; a speculation held to be dangerous, and indeed proved to be so in an eminent degree under adverse circumstances.

Of the modes of culture, the most marked differences are in relation to the implements used, the manner of preparing the ground, the manner of applying manure, and its quality, and the manner of trashing its surface or protecting it from drought.

During slavery, till the slave trade was prohibited, and a deficiency of labourers apprehended, the hoe was almost the sole tool employed in cane or other cultivation. Then, in aid of human labour, the plough and harrow were introduced, with other implemental aid, the efficacy of which had been tested, and approved at home. The success was marked and encouraging, till the arrival of unpropitious seasons, of drought, and short crops in consequence, extending through ten years commencing 1819—20. Then the plough fell into disuse, and the exclusive hoe was resumed, and strange to say, on the belief that it,—the plough—as much as the peculiarity of weather was to blame for the short comings, and scanty produce. This prejudice maintained its ground for about twenty years. The use of the plough was not resumed till 1839—40, and that only very partially,—gradually however increasing. At present it is far from generally employed, and the same remark applies to the other implements, and this even in situations and soil well adapted for their trial, with an almost certainty of success, and which will surely be obtained when there is an increase of confidence in their efficacy, and of knowledge and skill in their application and use.

The cane is commonly planted in holes, small square excavations, in rows, at regular intervals of four feet. Of late years a wider planting has been adopted and is most in use in the leeward district, allowing between each row six or eight feet instead of four, retaining the four feet distance between each plant in the row. Thus, affording more space, more soil, more air, more light, the cane is believed to grow with greater vigor, and acre for acre to be more productive of sugar, with greater economy and greater profit, than on the closer plan of planting; results which increasing experience seems strongly to confirm. In the same district flat tillage has to some extent been substituted for the old plan, above alluded to—that by holes and raised banks, —and it is believed also with advantage; a danger from stagnant water in the holes after heavy rain, being thus avoided; the risk from drought diminished in consequence of there being less surface exposed for evaporation; the roots, which tend to spread superficially and horizontally, being less in the way of injury; and at the same time facility in weeding and carting being promoted.

Manure for a long period has been in use in Barbados. Till of late years pen-manure with mould was almost exclusively used, prepared in the manner already briefly noticed. Other modes of manuring, such as by turning in a green crop, occasionally practised, were exceptions, and rarely made. Now, the plan of manuring is far less restricted; guano, nitrate of soda, soot, and some other imported manures have been pretty largely used, but with more or less of doubtful success. That which on the whole has appeared most promising is the first mentioned, guano, of which £50,000 worth was imported in one year—1846. It has been largely used in the leeward district. The most approved method of applying it has been by introducing it,—about a pint in a hole,—a few inches deep, just large enough to receive the quantity mentioned, in the line of plants at intervals, viz. between every two plants,—two intervening. This method by concentration is considered preferable to the opposite one by diffusion. The latter is thought to be wasteful, and to encourage the growth of weeds; the former to promote more the growth of the cane and that gradually and more lastingly; inferences confirmed also by experience.* Even when using pen

* In one instance that I examined a portion of soil thus manured, I detected guano in it, twelve months after it had been applied, and in a notable quantity close to the hole in which it had been put, and phosphate of lime in unusual proportion at the distance of several feet.

manure or stable dung, the cumulative method is always employed; a basket full is introduced into each cane hole, and that commonly some weeks before the planting of the cane.*

Trashing—the strewing of the cane fields with the leaves obtained from the former crop, or if cane-trash be deficient, with some substitute of an analagous kind—any kind of leaves—for the purpose of protecting the soil from drought, by checking evaporation,—is a process almost peculiar to Barbados. It is most largely used in the lower grounds, and especially in the leeward district, and is now getting into use in the higher grounds, and in the hilly portions of the island. Its effect appears to be excellent, answering not only the purpose for which it was principally intended, but others; acting for instance as a manure, promoting the growth of the cane, and tending to check and prevent the growth

---

* The basket in use in Barbados is very similar in form to that of Westmoreland, called a "swill," tolerably represented by the form of the hand when the fingers are closed for the purpose of scooping up water to drink ; it is of three sizes, large, middle, and small. The largest are used for holding grain of Indian corn, and the heads of guinea corn, of a capacity equal to about 80lbs. of these ; the smallest is chiefly employed as scales attached by cords to iron beams, which are to be purchased apart, imported for the purpose; that of medium size is principally used on the estates, a substitute for the wheelbarrow, for carrying manure, mould, stones, &c.; it will hold about 25lbs. of pen manure, when fully charged. It is carried on the head.

During the time of slavery when there was a command of labour, a large allowance of manure was apportioned to each hole, a basket-full of 80lbs. amounting per acre to 64 tons ; when the planting was 4 by 6, or 5 feet by 5, exceeding even the proportion used by market gardeners. Well has cane cultivation been called garden cultivation. Now, that there is not the same command of labour, and that the labourer cannot be coerced to carry so heavy a load, and since first crop canes have been more generally planted, this allowance per hole has been reduced to the smaller basket-full, and this at present I am assured is deemed high-manuring.

of weeds.  This, its double or triple function is very admirable,—the manner namely in which in dry weather when the winds are parching, and vegetation is arrested, it acts as clothing, and that in which in wet weather, when vegetation is active, it acts as food ; drought suspending its decomposition at the same time that it arrests vegetable growth, and rain promoting its decomposition at the same time that it advances that growth.  The earlier the trash is applied the better is the effect.

The cane requires about twelve months to come to maturity, liable however to be retarded or accelerated in its progress by circumstances, such as quality of soil, kind of manuring, situation as to exposure to wind and sun, and especially by the nature of the seasons, whether dry and unfavorable to vegetation, or rainy and favorable.  The month considered best adapted for planting, and that in which it is most actively carried on, is December, preparations having been made for it during the two or three preceding months, in the way of tillage and manuring.  In Barbados invariably, the cuttings, (by cuttings the cane is always propagated), are not placed in the ground horizontally, as is, or was practised in Louisiana and some other colonies, but perpendicularly or slightly inclined, in holes made with a crow bar,* one in each hole, each cutting being about one foot in length, and containing two or three eyes or buds at the joints, the embryo plants.  The

* The implement is commonly called a crow bar ; it differs only in its end, viz. in being conically pointed instead of divided and shaped somewhat after the manner of the crow's foot.

depth at which they are inserted is a variable one, tasking the judgment of the planter. Where the subsoil is marl or loam, and the surface soil of inferior quality, deep planting is found to be most successful. On the contrary where the upper soil is good and the subsoil not so, for instance a bluish clay, shallow planting has been found to answer best.* Two men, the holes being prepared, can plant about an acre a day. The process of trashing follows two or three months after, or earlier if the young shoots are tolerably advanced; they appear above ground in about a month. Weeding is contemporaneous, or rather, as in the best conducted estates, an unceasing operation, on the great principle in all farming, that weeds are as thieves or parasites, taking that nourishing matter from the soil on which the plants are fundamentally dependent for their growth.†

In the old mode of planting, that by square holes,

---

* Some remarkable examples have been mentioned to me in accordance with the above; in one instance how a proprietor on the verge of ruin owing to unremunerating crops, planting his canes, at the ordinary depth, more than recovered, acquired wealth, by planting deeply; in others how disappointment and loss ensued from deep planting. In the successful instances, from the account given, it may be inferred that an impervious layer consisting of carbonate of lime (the pan as the concrete is called in the North of England) had formed between the soil and subsoil, by breaking through which fertility was reached.

† Weeding on some of the best conducted estates is now performed as job-work; for instance after the planting has been effected, and the ground cleared of weeds, it is let to a party at a certain agreed rate to keep it clear of weeds, without restriction as to time or persons employed; so it becomes chiefly the occupation of the old, the young and infirm, and at leisure hours; and the work is commonly well done, because it is for the interest of the party so to do it.

the levelling the walls of these holes, or the banks, is the only additional operation required before harvest time, or time of cutting the cane. This, under the most favorable circumstances, commences towards the end of November or beginning of December; it is commonly at its height in January and March, most so in the former, and is generally completed throughout the island before the end of June. The commencement of harvest, of crop-time, and its ending, are indicated by the appearance of the windmills, whether their sails, their points, as the sails are briefly called, are up or down, a conspicuous mark even at a distance, and carefully looked for by those interested when nearing the island, arriving from England either in December, or in June and July. The harvest is necessarily a busy and anxious time to the planter, the process of sugar making being carried on simultaneously with that of reaping ; and the success of the former depending very much on the wind and weather, prospering when there is fine weather favorable for cutting, and a steady breeze favorable for working the mill, crushing the canes and expressing the juice, and the reverse of prospering when there is much rain or little wind.

The cutting of the canes is effected by the bill-hook, in form like that used in England from whence it is obtained. The canes as they are cut and clipt of their leaves and green tops are conveyed to the mill in carts drawn either by oxen, or by mules and horses, and the sooner they are subjected to its pressure the better.

As the juice flows out, it is conducted by a gutter or pipe to the boiling house, and forthwith exposed to the fire, in a set of boilers called 'taiches,' which, excepting the first, the racking copper, are heated by a common fire, the fuel of which is commonly the crushed cane dried, designated 'megass.' Under the most favorable circumstances, rarely occurring, the fuel derived from one crop is sufficient for making the sugar of the next. Where the megass fails, coal, or in the Scotland district mineral tar is the ordinary substitute. In the first boiler, the racking copper, an important part is performed, that of depuration, which is effected at a temperature little exceeding 160 Fahrenheit, with the addition of lime. Acids in the juice, are hereby neutralized, and albuminous matter coagulated. The juice received in this boiler in a turbid state is drawn off clear into the first taich of the set or plant, where under a stronger fire it is made to boil, a workman standing by, constantly stirring and skimming, and so in succession, conveyed by ladling, till it arrives at the last called the skipping taich, in which it is kept till sufficiently concentrated to crystallize or form sugar in cooling.* Arrived at this point, it is rapidly transferred, 'skipped' into shallow coolers made of wood, and from thence when sufficiently cool, and of a proper degree of consistence, into barrels in the curing house, with

---

* Depending on the principle, that the solubility of sugar in water increases in a high ratio with the temperature; thus, whilst water at 60° dissolves twice its weight of cane-sugar, at the boiling point it dissolves five times its weight. See Agricultural Reporter, for June, 1853, for some valuable practical remarks on the subject.

perforated bottoms standing on supports of wood, over a floor with a raised border, on which the leakage, the molasses, falls, and is retained, and from whence it is either transferred to the fermenting vat for the purpose of making rum, or to barrels, to be reserved for that purpose, or to be kept for sale in the island, or for exportation. During the period of slavery, as is now the practice in the slave colonies, sugar making was carried on by night, as well as by day, almost uninterruptedly, and the labourers severely tasked. Now, night work is in a great measure discontinued. I have heard a planter say, he never wished to see a light in his boiling house, the night-made sugar being commonly of inferior quality, from diminished care and attention, and the time favoring pilfering and other abuses.

The harvest season, though a hard working time, is a welcome one to the labourers ; it was so even to the slaves owing to the privileges then enjoyed, especially the being allowed a free indulgence in the field, at the mill, and in the boiling house, in partaking of the produce, of which they are excessively fond. In addition now, during crop time, the labourers have higher wages than usual, or if employed on task or job work, have an opportunity, by their diligence, circumstances favoring, to earn much more than their ordinary wages. This latter plan of labour has been successfully tried on some well managed estates, especially in the Leeward district, equally to the satisfaction of the proprietor and those employed, the remuneration of the latter being determined by the quantity of sugar made, so holding out

R

a constant and powerful motive to watchful exertion.
When, as is more commonly the case, this plan is not
pursued, when the labourers are hired by the day, the
risks and anxieties of the planter are increased; all the
losses are his, whether resulting from failure of wind
to work the mill, from negligence in supplying it, or
from various other negligences for which there is so
large a scope, even to the bracing of the mill and the
feeding of the fires.

The preceding is a brief summary of the principal
operations carried on, from the planting of the cane to
the harvest and the completion of the sugar making.

The sugar cane of which there are many varieties,
is in a manner perennial, once planted it is renewed
by shoots, requiring about the same time for their
ripening as the parent stock. These shoots are called
"ratoons" and second, third, and fourth year ratoons,
according to their age. This method of growing the
cane is partially practised in Barbados, but less than
formerly, and less than in most of our other sugar
colonies. It is most followed in those parts of the
island, where there is the greatest depth of soil, and least
risk of the soil being parched by drought. The small
expenditure attending it, whether in the way of labour
or manure, is its chief recommendation. Weeding, and
little else as regards labour is needed, from the end of
one harvest to the commencement of the next; commonly no manure is applied, though no doubt benefit
would result from its application: in one or two instances that a portion of cubic nitre,—nitrate of soda

was used, the beneficial effect I have been informed was well marked. The objection to ratooning is the small yield of sugar, seldom exceeding half a hogshead to the acre, or under the most favorable circumstances one hogshead, instead of two or three hogsheads, the produce of plant canes; and though the sugar from ratoons is commonly of superior quality, it is no adequate compensation for the smaller quantity: compensation can only be sought in the diminished risk.

Subordinate to cane cultivation is the growing of provisions and forage, such as Indian corn, (*zea mays*) Guinea corn, (*sorgum vulgare*) sweet potatoes (*batatas edulis*) yams, (*diascorea*) and eddoes, (*caladium*) Guinea grass, sour grass, woolly pyrol, pigeon pea, omitting mention of others of minor importance. Owing to their rapidity of growth, most of the provision crops are intermediate ones, that is capable of coming to maturity between the harvesting of one crop of canes and the planting of another; and in consequence the cultivation of them, of such as are not incompatible, may form a part of the preparation for the ensuing cane crop.

Of these secondary crops the more important are the sweet potatoe and yam, Indian corn and Guinea corn.

The sweet potatoe comes to maturity in about three months; it may be planted at any season; it is the most economical of crops, and one of the most productive and useful. It requires little tillage and no manure, though of course it is benefited by manure. By its wide spreading branches and leaves it tends to keep the ground moist and soft, and to check the growth of weeds. Its tuber is a good article of food, and for cattle and horses as well as man; and its leaves and stems are eaten freely by cattle, and are considered wholesome and nourishing. Containing little or no silica in its composition, and so differing from the sugar cane, and containing also a good deal of alkali, it is well adapted to precede the cane in a rotation system. The same remark applies to the yam. The sweet potatoe is often planted immediately after the reaping of the cane; in the short space of four months I have seen two heavy crops taken off the same ground, impressing one strongly with the great capabilities and productive powers of a good soil within the tropics in favorable seasons.*

The yam comes to maturity in about four months; the time of gathering it is commonly in November or December. It is a more delicate plant than the sweet potatoe, and requires more careful tillage, and also manure for its success. As an esculent it is in greater

---

* The sweet potatoe, of which there are many varieties, may be propagated by slips like the strawberry, or by small tubers; the former as the most economical, is the one most commonly employed. From 4000 to 5000 lbs. an acre, is said to be an average crop. Its price is very variable, from 1 dollar the 100 lbs. to 4 dollars or even more, as it may be abundant or scarce. The fresh tuber contains, I find, about 27·5 solid matter, and 72·5 water. Though in its ordinary state, it cannot be preserved with ordinary precautions more than a month or two, in the West Indies, (in a cool climate it may be preserved much longer), if cut into slices and dried, it may be kept for a very long time, and might in times of scarcity be exported advantageously. I have some in this state now quite sound which I have had more than two years. As it contains gluten or a principle analogous to that of the gluten of wheat, as well as starch, its nutritive power must be considerable. Its sweetness, it may be remarked, is not perceptible till it has been dressed; in boiling or roasting, it may be inferred, a portion of the starch is converted into sugar.

estimation than the sweet potatoe, though I am doubtful that it is equally nourishing; it has the advantage of keeping longer, the disadvantage of being more liable to blight, and failure from drought. The leaves and stems are eaten by cattle when hungry, but are less liked than those of the sweet potatoe.*

Indian corn, "the pride and boast of American husbandry," as it has been called by one American author, the cause of the indigenous civilization of many of the Indian tribes, as it has been considered by another, comes to maturity in about three months. It is often planted after canes; but injudiciously in the way of preparation for canes, the inorganic elements of both being very similar, especially their stalks and leaves. As an independent crop it is very productive, and commonly remunerative. The various uses to which it is applicable are not to be described in a few words, they are so many, whether for human food, as the grain,—for forage, the tops and leaves,—for fuel, the stems of the ripe plant,—for bedding, the spath, or the delicate, elastic leaves enveloping the grain in its head,—for manure, its ashes.

* The yam, of which there are also many varieties, is propagated like the common potatoe by the tuber cut into pieces, each about the size of an egg, with a portion of rind. 4000 lbs. an acre is considered a good crop; 3000 lbs. not a bad one. Its market price varies from two dollars the 100 lbs. (when abundant) to five or six dollars (when scarce.) It can be kept good twelve months even in the West Indies. One variety, the red yam, I found to contain 30·55 per cent. solid matter; another, a white yam yielding only 16·1 solid matter, the remainder water. The tuber often attains a very large size; Yams weighing 15lbs. are not uncommon, and occasionally they are met with of the weight of 20, or even 30 lbs. each. Each plant yields only one; the growth of the tuber under favorable circumstances is wonderfully rapid, three or four weeks sufficing, counting from its first appearance.

Guinea corn comes to maturity also in about three months. There is the same objection to its preceding the cane as that above mentioned in the instance of Indian corn. In the time of slavery it was very largely grown; about one-fourth of each estate was commonly devoted to it; its grain constituting the principal food of the slaves.* It is much less cultivated now than then, and the propriety of extending its culture seems doubtful, on account of its exhausting powers in relation to the soil. Indian corn, as an independent crop, appears to have the preference, and probably deservedly. Guinea corn, sown broadcast, to be used as forage, has been recommended; the yield has been said to be great and the cost of production small. It has been tried for the purpose of soiling, as has also Indian corn; the effect of turning them into the soil for manure in their green state has in some instances been reported on favorably, in others, especially of Guinea corn, not favorably,—even the reverse, without obvious reason; theoretically considered it might be expected that as green manures they would both prove powerful fertilizers to the cane.

The views of the planters at different times respecting these crops have varied in a remarkable manner, influencing, of course, their practice and system of agri-

* It may be considered as the peculiar grain of Africa, and the staff of life over a great part of that continent, being the same as the often-mentioned Dourra of African travellers, or millet. Under favorable circumstances, protected from insects, it may be kept many years,—as long as wheat. The plant is almost perennial, when cut down new shoots springing from the stock. The leaves and stalk in their green state constitute good fodder; the dried culms, after the panicles containing the seed have been gathered, are used as fuel in the boiling house.

culture. In the times of slavery, and in times of war, and before foreign concentrated manures were to be procured or had been thought of, when the island was mainly dependent on its own resources, then they were largely grown, even beyond the wants of the population; there was an excess to export. Then in favorable years there was abundance of forage; and thus likewise manure was largely obtained. Since emancipation, since the conclusion of the war, when commercial intercourse became easy, and supplies of all kinds were easily procured from abroad, with increasing pecuniary facilities by means of colonial banks, tempting to speculation and enterprise, the growing of these provisions diminished. It was thought by many, if not more prudent, at least more profitable to extend the cultivation of the cane, and to supply the deficiencies resulting, by the purchase of the imported articles. Accordingly the production of sugar became greatly increased, that of the provision crops greatly diminished, and the varied use of them and the importation of substitutes augmented, a large portion of which came from the United States, paid for, not in produce, but in dollars.*

So long as the market prices of sugars were remunerative, and the favoring circumstances unchanged, this

---

* An instance may be given. During the time of slavery it is estimated that of the 60,000 acres, or 200 estates of 300 acres, each produced 1,500 bushels of corn, or an equivalent, to feed 150 slaves, making for the whole an amount of corn, of the value of £300,000 besides some to export. Since then great has been the change,—for instance between the 5th of January, 1843, and the 5th of January, 1844, (not a period of drought) there were imported into the island 85,000 bushels of oats, 75,180 bushels of Indian corn, and 58,328 barrels of meal.

system of high and speculative agriculture appeared to
proceed prosperously. Wages were high, trade brisk, a
great activity prevailed. But when prices fell, as they
did in a remarkable manner on the admission of slave
grown sugar into the English market, all this apparent prosperity was at an end; disaster followed on disaster, insolvencies, scarcity of the circulating medium,
suspension of credit, reduction of wages, suspension of
cash payments, with threatened scarcity even of the
necessaries of life. The whole, such as was witnessed in
1847—48, was a concatenation of events as distressing
as instructive, and affording a lesson which should,
though it is doubtful that it will, be long remembered.
Since then the tendency has been to return to the old
plan; and what is remarkable, and very creditable to
the energies of the planters, whilst more provisions
have been grown, the culture of the cane has also
increased.*

Of the plants grown for forage, the principal are Guinea grass, (*panicum jumentorum*) sour grass, (*paspalum
conjugatum*) woolly pyrol, (*phaseolus mungo*) and two or
three more of the leguminous kind, such as the bonavis,
(*lablab cultratus*) archer bean, and pigeon pea, (*cajanus
indicus.*) Each of these in the system of agriculture
has more or less distinctive properties of a recommenda-
tory kind. The woolly pyrol, the bonavis, the archer
bean yield an abundant forage, protect well the ground,

* According to the latest accounts there is already a relapsing into a
neglect of growing provisions, in connection with the extension of forced cane
cultivation, even without the certainty of profit and with no inconsiderable
risk of loss.

and without exhausting the soil, render the tillage easy for cane cultivation; and if ploughed in, in their green state, enrich it greatly. The latter remark applies to the pigeon pea.

Guinea corn and sour grass are grown solely for forage; the former cultivated, the latter with little or no cultivation. Both are used commonly in their green state; they are rarely made into hay, for which both are well adapted; in consequence of which neglect, fodder in times of drought is scarce and expensive, and the stock and all the interests of the farm suffer. This is a subject to which I shall return; it affords ample scope for suggestion and animadversion.

The following is an exposition of the manner in which an estate in the time of slavery was apportioned, and is adduced in illustration of some of the preceding remarks. It is from the minutes of the agricultural society, the first that was formed in Barbados; they are dated July 15, 1815.

20 Acres under first crop canes.
23 ,,   ,,   second ditto.
5 ,,   ,,   yams, to be followed in canes.
13 ,,   ,,   potatoes, eight of which to be planted in canes.
7½ ,,   ,,   cane stumps and for potatoes, in September.
11 ,,   ,,   young Guinea corn.
21 ,,   ,,   corn-stumps for fodder, eleven of which are rented, five of which are to be followed in canes, the remainder in Guinea corn.
4½ ,,   ,,   Indian corn, to be followed in canes.
5 ,,   ,,   fallow, two of which are to be in canes.

s

12 Acres under sour grass.

15  ,,    ,,    yards, negro gardens and roads.

137 Total.

"The estate," it is added, "consists of 126 acres besides 11 acres rented, and is cultivated by 68 slaves, of whom 30 work with hoes; 6 horses, 39 head of stake cattle. The last crop was 82,500 lbs. of sugar. Hired labour for the year, £12 6s. 3d."

"The Society were of opinion" these are their words, "that the very flourishing condition, and good order of the estate exhibit proofs of the most judicious application of the labour, and merit their highest approbation."

I shall give one more example derived from the same source, bearing the date of the 3rd of August, 1811.

31 Acres in first crop canes.

31  ,,    ,, second ditto.

34  ,,    under preparation for the ensuing crop, in four fields, each 8½ acres, of which one is holed and planted with Indian corn on the banks, another is planted with red and white yams; the third will be holed immediately after the second crop canes now on it are reaped, and then planted on the banks with potatoes; the fourth is a Guinea grass field, which is to be destroyed in the month of November, and planted in the next spring.

68  ,,    ,, Guinea corn; 16 acres of which are planted in seed to give suckers to plant 36 acres, and 16 acres are in second crop.

24  ,,    ,, Guinea grass.

1  ,,    ,, sour grass.

2  ,,    ,, ocroes.

5  ,,    ,, potatoes.

3  ,,    ,, holed for eddoes.

8 Acres under young pigeon peas.
29  „      „     tenements, yards, negro grounds, &c.

236 Total acres.
162 Negroes, 60 of which are effective labourers.—
     17 Slaves were added last year.
43 Head of stake cattle.
5 Horses.
7 Mules.
6 Cows.
11 Calves.

"The society, after viewing the estate were unanimously of opinion that it was in fair condition; but they think Mr. —— (the proprietor) plants too large a proportion of canes."

For the sake of comparison, and for further illustration, I could wish to give examples of the apportioning of an estate, before the crisis of 1847—48, and subsequently, since there has been a wholesome retrograde movement in relation to the growing of provisions; but though I have applied to more than one friend residing in the island, and well acquainted with the manner in which its cultivation is conducted, I have not yet been able to obtain any very precise details on the subject, seeming to indicate that no general system is followed. *

* A friend writing to me in 1852 states that "the present mode of management adopted by a large number of planters is, to divide the cane land into two parts, one of which contains the growing crops, the other is in preparation to succeed it. This is the case in all the black soil estates, as in the parishes of St. Peter, St. Lucy, St. James, St. Michael, Christ Church, St. Philip, and St. George. This plan has been gradually gaining ground during the last ten or fifteen years, and seems to be the most successful plan of managing these estates. Along with this system was introduced that of wide planting, which has gradually increased from 3 or 4 feet to 6 or 8, and with marked improvement in the size, vigor, health, and yield of the cane. On the higher lands, and red soil estates, as St. John's, St. Joseph's, St.

The expenses of an estate, the cost of production, the profits of the planter, are subjects which I enter on with hesitation, and would willingly avoid, did I not think that they require to be noticed, and that some advantage is likely to result from directing attention to them, however imperfectly.

The establishment on an estate of average size,—one of about three hundred acres, is nearly as follows;—an attorney in the absence of the proprietor, who has the general superintendence of it, and who is paid by a per centage of about 5 per cent. on the net produce: a mercantile agent, through whom the sugar of the estate is shipped, and its sale conducted, and supplies of articles needed by the planter are received at a charge of about 2½ per cent.: a manager, resident on the property, who takes immediate charge of its culture in all its details, with a salary of about £125 a year, a house rent free, servants, and commonly a horse, with various privileges, such as keeping poultry and pigs as many as he pleases, feeding them on the produce of the estate, and the growing of vegetables for the use of his family: a book-keeper with a pay of from 6 to 8 dollars a month: a book-poster periodically employed in examining the books and estate accounts, whose pay may be about from 44 to 100 dollars a year: one or more watchmen, and a ranger, the former at 5, the latter at 6 dollars a month, whose daily duty it is

Thomas's, we think it a better plan to divide the land into three parts; one is planted in canes, another kept in ratoons, and the third in preparation." It is on the land in preparation that the provision and certain forage crops are mostly grown.

to take care of the property, the watchman being employed by night as well as by day, often armed with a gun:* lastly, a body of labourers, varying in number chiefly according to the extent of land in cane cultivation, or about one to each acre so cultivated, on an average about 120, comprised in about 50 families, constantly employed, receiving wages varying in amount from about 20 cents to one shilling a day; the latter in crop time; and varying also according to the mann'er in which they are employed. These labourers when residing on the estate have cottages provided them, one for each family, with about $\frac{1}{4}$ of an acre of ground, for which is paid commonly $\frac{1}{4}$ of a dollar a week.† They have to provide themselves with the ordinary simple implements in use, such as hoes, knives for cutting grass, bill-hooks for cutting canes and wood, and baskets for carrying manure. The cattle and horses required, it is difficult to estimate, some estates needing more than others, according to the distance from the port to which the produce has to be taken, and the mode of culture pursued, whether in part with the plough or entirely with the hoe. Referring to the minutes of the first Agricultural Society, I find that in 1812 on a well-conducted estate about five miles distant from Bridgetown, of 213 acres, with 131 negroes, 52 of whom were

---

* He is subject to a fine for negligence, or to make good the property that may be stolen.

† The terms somewhat vary in different properties; generally they are not so favorable to the labourer as justice requires, the holding being only from week to week.

labourers belonging to the 1st and 2nd gang, there were 66 head of cattle and two horses and mules.

To the English farmer the proportion both of labourers and of cattle to the land must appear very large.* What is the amount of labour he will naturally ask? a question not easily answered in a satisfactory manner. As a help to form an estimate of both, I shall enter into a few particulars.

And first, of the labourers,—in relation to the quality and amount of their labour. In their ordinary way of proceeding, they commence at six, a.m.; at that time they are expected to be in the field, and excepting an hour for rest and refreshment from ten to eleven, they should keep at their work uninterruptedly till four, p.m. when their day's labour is finished. Nominally in this, nine hours are comprised, but seven, allowing for snatches of rest, are supposed to be a more correct estimate.

Now, as in time of slavery, they work in parties, not classed according to sex, age, and strength, but men and women intermixed, strong and weak, and when using the hoe advancing in regular line, each

---

* Compared with the proportion of labourers in England their amount in Barbados is large, about three in the latter to one in the former. In Barbados 100 acres (judging from the last census and the amount of land under culture) appear to employ 30 labourers, whilst in England it has been estimated every hundred acres employ ten labourers; in France thirty six; in Ireland sixty; the agricultural skill and success in these latter countries, and so, generally, being very much in the inverse ratio of the numbers of labourers employed. Whether Barbados is an exception is doubtful; in some branches of agriculture that island certainly is not an exception, especially in the treatment of cattle and pastures.

party of thirty under the direction of an overseer; "a driver," who formerly was armed with a whip, and when making cane holes aided by a ranger, whose duty it is to see that regularity is observed in the work. If manure is to be carried and applied, the cart or wheelbarrow is not commonly used, or the spade or fork; a basket carried on the head, such as that already described, is the substitute for the former, and the hands or hoe for the latter. If grass is to be cut, the scythe is not used, or even the hook, but the hoe or knife; the hoe, the same as that employed in the ordinary operations of the estate; the knife, a short one, the blade about four inches in length and commonly blunt. Unfit as the former is for the purpose to which it is applied, it has rather the preference over the knife in the instance of sour grass; because when using it, the ground is broken and moved a little at the same time that the grass is cut.*

Next, of the cattle;—their case is very different from that of the labourers. Many years ago they

---

* For the sake of comparison, on one occasion I had the scythe tried against the knife in cutting Guinea grass, the former wielded by an Irishman accustomed to its use, the latter by a field labourer, a negro. The Irishman I found with three strokes of his scythe (he made 38 in a minute) cut about the same quantity of grass as the negro did with his knife in five minutes, denoting that the work done with the one was sixty three times as much as with the other. Another day the same mower with his scythe cut in three hours nearly twice as much grass as the negro cut, working a whole day.

For the scythe to be introduced and labour economized, and the pastures made productive, they require to be treated after the manner of meadow land in this country, instead of being left almost in a state of nature.

were designated as "the worst, because the most abused cattle in the world;"* a stigma I believe it must be acknowledged hardly less deserved now. They are commonly ill bred, ill fed, and ill taken care of, and consequently weak and unequal to hard labour,—numbers having to supply the place of individual power, and many working irregularly,—as is unavoidable,—and consequently with a great loss of power. In ploughing, eight or ten are used with two drivers besides the ploughman; in drawing a cart with no very heavy load—for instance two hogsheads of sugar, which two good horses will draw without being distressed,—about the same number, and with as many drivers, are employed; occasionally I have seen twelve in harness urged on by three drivers.

In the ordinary expenses of an estate a considerable amount is incurred for work done by artificers and tradesmen. The wages of artificers now are commonly about half-a-dollar a day, or a little more; but some are paid at a much higher rate, as coppersmiths; and of

* These are the words used in a valuable paper on the subject of the cattle of the island by John Rycroft Best, Esquire, read at a meeting of the first agricultural society in 1812,—a gentleman who will long be remembered in Barbados for his worth and ability, tried on various occasions, especially in the capacity of President of the Council. He was one of the hapless passengers in the Amazon, which was destroyed by fire on the 4th of January, 1852. His son and successor I perceive, following the example of his respected father, is exerting himself to check the prevailing evil, the bad treatment of cattle: he has proposed the formation of a society for the purpose,—"a society for preventing cruelty to animals," offering a donation of 50 dollars and an annual subscription of 10 dollars in furtherance of the object. See Agricultural Reporter, January 1853.

some the charges are said to be exorbitant, especially if long credit be allowed. During the time of slavery these expenses, it may be inferred, were very much less, each estate being provided with artificers from amongst the slaves.

In addition to all the above sources of expense there are two which must not be omitted, viz. the cost of freight of produce, and of the interest of money, should the necessities of the planter oblige him to have advances made to him. It is to be kept in mind that the freight charges are little variable, whether the market price of sugar is high or low; so too in regard to the interest on loans, which commonly is between six and seven per cent., and never less than five. The risks and losses to which the planter is exposed should not be left out of the account. They are not few, or inconsiderable. Besides the warring elements, from time to time so desolating in their effects, he has other destructive agencies to contend with, blights withering his crops, insects devouring them, and vermin destroying them. In one night an extent of many acres of sweet potatoes have been laid bare by caterpillars: year after year an estate has been rendered almost barren by an insect of the coccus kind attacking the cane, and exhausting its juices: the yam, season after season has suffered from a blight not unlike that in its effects, which of late years has so widely and ruinously attacked the potatoe in Europe: a beetle, the monkey, the rat often commit havoc to no inconsiderable amount, amongst the canes; the one by boring

T

into their substance, the other by gnawing into them, and occasioning their fall;* the third by pulling up the young plants.

What are the profits of the planter? This is a problem most difficult of solution, so many uncertain circumstances being involved in it. The particulars mentioned above show that his expenses both in growing, manufacturing and exporting sugar, the staple produce of his estate, are many and great, so many agents being employed, so much human labour being used, so little regard, in brief, to economy being observed. It was a kind of axiom till lately that the molasses and rum made, sufficed to pay the cost of cultivation, leaving the proceeds of the sugar sold as entire gain. This was the estimate when the price of sugar was held to be remunerative, and was held to be sound even as late as 1846, for so in that year I often heard it announced. Shortly after, with falling prices, the calculation was altered; it was thought necessary to include one third of the price of the sugar in addition to that of the rum and molasses, to cover the cost of production, leaving two-thirds of the selling price as the planter's portion of profit. Later, the price of sugar falling still lower, as after the introduction of slave-grown sugar into the British market, the rate of profit further declined, so much so as to render

* In one year, and that recently, 15,000 rats were killed on Turner's Hall estate, the head-money due from the parish for which amounted to £100. This animal is widely spread over the island; the monkey is now found only in two or three of the larger and least accessible gullies.

it almost questionable whether the cultivation could be continued without loss, and leading to the conclusion approved by common sense and reason, that "unless our staple article be produced cheaper, or more remunerative prices be obtained, we must cease altogether to grow the cane." Such are the words used in the Annual Report of the Leeward district Agricultural Society of Barbados, for 1847, a society consisting of some of the ablest and hitherto most successful agriculturists in the island, and with a very few exceptions, all resident proprietors.

The following is this society's estimate of the cost of producing a hogshead of sugar, weighing 15 cwt. of the crop, an abundant one, reaped in that year, made from the books of four estates of the district which were considered well managed, and by a committee appointed for the purpose.

| | |
|---|---:|
| Amount expended on the cultivation .. .. .. | $3,000 |
| Workmens' accounts, viz. blacksmith, coppersmith, wheelwright, plumber, cooper, repairs of buildings .. .. .. .. .. .. .. .. | 840 |
| Plantation supplies, drogherage, freight, and individual expenses .. .. .. .. .. .. | 1,282 |
| Cost of foreign manures .. .. .. .. .. | 400 |
| Colonial taxes and export duty .. .. .. .. | 150 |
| Keeping up the number of stock.. .. .. .. | 300 |
| Salaries for attorney, manager, book-keeper, and agent .. .. .. .. .. .. .. .. .. | 700 |
| Cost of producing 100 hogsheads of sugar.. .. | $6,672 |

Equivalent this to 66 dollars 72 cents the hogshead ; or making a reduction of $2,000 for the molasses and

rum in addition, equivalent to $46 72 cents, which
in English money, rating the dollar at 4s. 2d. is £9
14s. 8d., or per cwt. 14s.

1847 was a favorable year; its crop, (32,500 hogsheads were exported) exceeding in amount of produce,
that of any preceding year. In the report this is
pointed out, with the remark that were the estimate
made on the crop of the following year, that of 1848,
one less profitable to the planter owing to drought, the
cost of a hogshead of sugar would be $75, 72 cents
(an increase of $29,) or 21s. the cwt.

At the time sugar was selling for ready money
prices, at 4 dollars, or 16s. 8d. the cwt. in Bridgetown
bought up by merchants, and even lower;—its average
for several years previously having been about 7 dollars
or 29s. the cwt.

The following is another estimate made in the same
year, given in the third annual report of the St.
Thomas's district Agricultural Society.

Expenses of an Estate capable of making one hundred hogsheads of sugar on an average of ten years:—

| | |
|---|---:|
| Cultivation, .. .. .. .. .. .. .. .. | $3,000 |
| Manager's salary .. .. .. .. .. .. .. | 480 |
| Book-keeper's ditto .. .. .. .. .. .. | 96 |
| Apprentice .. .. .. .. .. .. .. .. | 32 |
| Book poster .. .. .. .. .. .. .. .. | 64 |
| Workmen's accounts, viz:— | |
| Coppersmith .. .. .. .. .. .. .. .. | 100 |
| Blacksmith .. .. .. .. .. .. .. .. | 60 |
| Plumber .. .. .. .. .. .. .. .. | 30 |
| Millwright .. .. .. .. .. .. .. .. | 50 |

| | |
|---|---:|
| Wheelwright.. .. .. .. .. .. .. .. | 60 |
| Carpenter  .. .. .. .. .. .. .. .. | 60 |
| Mason  .. .. .. .. .. .. .. .. .. | 30 |
| Making 100 hogsheads, including staves, hoops, and nails .. .. .. .. .. .. .. .. .. | 400 |
| Lumber, plank, boards, &c... .. .. .. .. | 200 |
| Stores :—oats, oil, meal, provisions, &c  .. .. | 200 |
| Foreign manures .. .. .. .. .. .. .. | 500 |
| Freight of produce to shipping port .. .. .. | 150 |
| Taxes and export duties  .. .. .. .. .. | 150 |
| Loss upon stock .. .. .. .. .. .. .. | 300 |
| Wear and tear of machinery  .. .. .. .. | 650 |
| | |
| Total expenses .. .. .. .. .. .. .. .. | $6,612 |
| Rum crop, 40 gallons to the hogshead, at 40 cents. | 1,600 |
| Molasses crop, 20 gallons to the hogshead ; 2000 gallons at 20 cents... .. .. .. .. .. .. | 400 |
| Provision crop  .. .. .. .. .. .. .. | 300 |
| | |
| | $2,300 |

Deduct $2,300 from $6,612, and there remain $4,312, the actual outlay for making 100 hogsheads of sugar.

Taking each hogshead at 15 cwt. net, we have, as the actual cost of making 1 cwt. of sugar, $2 80c.=11s. 8d. sterling.

To this add the following charges,—

| | | |
|---|---:|---:|
| Freight  .. .. .. .. .. .. | 4 | 0 |
| Imperial duty .. .. .. .. .. | 14 | 0 |
| Merchant's charges .. .. .. .. | 3 | 0 |
| | Total 32s. | 8d. |

The average price, duty included, appears by the *Gazette* to be 40s. sterling.

" Hence (it is added,) it must be evident that culti-vation, carried on at such prices, must soon cease, as it appears from the above that there is only a surplus of

£5 per hogshead over and above the mere cost of production, which falls far short of the interest of the invested capital.*"

Well might the planters be filled with apprehension at the time and fear utter ruin. "Great has been the distress, (the report of the first-mentioned society states,) which has overtaken many of them on finding that the cost of production had absorbed the returns of their estates and left them without the means of subsistence."

The time was truly a critical one, but the crisis, I trust and believe, was of a favorable and restorative kind, conducing to a sounder state of things, and not without promise of renewed prosperity. Dire necessity compelled certain changes; wages were greatly reduced, on field labour from one third to even one half; salaries were diminished; in many instances proprietors took upon themselves the management of their own estates; a more economical system throughout was adopted with a great reduction of expenditure, even in some instances to the amount of 40 per cent.; and what is creditable to all concerned, especially to the labouring class, with a feeling, it may be said, of general content, arising out of the conviction

---

* According to a statement made by a friend, (it was in March, 1852,) the cost of production of a cwt. of sugar was about seven or eight shillings. The merchant's charge for freight and commission six shillings more, and in consequence, with ten shillings duty it was not landed under 23s. so that when sold,—sold for 28 or 30s. as it commonly then was, little was left for the interest of capital and to compensate for losses by drought and hurricanes and other disasters.

of the imperative necessity of the circumstances. Anecdotes may be mentioned in illustration; I shall relate only one, and no one better deserves to be recorded. A proprietor in the Leeward district who by the liberality and justice of his dealings with his labourers, had gained their confidence and good will, in the midst of his difficulties continued for a time to pay them at the old rate of wages: when he found that he could do this no longer, he called them together, stated his case, and how he was compelled to make a reduction in their daily pay, or to cease to employ them. He was proprietor of three estates one of which he had recently become possessed of. All the labourers excepting those employed on this estate, who were not well acquainted with him, assented to his terms, these declining them, struck work: a conduct that called forth the good feeling of the others, marked by the offer which they presently made, unsolicited, to keep the third estate in cultivation by extra labour and on any pay he could afford to give them. This gentleman who had become what he was from a humble beginning by his industry and skill, and had so gained the hearts of his labourers, was in the habit of paying them daily, and with exactness attainable only by this method, according to the portion of the day that each was employed. Much also of the work on his estates, such as the weeding, cutting of the canes, and the making of sugar was paid for by the job, creating a common interest.

That the crisis of 1847—48 will be salutary, as I

have expressed the hope, and eventually beneficial, there are many circumstances, I cannot but think, of good omen. Besides those already mentioned connected with increased economy, there are others. There is a growing disposition to view agriculture more in the light of a science, and to consider the making of sugar more as a chemical process to be directed by science. There is an increasing demand for skilled labour, and a firmer confidence in the advantages of implemental husbandry, and a stronger conviction of the necessity of extending education to the labouring class. The advantages resulting from agricultural societies are becoming more appreciated; in brief, throughout the community there appears to be an advance, and that founded not on what is specious and may be deceptive, but on what we are sure is sound and should be enduring, viz. greater enlightenment, greater exertion, greater economy.*

* In the Agricultural Reporter, the number for April 1853, mention is made of the more remarkable improvements recently effected in Barbados, in answer to a statement that the planters of Barbados instead of exerting themselves, "sat like a spoiled child crying for the moon"—Protection. It would be unjust not to quote the writer's spirited reply. "What then, and how great are our improvements? To begin with the field; ten years ago the system of farming, or jobbing out fields to the labourers to weed by the week was unknown; now it is almost universally practised; ten years ago the first ploughing match had not come off; now there is scarcely an estate that will admit of their use, in which the plough, grubber, and horse hoe are not daily at work; and to these two improvements conjointly we owe the comparative steadiness of our labour market, the destruction of devil's grass, the beautiful thyme-bed appearance of our fields, and under providence the unprecedentedly large crops which have crowned our efforts. Let us pass to the mill; ten years ago there were not a dozen horizontal mills in the whole island, now it is hardly too much to say that there are as many in every parish, and more are to follow every day. Next, look in at the boiling house; ten years ago there was scarcely a planter in Barbados who knew what a

Particular instances are not wanting, which may be adduced in confirmation.

I shall mention a few, that came to my knowledge. By a change in the system of culture, the adoption of that improved system in use as already mentioned in the Leeward district, estates which for nine years previously averaged a produce not exceeding 37 hogsheads, in 1848 yielded 207 hogsheads.

Even by a more careful culture, a better tillage, better weeding, without any change of system the produce has been greatly increased. On one estate, the proprietor by adopting this method obtained from 50 acres, 180 hogsheads of sugar, 112 more than he had previously procured from 100 acres cultivated with less care, the yield from them not exceeding 68 hogsheads. There was no marked difference in the seasons.*

vacuum pan was, or had any idea of the possibility of evaporating cane juice at a lower temperature than that produced by a roaring fire under an open taiche; now there are four vacuum pans, besides the plant at the Refinery; Gadesden-pans innumerable, and other means and appliances which have been more partially adopted; above all, ten years ago we were unacquainted with those valuable adjuncts to the production of good sugar, Precipitators, and Centrifugal Desiccators; now they are coming so rapidly into fashion, that we shall not be surprised if the man who is unprovided with them next crop is accounted a very slow coach indeed."

* The old maxims, " fæcundior, est culta exiguitas, quam magnitudo neglecta."—" laudato ingentia rura, exiguum colito," are gaining favor, and even in the United States; there we are told that the farmers have come to the conclusion, forced upon them by dear experience, " that superior cultivation on an old soil is an overmatch for the natural resources of virgin land, or of good soil with slight or careless tillage." See " The Journal of Agriculture, and the Transactions of the Highland Agricultural Society of Scotland, No. 28, new series," for some valuable observations on this subject, —the careless, ignorant, exhaustive system compared with the careful, intelligent, and supporting system of farming.

A planter who has distinguished himself by advocating the use of the plough and other efficient implements, and has set the example on his own property, has stated that by substituting the plough for the hoe, work, which with the former cost 30 dollars, with the latter was reduced to seven, and was as well done ; land so tilled had yielded him four hogsheads of sugar. *

On a property on which sugar has been made by the vacuum pan, under careful and skilled management there has been a gain of 25 per cent. The ordinary proportion of juice obtained from the canes is about 50 per cent., by improvements in the mill, the quantity has been increased to 60 without injury to the megass as fuel ; and where steam power has been used, even to 70.

The scope for improvement, of which these are a few examples, seems to be almost unlimited, increasing always with the advance of knowledge, and making progress always if there be energy and industry ready to bring that knowledge into action; and very recently we have had proof corroborative that this influence is not imaginary or overstrained, in the great and increasing produce of sugar that has been obtained season after season.†

* Long, in his History of Jamaica, written nearly a century ago, states that one plough in that island turned up as much ground in one day, and in a much better manner than 100 negroes could accomplish with their hoes in the same time ; the ploughed land yielding 3hhds. per acre, the same hoed, 2hhds.

† During the time of slavery the crop never reached 30,000 (the average of sugar shipped from 1805 to 1833, both years included, was 17,409 hogsheads, 4,740 tierces, 603 barrels.) In the year 1849 and 1850 it exceeded

Let us imagine Barbados an example in point
and the improvements commenced carried further;
science and skill brought to the aid of industry as much
as possible; an educated peasantry, an enlightened
proprietary; no means wasted, no resources neglected!
then, we apprehend, more than existing difficulties
would be got over; her condition would be more secure
and prosperous than at any former period; and what is
more, she would be able to compete with, and prove
the superiority of free over slave labour, and thereby
afford a demonstration of a great truth, viz., that what
is right in principle is right in practice. Even at present indeed, it is a question whether the superiority is
not already on her side. Before I left the West Indies,
I heard in conversation an opinion expressed to this
effect by intelligent planters, and I have now before
me calculations in proof of it, and these made before
the rate of field labour was materially reduced. The
price of land would seem also to lead to the same conclusion. Up to 1846, when their returns were estimated at from ten to thirty per cent., estates sold for
as much in most instances as before emancipation, and
sometimes even at a higher rate; that is when the
slaves on the property were included in the purchase,—
an occurrence this, which formerly would have been
pronounced a thing impossible. That the capital required and invested in time of slavery was greater,

this amount, in 1851 it reached about 40,725, (38,725 hogsheads were exported;) in 1852 the surprising quantity of about 50,000; 48,000 hogsheads
were shipped, and at least 2,000 it is supposed were retained for island use.

cannot be questioned, nor that the risks then were greater; risks of every kind, even of life as well as of property; and this moreover we are sure of, that were slavery to be offered now as a boon to the planters, it would be rejected as a curse.

From the large I shall now proceed to the consideration of the smaller landed properties and their culture. It has been already mentioned that these, ranging from one acre to nine in 1840, were estimated to amount to 1367. Since that time no doubt the number has increased, and it is believed considerably, for the colored race have a great desire to possess land, are ready to pay high prices for small portions, have a peculiar facility in locating themselves on their purchases, and are not impeded by expensive legal processes.* The number, moreover, of small farmers is greatly increased by the system of letting portions of the larger estates, some proprietors being tempted so to do by the high rents they can obtain, varying from twelve to twenty dollars an acre, and occasionally reaching even thirty or forty dollars. They also, but more rarely, let land, not for money, but for labour rent, at the rate of one day's labour per week for a quarter of an acre, or two for half an acre, according to agreement. In this in-

---

* I have known thirty pounds currency, £19 7s. 0d. paid for the one eighth of an acre of shallow rocky ground, which could be made productive only by great labour, viz. by breaking up the rock, and collecting and forming a soil. The houses of the natives being altogether of wood, they are easily taken to pieces and removed from place to place. It is a common practice to rent a small portion of land, on which to place a house, the property of its inhabitants. The registering of the transaction in the Treasurer's office suffices in the instance by sale of small landed properties,

stance a substitute is commonly sent to the planter, the tenant devoting his own labour as more profitable to the land given up to him. The former is only from week to week. The culture of these small properties is remarkable, but more for the industry than for the skill displayed. The hoe is almost the only implement used. The crops grown are extremely various, and the produce commonly large, affording striking examples of the fertility of the soil, under the genial influences of a tropical climate. On one little property, whether its extent be limited to a quarter of an acre as in the instance of the labourer attached to an estate, or is greater, as much as an acre or two, there may be seen growing side by side, or intermixed, almost all the different vegetables which are in request in the island,— the sugar-cane, yam, sweet potatoe, eddoe, cassava, ground nut, and in some parts of the island in addition the cotton plant, ginger, arrow root and the aloe.*

The amount of the produce, and its value, it may be difficult to estimate with any exactness : that it is large, however, and remunerative is certain. The land letting for 20, or 30 dollars an acre, is expected, when cultivated in canes, to yield three hogsheads of sugar, and two or three crops of provisions ; little manure being used, and sometimes none, but a good deal of tillage and

* In 1847, 639 packages of arrow root, 333 packages of ginger, 346 bales of cotton, 559 gourds of aloes were exported in addition to 32,500 hogsheads of sugar;—all but the sugar, the produce of small portions of land.

careful weeding.* Even the small allotments to the labourers yield no inconsiderable quantity of sugar; that of the year 1846—47 amounted to 7000 hogsheads, and from one estate to 25 hogsheads, on which there were 40 holdings, each of a quarter of an acre. The canes thus grown are cut by the labourers and carted by the planter, who undertakes the manufacturing process in all its details returning to the labourers two-thirds of the sugar and molasses made, retaining one-third of each as a remuneration, and the whole of the megass or crushed canes, he supplying fuel for the boiling of the juice. The culture of the aloe, which is confined to the small farmer entirely, and carried on chiefly in the parish of St. Philip, towards the sea shore where the soil is scanty and dry, at times has been very profitable, for instance when the inspissated juice has sold for £30 the cwt. Even now that it is reduced to about 24 dollars, or a little more than five pounds, it makes a good return, the produce of an acre of land being about 140 lbs. This information I had from an intelligent man who said that in addition to this yield, he had from his ground in aloes, three crops of Indian corn in the year, besides yams and potatoes. The aloe plants require to be renewed, about every fourth year. Cotton, which was once largely grown in Barbados,

---

* An instance was mentioned to me of a portion of land in the Scotland district that had for thirty years afforded good crops without manure, so good that the proprietor did not think it worth his while to be at the trouble of using manure, which, close at hand, was offered him, for the carrying it away. The soil of this little property was a calcareous marl.

when its selling price was occasionally so high as
1s. 8d. per lb., is now very partially cultivated, even
by the small farmers, and that principally in the poorer
soils of St. Philip's and St. Lucy's. Probably the time
will come when its culture will be extended, and be
resumed in those parts of the larger estates, which
from the nature of their soil and situation, are better
adapted for it than for the cane. Were the finer varieties, those that fetch the highest prices carefully grown
and the most made of its seed and cuttings in the feeding of cattle, and the making of manure, it would
probably prove a fairly remunerative and useful crop.
It is not an exhausting one, provided all but the cotton
wool, in which are no inorganic elements, be restored
to the soil.*

The small farmers, and the labourers who have
a portion of ground, have in addition to their varied
crops, stock of some kind, such as a cow, a bullock,
one or two goats or sheep, a pig or two, not unfrequently a horse, and are rarely without poultry.
The number of animals they keep is often beyond
their means, tempting them to pilfer for their support,

---

* A friend writing from Barbados informs me that when cotton was
largely cultivated in Barbados, three kinds were usually grown, the Persian,
the Vine, and the Great Lock or coarse cotton. He adds, " The two former
are very superior in quality. May and June are the planting months;
January, February, and March the reaping. The Persian, and the Vine,
yield about 300 lbs. per acre of clean cotton, the coarse about 500 lbs. The
plant" (in accordance with what is stated in the text and with general experience ;—seeming to forbid a far inland culture of this crop) " thrives best in
land near the sea coast, where the cane does not grow well; in the interior
of the island where the cane grows best, cotton does not thrive."

and commit depredations on the adjoining estates. It is not even uncommon for persons without land, not only to keep poultry, but also a sheep, or a goat, which by day are tethered by the road side, and at night too frequently let loose in an adjoining field. The tethering plan here is very much in use, it is even applied to poultry to prevent their wandering; a tethered hen with a young brood of Guinea fowl, is no unusual sight. The goat and sheep are more in request than the cow, feeding more on the leaves of shrubs, and more manageable and prolific. Both are said to breed in Barbados, twice in the twelve months. This I cannot confirm from my own observations. A goat that belonged to me, had young twice in eighteen months. The milk of the sheep as well as of the goat, is used, and the ewe by management may be kept in milk eight or ten months.*

* It is interesting to see the effects of climate on the former, how in two or three generations, its woolly coat disappears for one of hair like that of the goat, more suitable to a tropical climate.

The effect of climate is supposed also to be witnessed in the toad of Barbados, and undoubtedly would be, were we sure that it is identical with our common English toad, (bufo vulgaris) and was originally imported from Europe. But though there is a considerable resemblance, so much; that they have been pronounced by a naturalist, (to whom specimens were sent,) to be identical; the probability is, that they are different species; all the reptiles of the American continent hitherto examined by competent persons having been found distinct from their European congeners. The Barbados toad is far more active than an English toad, is of a less robust make, of a lighter and brighter color, and the viscid matter, the secretion of its glandular skin, is I believe more copious and a more powerful poison than that of our species. It is worthy of remark that this toad was unknown in Barbados twenty years ago: it was imported from one of the French islands, and so rapidly has it spread, that now it is everywhere common; feeding entirely on insects, it is held to be, and probably justly, beneficial to the planters.

The subject of small farms and the letting of portions of estates to labourers has received a good deal of attention and has been much discussed by the planters. They have commonly expressed themselves opposed to both, especially the latter, and more than one of the agricultural societies have come to resolutions declaratory of this feeling, founded chiefly on the belief that such farms and lettings have a tendency to withdraw labour from the larger estates and to deteriorate agriculture. The fears entertained by them it may be admitted would be well founded, were the system to be extended greatly, as that of sub-letting was in Ireland ; at present, however, there seems to be little occasion for alarm. In moderation, small farms, it may be held, are not without their use, and might be made useful in a high degree were they judiciously cultivated, chiefly for growing vegetables, provisions and fruits, of which, especially fruits, there is a scarcity and want in Barbados, as well as of all but the most common roots. Under garden cultivation the fertility of the soil is tested, the industry of the holder is stimulated ; an opportunity is offered to the industrious and intelligent to better their condition, thus introducing an element, the tendency of which in operation, is to improve and elevate the peasantry, and create grades in the social body, in itself favorable, to the interest and strength of the whole. "Wherever you find hereditary farmers, or small proprietors, there you also find industry and honesty;"* so writes the great historian Niebuhr, reflect-

* Life and Letters of G. B. Niebuhr, Vol. ii. p. 149.

X

ing on the wretched condition of the Roman peasantry, and the evils arising from vast landed possessions ; and here—in Westmoreland,—we can most fully adopt his conclusion,—a county once abounding in small proprietors, " statesmen," distinguished for the good qualities referred to, now every year becoming fewer, to the deterioration, it is believed, of the character of the people.

The same is witnessed in Scotland. " By the absorption of the smaller farms the labourer has been deprived of the means of securing that humble independence in old age to which he had been formerly accustomed to look. In some cases his spirit has sunk, and he has become an unreasoning serf; in others, that ambition which is the failing of noble minds, whether they be found in a palace or in a hovel, curbed and crushed, but unsubdued, has filled him with jealousies of those above him, and made him lend an easy ear to the discontented demagogue. Throughout the mass of the population a feeling of estrangement between the upper and lower classes has at the same time been engendered, injurious to the interests and happiness of both."*

---

* See "a memoir of Annie M'Donald Christie, a self-taught cottager," the grandmother of those remarkable men John and Alexander Bethune, and appendix, in which are some notices of them and extracts from their writings.—

Is not the same absorption of the smaller into the larger farms to be apprehended in Ireland, if it have not already taken place? It is worthy of remark that neither in the West Indies, nor in Scotland, nor Ireland is there any "common," common-land, as in England, so favorable to the comfort and well-being of the labouring class!

Besides the subjects hitherto treated of, there are others, and not a few, even in this small island, which in an account of it should not be altogether passed over, so rich is it in objects of interest. A notice of some of its principal institutions and of its chief town, I shall reserve for an after-part of the work, on the towns and institutions of the colonies generally. At present I shall restrict myself to the pointing out a few of the objects and scenes most worthy of the attention of the traveller, and in visiting which he will have an opportunity of becoming acquainted with those parts of the country also, which are most interesting.

"The burning spring."—On Turner's Hall estate in the parish of St. Andrew, in a lofty situation there is a small portion, a solitary instance, of primeval forest remaining.—Adjoining it, just on its skirts in a water course, is a jet of gas, of the nature of coal gas, as I have ascertained, derived probably from a bed of coal or asphaltum, which from its property of kindling when a lighted candle is applied to it, and burning with a flame, is known by the name of "the burning spring." It is considered one of the sights of the island, and is shewn in an amusing manner by one or other of a negro family residing in a cottage hard by. The excursion to it on horseback through a considerable part of the interior,—its central, hilly, and table lands,—by way of Vaucluse, Pory Spring, Dunscomb, Mount Hillaby, Greg's Farm, will amply repay.

Pory Spring is a charming little region. There besides the never failing source of excellent water, whence the name,—there is a luxuriancy of vegetation and abundance of fruit trees,—clumps of the graceful bamboo, and almost groves of the orange and lime,—which one could wish were more common in Barbados.

Mount Hillaby.—This, the highest hill in the island, and from whence a large portion of it may be seen, especially of the Scotland district, affords an instance of the junction of the two principal classes of rocks or geological formations of which Barbados consists, coral rock being found within a short distance of its summit, and the summit itself being composed of the cretaceous deposit, rich in infusorial remains. It affords also a happy instance of productive agriculture, its slopes, even to its very top being under cultivation, let to labourers, when I was there, at the rate of twenty-four dollars an acre, and yielding rich crops both of the sugar cane and ground provisions.

Bissex Hill.—This hill, one of the stations of the mounted Police,* situated in the parish of St. Joseph, and almost in the centre of the Scotland district, though lower than Mount Hillaby, commands even finer views of the adjoining country, and is interesting in itself from its geological structure already adverted to. Approached by way of Blackman's, Castle Grant, and Sugar Hill, some of the finest parts of the interior are traversed. It is a peculiarly agreeable ride or walk

---

* This useful body is composed entirely of natives, chiefly half castes, in every way adapted to the service, being commonly intelligent, sober, active, and healthy, exempt in a great measure from those diseases of climate which have so often decimated white troops even in Barbados.

from the variety and pleasing character of the scenery through which the road leads in ascending, and from the refreshing coolness of the air; and this (if using an umbrella as a protection from the direct rays of the sun) even in the hottest weather.

The Animal Flower Cave.—The cave so called from an elegant species of polypus, one of the membranous kind, an *actinia*, (sea anemone) which is to be seen in it, inhabiting a pool of water communicating with the sea, is situated in the bold rocky shore of St. Lucy's, beyond Spight's Town. The whole of this portion of the coast has a peculiar character, as has also the country bordering on it, lying between it and the first terrace elevation, the one being so worn by the sea, and cavernous, the other so low, flat, rocky and naked, and in its barrenness remarkably contrasted with the adjoining part of the same parish which is distinguished for its fertility and its advanced agriculture. It is worthy of note that whilst at the foot of the inland terrace there are indications, as already noticed, of an ancient sea beach, there are on the present shore not less clear marks of the sea encroaching on the land, suggesting the idea that at some remote future the old beach may be again washed by its waves.

Cole's Cave.—The cave so called is situated inland in the parish of St. Thomas, between seven and eight hundred feet above the level of the sea, and is the only one in the island anywise remarkable. The entrance to it is on the estate called the Spring, by a chasm like a shaft, or the opening of one of the great natural drains, designated here a "suc." Its recesses are of vast extent, some of them have never yet been explored. Its walls in some places lofty and wide asunder,—oftener low, and contracted,—are naked except where ornamented with stalactites. Its floor everywhere rugged is in part dry, and in part the bed of a noisy and rapid stream. Its atmosphere, even in its innermost recesses is cool and fresh; its temperature when I visited it, was 77°, which was that also of the running water. A region of darkness, though in other respects favorable to vegetation, it affords no traces of vegetable life, nor, I may add, of animal life, with the exception of innumerable bats which make its roof their resting place, and a few crustacea said to be found in its stream. *

Whilst there is an almost entire suspension here of organic activity, the mineralizing processes are in constant operation. Not only are stalactites and stalagmites in continued course of formation, but also it would appear of change from molecular action, some specimens, when broken, exhibiting the form of Iceland spar, some

* In some of the great caves of North America, and of Southern Austria, eyeless animals have been found, and of two or three different classes. Diligent search in this cave, it is not improbable, might lead to the discovery of some insects at least; and whether eyeless or not, the results would not be without interest, especially should extended enquiry prove that the blind species are not otherwise peculiar, and that duration of time alone is required to produce a wasting and finally a loss from disuse of the visual organs; and if so, the inmates of dark caves would become in a manner geological chronometers. The crustacea, as well as the bats, may be only part of their time in the cave, and consequently, little is to be inferred from the state of their eyes.

of granular marble.* Those who visit this cave, especially if they wish to explore its depths, should be well
provided with wax lights, and should forbid the burning of wood and cane leaves, which the attendants
commonly provide, and which unless thoroughly dried
yield ordinarily more smoke than light, and burn only
for a short time.

Skeet's Bay.—This bay is in the parish of St. Philip. It is interesting on account of its geological phenomena.  There are to be seen some of the strongest
indications of the powers which have acted in the
formation and elevation of the island ; for instance,
coral rock, under water, in the act of being produced
by organic growth ; the same rock with sandstone containing sea shells, not yet fossilized, and many of them
even retaining their color, elevated a few feet, resting
on clay, forming the low boundary cliff, and in addition
a bed or stratum called rotten stone, which possesses
all the characters of volcanic ashes.   In going to this
bay from Bridgetown by the direct road by Oistin,
passing through the parish of Christ Church into St.
Philip, a large portion of the south-east quarter of
the island is traversed, the greater part of it rugged
and low, well fitted by the dryness of the soil and its
nearness to the sea, for the cultivation of cotton and
aloes, which formerly were the principal crops of the

---

* In the xl. vol. of the Edin. Phil. Journal, edited by Professor Jameson,
a fuller notice is to be found of the changes, and to Sir Robert Schomburgk's
History of Barbados, I may refer for an ample account of the cave generally.

district. A village called "the Three Houses," close to
the road, is the reputed place of abode of the earliest
settlers. A fine spring of water gushes out there
which in seasons of drought applied to the irrigation
of the adjoining cane fields, has preserved their fertility.
The soil in the neighbourhood is calcareous marl of the
best quality, different from that of the parish generally.

Pico Teneriffe.—This singularly formed hill is in the
parish of St. Lucy close to the sea shore. Its summit
is a mass of coral limestone, resting on chalk, tapering
almost to a point like a capped pyramid, the former
projecting over, and sheltering and defending the latter.
The cretaceous matter which forms the principal portion
of the hill, indeed all but its head, is peculiarly rich in
various forms of microscopic organisms. In visiting
the pike, if the road be taken by Hole and Spight's
town, along the shore diverging beyond the latter, and
crossing by Lambert's, the greater portion of the Leeward district will be seen,—a district for most part
lying low, with few inequalities, except its deep transverse gullies, breaking its inland terraced heights, and
distinguished now for its productiveness, the result
mainly of good culture. At Lambert's may be witnessed, the first attempts made in the island to introduce thorough draining by the enterprising and enlightened proprietor of that estate.

Codrington College.—Of this nobly endowed college,
the only one in the West Indies, I shall at present
merely advert to its site. By nature, it seems as it
were designed for a studious retreat, so retired is it and

secluded, lying apart in a little hilly district of much
and peculiar beauty, sheltered by an inland cliff, and
yet open to the sea, with the advantage of good air
and abundance of water, even in excess, so as to be
applicable for ornament as well as use, the spring, a
never failing one, supplying two little lakes or ponds
in front of the college building, and afterwards in its
descent, when needed, being made to irrigate the lower
cane fields and provision grounds. The way to the
college by St. John's Parsonage is through the most
charming part of the island, the greater part of the
road being along an elevated ridge, little less than
eight hundred feet above the level of the sea, skirting
and looking down upon the fertile upland valley deservedly called the "Sweet Bottom." From the church of
St. John's standing almost on the verge of the great
coral inland cliff, the prospect is of the finest kind,
marked equally by wildness and beauty. Not far
below, on the Newcastle estate, is a small stream called
the Collington, the water of which has a petrifying
quality from the large proportion of carbonate of lime
held in solution in it. In an umbrageous spot, where
it is pent up between steep banks, and where it forms a
little cascade with a fall of about twelve feet, the only
one I believe in the island, there the loose roots of the
shrubs and trees washed by it or exposed to its spray,
have become incrusted, and occasionally exhibit singular forms; one specimen now before me, very much
resembles a vertebral column, being jointed and tapering—a jointed stalactite,—its joints loose, connected

Y

only by the root, round which the incrustation has formed in this singular manner.

These notices might be greatly extended, for I know no island of the same dimensions so abounding in natural objects deserving of note.  Those who visit Barbados,—to others they can be of little interest,—will find a good directory in Sir Robert Schomburgk's ample history.

# CHAPTER V.

## ST. VINCENT.

THIS island, the nearest of the Antilles to Barbados, about 78 miles distant due west, and within sight of it when the atmosphere is clear, is yet in most respects remarkably different, so as to afford an example of striking contrast.

In structure, geologically considered, it exhibits none of those peculiarities which are so conspicuous in the greater portion of Barbados. We in vain look for any traces of organic remains,* such as rocks of coral, and shell limestone, and of infusorial chalk; and equally in vain for any marks of the modelling and transporting agency of submarine currents. Even the coasts and shores are not an exception. The former are not like

---

* It is said that petrified wood is occasionally met with in the northern part of the island, and that there a tufa occurs in which there are casts, or impressions of leaves, but if true, these can hardly be considered as exceptions.

those of the sister island, guarded by outstanding coralline reefs,* nor are the latter, except in one instance,
and that very partial, strengthened by sandstone now
in the act of forming.†

St. Vincent may be pronounced to be essentially volcanic. To this its origin, it owes all its peculiarities;
the forms of its hills and mountains, often rising into
cones; the shapes of its vallies, whether circular or crater-like, opening by an abrupt gorge, or narrow and
contracted fissure-like, gradually descending and expanding; the quality of its rocks, all igneous, whether
crystalline, such as its basalts, or aggregates of sand-
stone-like formation, such as its tufas.

It has been called, and justly so, the most beautiful
of the West India Islands. To this its origin, likewise,
it principally owes its beauty. Rising out of a deep sea,
remote from any great river, it is surrounded by a surface of water of the pure blue of the ocean; this is one of
the elements of its beauty. The others can easily be
imagined, however inadequately, in the features of its
scenery; mountains clad with native forests, sufficiently
high to reach the region of the clouds;‡ hills and val-

---

* At the mouth of the little bay of Calliaqua it is said there is a coral
reef, if this be an exception, I heard of no other.

† This solitary example I observed on the shore a little to the eastward
of Old Woman's point; it consisted of black volcanic sand, cemented by
carbonate of lime.

‡ The Morne, or Geru group or chain of which the Soufriere forms a
part, about 3000 feet high, the grande et Petite Bonne Homme, of a fine
conical form, wooded to their summits, little inferior in height,—Mount St.
Andrew's of the same form and similarly wooded, are some of the most
conspicuous.

lies whether wooded or cultivated, ever verdant; with variety in all, whether mountain, hill, or valley, in form and coloring more than sufficient for picturesque effect. The points of difference do not end with its physical condition; they are hardly less remarkable in its civil and political history, in its population and agriculture. When Barbados was most flourishing, about the end of the 17th century, St. Vincent was nearly in the same state as when first seen and named by Columbus, almost uncultivated, and inhabited almost exclusively by wild Caribs. Little beyond the commencement of the present century have we had peaceable possession of it. For a series of years, viz. from the middle of the preceding century almost to its termination, it was the scene of war; first between the French and the English, and last between the English and the Caribs, aided by the French. So late as 1793* it was formally and finally ceded to Great Britain:† as late as 1795—96 was the great struggle with the natives, that which terminated in their subjection, and the removal of the greater number of them to a distance, to Rattan or Ruattan, an island in the bay of Honduras.‡

* At this time, according to Leblond, when the English became masters of the island, the Leeward portion of it alone was known to Europeans; he describes it as the only part inhabited, and at all cultivated, its population consisting of 800 whites, of more than 200 free men of colour, and about 3000 slaves. *Voyage aux Antilles, etc. commencé en 1767 et fini en 1802 par J. B. Leblond, Paris,* 1813.

† It had been ceded before in 1763.

‡ See for the details of this war, its beginning and end " An historical account of the island of St. Vincent, by C. Shepherd, Esq.," London, 1831. The number of natives removed of all ages and both sexes was 5080; they were chiefly those called black Caribs of African or mixed descent. The few

This period, after the destruction of the sugar estates in St. Domingo, was, as has been already pointed out, an auspicious time for our West Indian Colonies, and especially St. Vincent. Till then, some of its finest land,—a considerable portion of the windward district, the Carib country,—had not been appropriated.* So rapid was the progress then made in cultivation, that in 1800 the sugar produced in that year amounted to 16,518 hogsheads, a quantity exceeding considerably that which of late years has been obtained since slave emancipation, and was inferior only by three or four thousand hogsheads to the maximum yielded in the time of slavery. To have a just idea of the exertions made, and the capital expended during this period, it is almost necessary to visit the island. Many of the works,—they may be called enterprises,—then accomplished, are of a magnitude, and were effected at a cost to excite one's surprise: I allude especially to the manner in which water has been conveyed to some estates

allowed to remain had a portion of land conditionally assigned them, on the Windward and Leeward coast towards the northern extremity of the island, where the small remnant of the race is still located.

Mr. Shepherd in a note remarks, Baron Humboldt in his personal narrative, vol. vi. p. 32 says, " These unhappy remains of a people heretofore powerful were banished because they were accused by the English government of having connexions with the French."—justly adding, " should the learned foreigner ever meet with this (Mr. Shepherd's) humble narrative, he will be enabled to give a better reason for the punishment inflicted on them."

* The land taken possession of by the crown after the expulsion of the Caribs was divided into lots, and either granted to individuals as a reward for services, or sold at a fixed price per acre. At the last sales made in 1809 the price of cleared land was £22 10s. per acre. Similar sales had been made previous to 1777 when the island was first ceded.

by aqueducts of solid masonry; or in which roads have
been carried along the face of precipices by laborious
cuttings, or through them by tunnels; and embarking
places have been formed by jetties run out in a bois-
terous sea.

Even in these particulars, the works referred to being
partial and limited, the contrast does not fail between
the two islands.   Whilst in Barbados, the greater portion of the country is intersected by good carriage roads,
and cultivation is carried to the summits of the highest
hills, in St. Vincent it may be said there is only one
line of road passable for carriages and carts, and that
confined to the coast and yet not surrounding it;—on
one coast, the windward, not exceeding 27 miles in
length; on the other, the leeward, extending only to
about 24, and this latter little better than a bridle path.
And its cultivation is so restricted, as with one or two
exceptions, to be confined to a belt also bordering the
coast, of variable depth, on an average not exceeding
one mile and a half, and limited to about three-fourths
of the circumference of the island.

There is one more point of difference between
the two islands not unworthy of mention,—that which
relates to tradition and remote times.   In Barbados
there are no traditions, no vestiges of the remote
past; its history is entirely recent.   There, there
is not a spot which bears other than a modern
name.   In St. Vincent, not only is there a remnant of the ancient race, but likewise the majority
of the more conspicuous spots, and those most resorted

to, bear the names which were assigned them before the island was known to Europeans, mixed indeed with French and English names in a very characteristic manner, in themselves as it were a history; and there are objects of like significance, such as rocks, on the face of which rude figures are cut of an unknown antiquity. Even the recent history of the island has a superiority of interest, arising out of the many hard-fought battles and incidents of war that occurred before the island became an exclusive possession by right of conquest. The traditionary record of the more memorable actions, memorable in a small way as feats of arms, is not yet lost, nor are the localities, the scenes of them, yet forgotten.

The island is commonly estimated as being about eighteen and a half miles in length, eleven miles in width, where widest, and as comprising an area of about one hundred and fifty square miles.*

The brief information already given of its geology, may almost suffice. It may be remarked further, that though the rocks, so far as I had an opportunity of examining them, may all be considered either igneous or volcanic, there is a great variety amongst them; and that in consequence of this variety, and the resem-

---

* Of the total area of the island, supposed to be about 150 square miles, 100 have been considered as waste, and for the most part untenanted,— crown lands: the right to these was acquired from Mrs. Martha Swinburne, in 1786 by a payment of £6000 in accordance with the vote of parliament. In 1780, when the island was in possession of the French they, "*les terres incultes, vagues, et non concedeés,*" had been granted to this lady by a royal ordinance.

blance of many of them to our green stones, and others of the trap formation, they are very deserving of being studied. Fine examples of columnar basalt occur on the leeward coast, especially in Cumberland valley. Lava, sometimes scoriform, sometimes basaltic, the one passing into the other, is frequently to be met with. Tufas, resembling sandstones, and conglomerates, and commonly called by these names, are even more common. Many of the inland precipices and sea cliffs, " bluffs " as they are named, consist of them, either alone or alternating with lava, and most frequently stratified. The finer tufas are composed of volcanic ashes; the coarser of lapilli, of ashes and fragmentary masses. Some have considerable compactness; others little coherence. One kind called " terras " of a red color, has the character and properties of pozzolana, and has been found to answer well as a substitute for the Neapolitan in the public works carrying on in Bermuda.

On the leeward coast, where there is the greatest boldness as well as beauty of scenery, the igneous rocks, basalt, and basaltic lavas and other varieties, most abound. On the opposite coast, where the scenery is tamer, where gentle slopes and extended declivities gradually descend to the sea, there the tufa and " terras " formation is most prevalent. The aspect of the one is such, as if formed under more violent agencies than the other; the leeward side, for instance, as if the result of the outpouring chiefly of volcanic matter, in overwhelming quantities, in a state of fusion;

the windward, as if, on the contrary, from the fall of ashes and of the finer lapilli, and that at a time of comparative atmospheric calm. Such were the ideas that occurred to me on the spot, where it was almost impossible not to indulge in speculation relative to the causes in operation to which such diversity of effect was owing, and the more so in conjunction with the feeling of wonder and admiration that beauty of the highest order, and fertility, should be the result of agencies, at the instant only destructive.

Of minerals,—distinct mineral species, there is a remarkable scarcity; the only ones I have seen have been hornblende, augite, and olivine, and a glassy felspar. The best examples I met with of them were in some fragments of rock lying high up on the side of the Soufriere mountain, and near the edge of the old crater. The absence of quartz, whether crystallized or amorphous, is remarkable, especially considering that silica in a chalk-like and marly form is common, occurring often in nodules, and thin layers in fissures. The absence too of calc-spar, and of carbonate of lime, in all its various forms, is also worthy of note.

Of mineral waters, the only kind hitherto found in the island, is an effervescing chalybeate, of which there are several springs, most of them very weak. Their principal constituents are carbonic acid, carbonate of iron, and carbonate of potash and silica; all with the exception of the acid gas in very minute proportions. I state these results from the examination I made of one that is most used, and is in estimation as an agreeable and wholesome beverage. So weak was it in its mineral contents, that its specific gravity was nearly the same as that of distilled water. It may appear singular that in this so distinctly volcanic island, no hot springs are known to exist.

The soils of the island are in accordance with its geological structure, offering additional contrasts compared with those of Barbados. They may, I believe, be described as to origin generally, under two heads, either as composed principally of volcanic ashes, such as are the lighter soils, especially those of the windward district; or of clay or loam derived from the decomposition of the igneous rocks, mixed more or less with ashes, such as are the stiffer and more retentive soils which are common on the leeward coast. From the partial examination I made of some specimens, I was led to infer that they contain very little vegetable matter,—even the virgin soil of the forest land that had never been broken up, contained hardly one per cent.,—that there is a notable portion of alkali in them, and, with the exception of but little lime, those other ingredients common to good soils, such as silica and alumina, with a proportion of magnesia, and traces more or less of the phosphates, with oxide of iron as the coloring matter. Were careful and proper enquiry made, such as the importance of the subject demands, probably great variety of soil would be found, depending on difference of composition, the knowledge of which might be of essential service to the planter.

The climate of St. Vincent, as a tropical climate, is

good; indeed it has marked excellencies, rendering it
equally favorable to agricultural fertility and success,
and with a few exceptions, not less so to the health
and comfort of its inhabitants. It may be pronounced
generally to be cool and equable; the thermometer
even on the leeward coast seldom by day rising above
85° or 86°, and scarcely at night falling below 70°
On the windward coast the atmosphere is cooler by
several degrees, even at the level of the sea, and not
less equable. Of course at various heights different
degrees of coolness may be obtained. At the Soufriere,
when I ascended that mountain in the month of April,
the thermometer at 3 p.m. on the edge of the crater
was 63°, and at 1 p.m., when about half way up the
ascent, 67°. A further excellence of its climate, and
no doubt its greatest in relation to agriculture, is its
exemption from protracted droughts, owing, it may be
inferred, to its mountainous character. I have before
me a register of the rain guage kept for six years,
from 1824 to 1829, both years included; and for the
longer period of 17 years, viz. from 1835 to 1851, in
Kingston, and part of the time at Fort Charlotte in the
neighbourhood. It is remarkable that in the whole of
this period, extending to twenty three years, not a
single month is to be found without rain. The
agreement of the two registers is also remarkable both
as to monthly, and of course, the average yearly fall of
rain. In the first of the two following tables, this
agreement is shewn; in the second, the monthly fall,
in each of the last six years.

FIRST.

| *From 1824 to 1829.* Monthly mean. | | *From 1835 to 1852.* Monthly mean. |
|---|---|---|
| January .. .. | 4.08 inches | .. .. .. 4.08 inches |
| February .. .. | 3.67 ,, ,, | .. .. .. 3.01 ,, ,, |
| March .. .. | 3.09 ,, ,, | .. .. .. 2.51 ,, ,, |
| April .. .. | 2.32 ,, ,, | .. .. .. 3.54 ,, ,, |
| May .. .. .. | 5.20 ,, ,, | .. .. .. 4.99 ,, ,, |
| June.. .. .. | 10.17 ,, ,, | .. .. .. 9.74 ,, ,, |
| July .. .. .. | 8.94 ,, ,, | .. .. .. 8.20 ,, ,, |
| August .. .. | 8.83 ,, ,, | .. .. .. 7.99 ,, ,, |
| September.. .. | 7.65 ,, ,, | .. .. .. 9.24 ,, ,, |
| October .. .. | 9.58 ,, ,, | .. .. .. 8.82 ,, ,, |
| November.. .. | 8.02 ,, ,, | .. .. .. 8.74 ,, ,, |
| December .. .. | 6.59 ,, ,, | .. .. .. 8.08 ,, ,, |
| | 78.14 inches. | 78.94 inches. |

SECOND.

*Monthly fall of rain.*

| | 1846. | 1847. | 1848. | 1849. | 1850. | 1851. | 1852. |
|---|---|---|---|---|---|---|---|
| January .... | 4.45 | 6.24 | 2.54 | 2.75 | 2.43 | 4.67 | 5.30 |
| February.... | 1.19 | 1.12 | 2.75 | 1.82 | 3.90 | 2.52 | 1.85 |
| March ...... | 1.74 | 2.78 | 2.00 | 2.20 | 1.61 | 2.00 | 1.95 |
| April ...... | 3.59 | 3.03 | 2.00 | 4.68 | 3.21 | 1.75 | 5.93 |
| May ...... | 2.01 | .87 | 6.86 | 6.39 | 4.44 | 4.90 | 3.22 |
| June ...... | 12.50 | 4.83 | 6.69 | 11.99 | 20.56 | 12.50 | 4.05 |
| July........ | 15.48 | 2.82 | 13.10 | 6.35 | 6.87 | 10.62 | 7.42 |
| August .... | 9.17 | 5.46 | 9.12 | 7.03 | 11.35 | 9.36 | 7.14 |
| September .. | 15.49 | 12.36 | 8.11 | 9.83 | 5.54 | 10.23 | 3.00 |
| October .... | 7.85 | 5.19 | 15.81 | 17.82 | 7.07 | 11.76 | 5.58 |
| November .. | 5.81 | 8.37 | 5.90 | 5.68 | 12.48 | 8.44 | 9.93 |
| December .. | 13.34 | 8.82 | 13.57 | 7.17 | 9.97 | 6.90 | 11.93 |
| | 92.62 | 61.89 | 88.45 | 83.71 | 89.43 | 85.65 | 67.30 |

On the windward side, it is believed that the quantity of rain that falls is greater than on the leeward, and on the higher than on the lower grounds, and most of all on the mountains, in accordance with what is best known on the subject of rain.

As regards salubrity, the windward coast is in greater estimation, and I believe deservedly so, than the opposite coast, especially the close and narrow vallies there situated. In them fevers are more or less prevalent, and there on some of the estates the Portuguese immigrants have suffered severely, at the same time that others of the same people and English labourers retained their health when located on estates on the windward coast.

Owing to the character of the country and the ample supply of rain, streams of rapidly running water abound, adding not a little to the beauty of its scenery, and affording the advantage of a moving power to the planters, of which they have availed themselves in many instances. There are twenty rivers of sufficient size and force to turn water mills. In the larger streams the pleasant recreation and exercise of fly-fishing may be enjoyed, a species of mullet being common in them, which, rising freely at the artificial fly, and being strong and active, affords good sport to the angler ; it is moreover delicate eating.

Except in the old crater of the Soufriere there is no lake in the island, or any piece of stagnant water worthy of mention. The lake of the Soufriere is peculiar from its situation and all its attending circumstances. It reminded me of the lake of Albano, and of some of the mountain tarns of Westmoreland. The circumference of the hollow of the crater is said to be about three miles, and that of the lake can be little less, as the descent to it is rapid, in some places perpendicular, and altogether only about five hundred feet. Its depth is reported to be about nine hundred feet. When I visited the mountain, though the air was nearly calm where we stood looking down upon the water, its surface we saw was agitated by a strong wind. Thus moved, its color was pure blue, finely contrasted with the bright green of its inclosing walls,—a hue which these owed to the small shrubs and plants which clad them wherever they could take root.

Like Barbados, St. Vincent is subject to hurricanes, but there is reason to infer in a less degree. In "a chronology of the most remarkable events relative to the West Indies," appended to Mr. Shepherd's work on this island, its name is not once mentioned in connexion with such a catastrophe, nor in the extended list of hurricanes given in the appendix to Sir Robert Schomburgk's history of Barbados, excepting in one instance, that of the 10th and 11th of August, 1831, from which it is stated that St. Vincent experienced great damage.*

Earthquakes have occasionally been felt in the island; but there is no instance on record of any damage to buildings resulting from them. The vol-

---

* An interesting account of this hurricane is to be found in Mr. Carlyle's Life of Stirling, contained in a letter from him, who was then residing there. The losses sustained were estimated at £163,420; nineteen vessels were driven on shore in Kingstown bay.

canic eruption which occurred on the 30th of April, 1812, finding vent by a new crater, little inferior in size to the old one, and contiguous to it, and which proved so destructive, not by its lava or fire, but by the ashes discharged, has been so often described, that it is unnecessary more than to advert to it.

With all its natural advantages the population of the island has increased slowly, and is scanty in relation to the extent of its surface and the land under cultivation. In 1844, according to the census then made, it amounted to 27,573. In 1851, the date of the last census, it did not exceed 30,122, notwithstanding that in the interval not fewer than 2,874 immigrants, natives of Madeira and liberated Africans, had been introduced ;* which is in the proportion of only about 208 to the square mile. Of the total in 1844, the number of whites was returned as not exceeding 1268 ; at the present time it is not probably greater ; this can only be conjectured, as the distinction of color is not preserved in the tables.

On the condition of the people and the state of society I shall offer but few remarks, many of those made respecting the several races in Barbados being hardly less applicable here.

First, of the people of color constituting the great

---

* The Lieut. Governor has expressed his opinion that the real amount is greater, not less than 32,000. Be this as it may, in considering the amount, allowance should be made for the loss by small pox, which broke out in 1849, brought by immigrants from the coast of Africa, and prevailed till the middle of the following year, proving fatal to from 900 or 1000, or to about ten per cent. of those attacked.

majority of the whole. Their condition unquestionably has vastly improved since the time of slavery, and is improving. Very many, even of the labourers, can read and write, and the number of such is daily increasing in consequence of the attention paid to their education. Compared with the same class in Barbados, perhaps on the whole they are better conducted and not less industrious. Fewer in number and more in request, they are more independent, and more disposed, being more able to enforce their own terms. The character given them by the late Lieut. Governor, Sir John Campbell, after a residence amongst them of several years, and his account of their advance is of a very satisfactory kind, and in justice I am induced to quote it. After noticing favorably the energy of the planters in their difficulties, and the merchants, contented with smaller profits, exercising greater economy, he continues,—"But when we turn to the great body of the native population, it is beyond all dispute that it has been the subject of progressive melioration both moral and physical. In treating of the negro people as here existing, it must never be forgotten that seventeen years only have now run their course, since they were emancipated from a state absolutely opposed to all improvement, and with this in view I record not only my satisfaction, but a feeling of joyful surprise at the advances made by them during the six years to which my observation and experience have extended. As a general rule, they possess, beyond all reasonable question most of the essential elements of progress, and in a preëminent degree natural intelligence and quickness of perception sharpened by a praiseworthy desire to better their condition, somewhat controlled in action indeed, by the indolence incident to a tropical climate, by the facilities for acquiring a comfortable subsistence and by dilatory habits thereby accruing.

Looking around this country, and considering the extent of land recently acquired by the labouring population, and the thriving villages reared by them, any unprejudiced observer must be convinced of the fact of an improved physical condition of the lower orders. Now I believe that although some blemishes or even vices may proceed from such advancement in material comfort, it will be found always on the whole attended with moral amendment. It is accordingly observed, that where a body of what may almost be termed yeomanry has associated in a village, they have sometimes acquired too free ideas of liberty and independence, which lead them occasionally to make some attempts at resistance to lawful authority. These attempts, it is however right to remark, are made against officers of the law, or constables of their own color, often inferior to themselves in intelligence and generally when employed in the distasteful business of enforcing the collection of taxes. On the other hand, I have found crime to be rare amongst these people ; and although it is not so easy to pronounce as to vice, owing to its more hidden nature, I am yet pretty confident that it too has sensibly abated under the influence of a better external condition, and that in the aggregate it is infinitely less than during the days of slavery."

Another important trait of improvement he notices. Adverting to a former feeling of aversion to labour, he remarks, "I do not believe that any symptoms of aversion to field labour are now to be observed, and feel nearly confident that wherever proper relations subsist between the employers of labour and those in authority under them, and the labourers, there is little difficulty in procuring people to work and in retaining them as long as required. No doubt the quantity of work done is frequently unsatisfactory. A labourer of any race is perhaps not able to work hard from morning to night, as is done in a temperate climate, but he may do more than is often done here without distressing himself." In relation to intelligence and trustworthiness he remarks. "There are several instances of black men being employed as overseers, or even as managers upon estates, and I believe they are found to do their duty well and faithfully, and the extension of education will, I hope, add gradually to the number of this respectable class of the native population."*

Of the condition of the higher class, the comparatively few, such as the very small number of resident proprietors, and of those who may be considered as belonging to professions, I need hardly remark that it is not prosperous. Absenteeism,—the low price of produce, doubtfully remunerative,—the high price of

* Report to accompany Blue book, 1851, p. 105. The whole of this report is deserving of perusal, especially as regards the colored population.

most of the necessaries of life, and of all the luxuries, *
are causes amply sufficient to account for a want of
prosperity in this class, compared with that of the
working class generally with their many advantages,
and in such a climate, with few necessary wants. Details
might be given in illustration were it necessary.

Whilst the friends of humanity may well derive
pleasure from the improved circumstances of one portion of the community, all right-minded persons must
feel regret at the fallen fortunes of the other portion,
and that in relation to the common interests of the
colony; it being obvious to reason that a society must
decline and fall lower and lower in point of intelligence
and knowledge, art and science, and may it not be
added in morals as well as manners, wherever no encouragement is given to professional excellence and
scientific attainments,—in brief whenever the animal
wants, whatever may be the cause, have become the
engrossing subjects of attention.    St. Vincent is more
in danger of such a decline from having been newly
settled, and from the manner in which it was settled.
Widely apart as the estates are, intercourse is not easy,

* In the West Indies generally, the prices of provisions are high,—indeed
most articles, whether of food or clothing or household furniture, are costly.
The prices of the following articles are from the reports of the Lieut. Governor of St. Vincent for 1845 and 1848 ; and they differ but little from the
charges for the same in the other colonies. Mutton, 1s. per lb. beef, 10d., of
bad quality, ' vile' is the term used, and deservedly, especially in the instance
of the beef supplied to the troops ; turkeys from 16s. to 20s. a piece ; eggs,
1s. to 1s. 4d. a dozen ; milk, 6½d. to 8d. per quart ; potatoes in hampers, imported of 25 lbs., from 4s. to 6s. " Native vegetables," it is added, " are more
expensive than would be expected, which is accounted for by the limited
cultivation, the negroes being the only persons who rear vegetables for sale."

except of a very limited kind; the proprietary body having always been composed more or less of absentees, absenteeism may be said to have become a habit, and no efforts have been, as probably would have been made, were the island their permanent home, to administer to the higher wants of their nature, whether in ornamental planting or gardening, or in the formation of libraries and societies, such, limited as they are, which have been established in Barbados. Even in the principal town, with a population, according to the last census, of 4,983, of whom at least 473 were whites, there was not, when I visited the island, a single library, and only one small book shop. An agricultural society it is true has been formed, and recently has been reported on favorably, as exerting itself, especially in the breeding and rearing of live stock; to encourage which shows have been held, and prizes offered;—£50 a year being allowed from the treasury in aid of subscriptions for the purpose. That there is a disposition towards higher objects, I hope may be inferred from this; and, as indicating the same, I may mention, that a book club has been organised in the town through the exertions of the enlightened Chief Justice; and that still later, a literary and scientific society has been instituted, at the meetings of which it is proposed that lectures should be given upon various subjects of general information and practical utility.*

The state of agriculture, and the manner in which

* See Report of Lieut. Governor, for 1851.

land is appropriated, tend further to show the condition of the people, and the changes which have occurred and are in progress.

In the time of slavery, the number of estates was 98, containing 37,842 acres, individually varying in size from 750 to 82 acres, averaging 386. Then there were no small independent holdings, or if any, so few as to escape notice and mention. Then the produce of the land was almost exclusively sugar; the average yield might be about 19,000 hogsheads per annum,— no provisions being grown, except by the slaves for their maintenance, on the ground allowed them for the purpose, or by the planters for the use of their own families.

At the present time whilst the number of large estates has continued much the same,* many small holdings have been created. According to the report referred to below, in 1845 they were not then fewer than 158, varying in extent from 11 acres to spots no more than sixty feet square, obtained by purchase at the ordinary rate of about £27 per acre. Now, the produce of the larger estates—still restricted to sugar, has fallen off almost one half; in 1847, a favorable year, it was estimated at 10,905 hogsheads. What the amount of produce of the smaller holdings is, cannot be specified, except in the article arrow root, that being the only one officially returned; it, in that year,

---

* In the Report of the Lieut. Governor accompanying the blue book for 1845, the number of estates is given as 100 and the land not altogether waste as between 34,000 and 35,000 acres.

amounted to 297,587 lbs. It is believed that there are now not less than from ten to twelve thousand acres under cultivation in this plant, and for growing provisions; land either the property in fee of the labouring class, or rented by them at the rate of about sixteen dollars the acre.

Another marked difference, in comparing the two periods, is in the number of labourers employed on the larger estates, and the produce obtained. In the former period, for instance in 1828, when the quantity of sugar was about 19,000 hogsheads, the number of slaves engaged in the cultivation of the cane was estimated at about 11,000, being at the rate of one and three quarters hogshead to each labourer, whilst in the latter period, for example, in 1845, the quantity obtained was no more than 9,000 hogsheads, but procured by the employment of no more than 4,500 free labourers, being at the rate of two hogsheads to each labourer.* Results certainly not unfavorable to free labour, even making allowance for less land of secondrate fertility being under culture in the latter than in the former period, and for more skill and care, granting it to be so on the part of the planters, being exercised in the one than in the other.

As regards the system of agriculture in use in this island compared with that followed in Barbados, there are some points of difference. I shall notice only those which appear to be most important.

* See the Report of Lieut. Govenor for 1845, from which the above statements were taken.

1. With a very few exceptions, the tillage is effected entirely by the hoe.

2. Close planting, about five feet by five, is almost invariably followed. The canes are planted not by means of the crow bar, one piece to each hole, placed nearly erect: less economy is observed; three or four cuttings are apportioned to each, and these are stuck in with the hand, and are slightly inclined; and to insure regularity in the arrangement, each cane hole is marked out by stakes, one at each angle.

3. The cane field is not trashed to preserve moisture in the soil; on the contrary, the canes are repeatedly stripped; their lower leaves are taken off to promote ventilation, evaporation, and the ripening of the crop.

4. Little attention is paid to moulding; little to the making of manure, such as farm manure or stable dung, whilst much imported manure is used, especially guano.

5. No green fallows are used, no rotation system is followed; nor have I heard of irrigation having ever been employed.

6. Ratooning is commonly practised, extended to three or four years.

Next as to management and labour:—as regards the first, there is this marked peculiarity, viz. that the majority of the proprietors of the large estates are absentees. When I was there in 1848, I was assured by a person competent to know, that out of a hundred, twelve only were resident, eighty-eight properties being under the direction of attornies.

As regards the second—labour; it is paid in a manner somewhat different from that followed in Barbados, and less to the advantage of the planter. The pay is seldom in money wages entirely; in 1848 that was 10d. a day. Commonly a money payment and allowances are made together; the former $7\frac{1}{2}d$. sterling per day, the latter a house free of rent, with forty square yards of cane land for growing yams, reckoned about equal to six or seven dollars a month, with leave to make charcoal on the woodland, and besides, in many instances, some other perquisites, such as allowances of sugar and rum, notwithstanding that the latter has been prohibited by law, with the intent of checking intemperance.

Whilst the cost of labour is materially greater than in Barbados, and the difficulty of obtaining continued labour is very much greater, the interests of the working class so much interfering, the other expenses can hardly be considered less, especially on the estates of absentees; thus the ordinary pay of an attorney varies from £50 to £100 sterling per annum, according to the extent of the property; that of the manager, who is resident, is on an average from £100 to £160, occasionally even £300, with house, provision grounds, domestics, the use of horses, with, in some cases, additional perquisites; that of the overseers averages from £70 to £80 per annum; with similar allowances, only on a more limited scale.

The cost of production I have heard estimated at 17s. the cwt of sugar, and that it may be so high, except in the most favored situations, can easily be imagined, taking into account the great expenses incurred. Such a cost is clearly incompatible with the then,—I speak of 1848,—and present market price of sugar. Either the price must be raised, or the cost of production must be diminished to allow of profit and the keeping of the estates in cane cultivation. The prevailing impression in the island seemed to be amongst well informed and reflecting persons, that a complete change must take place; that the estates must be conducted on a different system, either by being let out so as to be managed by resident agriculturists, like our farms at home; or be brought into the market and sold for as much as they will bring, to men competent and willing to undertake their management and conduct it according to the most approved methods and in the most economical manner.

Of the manufacturing processes in use, and of the condition of live stock I have little to remark. The cattle in St. Vincent are generally superior to those in Barbados, are better fed and better taken care of, and are of course capable of more work. Few windmills are in use; watermills are more employed in expressing the juice of the cane. On more than one of the larger estates, the vacuum pan has been introduced and the steam engine, at a great cost, and after the old enterprising manner, viz. by means of British capital; but with what success remains to be determined. In consequence of there being many non-resident proprietors with large means, there is a disposition, I may remark, to attempt here new manufacturing processes attainable by means merely of an outlay of capital, which for their success require knowledge and skill, not so easily attainable, and these failing, disappointment and loss follow.

Respecting the smaller holdings and their cultivation I may also be brief in my remarks. No sugar is obtained from them as in Barbados; provisions and arrow root are their principal produce; the former for the subsistence of the proprietors and their families, and for sale in the island; the latter for exportation. Since emancipation the culture of arrow root, (*Maranta arundiacea*) has come into great request, and from a very small beginning has become extensive. During the time of slavery no attention was paid to it; now the cultivation of the plant is second only in importance to that of the previously exclusive cane, and has even been introduced on one or more of the larger estates. The amount exported is now above one quarter of a million of pounds; in 1847 the quantity returned was 297,587 lbs.; in 1851, 490,837 lbs. or very nearly half a million. Many circumstances have promoted this increased culture. When commenced, the price of the article was high, the grower obtained a largely remunerative profit; its culture was little laborious, it was subject to few risks; it did not for its success require rich land, or much manure; there was a constant and increasing demand for it; and in consequence of the abundance of pure water great facilities were afforded for the manufacture, and that

by a process so simple, easy and cheap, as to require
little skill in conducting it, and scarcely any capital.*
Whether there will be much further extension of its
culture seems doubtful, inasmuch as in the instance of
the sugar cane, greatly increased production and that
widely in other countries stimulated by large profits,
has so much reduced its marketable value as to allow
now of little profit to the grower,—that which a few
years ago was in request at 3s. 6d. the lb. now selling
at from 6d. to 8d. How strong and instructive is the
analogy in relation to sugar! Is not the history of the
one the counterpart of that of the other, and, had
there been no slave labour employed in the cultivation
of the cane, and no differential duties encouraging,
would not the result have been much the same as that
we now witness, viz.—a large supply at a cheap rate
to the consumer and a small profit to the producer, or
none without skill, care, and economy? Indeed, is not
this the general history of all productions whether agricultural or manufactured, such as are not restricted in
amount by any marked speciality? Were the diamond
as easily obtained as coal, it would be cheaper probably; could guano be procured with as little difficulty

---

* When made by the labourers on a small scale, it is prepared much in
the same manner as potatoe starch in this country for domestic use; the
only implements required are a grater and wooden trough sand trays; when
on a larger scale, as on the estates of the proprietors, the crushing of the
root and the reducing it to a pulp, is effected by simple and cheap machinery, (a wheel and rollers,) worked by water. The arrow root is dried
under sheds.—Little or no use, it may be remarked, is at present made of
the pulp after the extraction of the starch by lixiviation: probably, a serviceable paper might be made of it, at a trifling cost.

as lime or gypsum, it would fall proportionally in marketable value. But these are truisms in political economy; were they not too much overlooked, especially in our colonies, I ought to apologise for introducing them.

In conclusion, I would remark, that the stranger visiting the West Indies, may spend a week or fortnight in this island very agreeably, especially in the cool season of the year in December and January when the weather is very delightful, the woods of their richest coloring, vocal with the songs of birds, especially of the mocking bird* and the cane fields are in their greatest beauty. In the time mentioned he may see a good deal of the island and explore the parts of it which are most interesting and best worth visiting. It will be a great advantage to come provided with letters of introduction, as, excepting in Kingstown, where there are one or two indifferent inns, he must be dependent on the hospitality of the planters, which, I believe, is never grudgingly given, but most willingly and kindly, the demands on it being 'few, and far between,'— and not unwelcome, as breaking the dull monotony of colonial life.

The mode of travelling on one side of the island, the windward, is commonly on horseback ; on the other

---

* Columbus in the first description of the new world ever given, writing of Cuba says, "the nightingales and various birds were singing in countless numbers, and that in November, the month in which I arrived there." *Select letters of Columbus, published by the Hakluyt Society.* His "nightingales" were probably mocking birds ; they have many notes in common or nearly so.

side in a boat. I hardly know which is most pleasant; both afford an excellent opportunity to see the country, and mark its peculiarities.

The windward coast being fully exposed to the prevailing winds which are often violent, bears the impress of them in the swell that constantly breaks on its shore, and in the stunted growth of the woods or rather thick copse, which here and there fringe it, kept low by the gale, shorn and inclined as if the result of art, an exaggeration even of what we witness of like effect on our south western coast.

On the leeward sheltered side, the opposite character of wood pleasantly meets the eye, unchecked and luxuriant almost to the water's edge. Proceeding along this coast some fine vallies come into view, in a high state of cultivation, and two or three pretty villages; the vallies of Buccamont, Layou, Chateaubelair, and others; and the villages Rutland Town, Barrawallie or Princes Town, and Richmond Town. The object of this excursion commonly is the ascent of the Soufriere mountain, which is not difficult of accomplishment, there being a rude kind of bridle path, almost to the summit; but even were it less easy, so extraordinary is the scenery near and distant, so varied, so contrasted, that the labour to most would, I think, be amply repaid by the enjoyment;—indeed excepting Etna I know no volcanic mountain more impressive, no one where are better displayed, on one hand, the destructive energies of nature in the effects of the volcano, or on the other her restorative and preservative influences, in the formation of fertile soils and the growth of a luxuriant vegetation.*

Passing to the windward side, and deviating a little from the line of coast, the fine circular valley of Marriaqua may be seen,—a little district within itself, about eight or ten miles in circumference, surrounded by high hills and mountains, the greater part of it laid out in sugar estates, forming a whole of peculiar beauty. This charming valley, perhaps once the enormous crater of a volcano, and afterwards, it may be, the bed of a lake, opens abruptly towards the coast by a remarkable gorge, through which one of the largest streams of the island finds its way to the sea by a succession of pools and rapids, often peculiarly picturesque from the accompaniments of rock and wood.

Other objects not without interest are passed,—such as the bed of the Rabacca, or " dry river," rendered so by the last volcanic eruption,—that of 1812, when, it is said a current of lava flowed through this channel, and produced the effect. Be this as it may—the actual lava stream seems very doubtful,—the bed itself is a very singular one ; where widest, about three quarters of a mile wide, it divides into two branches, leaving intermediate a spot of raised ground, covered with shrubs and trees,—an island of verdure. The dry bed is a surface of small masses of scoriæ, with which larger

* An amusing account of the ascent of the Soufriere, Morne Garou, and of the difficulties to be encountered, is to be found in the *Phil. Trans.* vol. xv. abridged, p. 634. The time was 1785, when, it appears from the description, that the volcano was active.

fragments are intermixed together with sand and pebbles of a dark hue,—a little desert plain furrowed longitudinally, from the effect of water, which, after heavy rains, I was informed, flows through it in small streams. So barren is the dry bed, that not a plant is to be seen in it,—not a blade even of grass, so entirely destitute is it of vegetation. This I was assured of by a gentleman residing near, who had made the search in vain;—a peculiarity, it may be, owing to the condition of the surface, nowise retentive of moisture, and highly heated and parched, from exposure to the sun and wind. The Rabacca before it is thus lost sinking into the fissured earth, a few miles higher up, presents, compared with this, a scene of perfect contrast, and one of great beauty for picturesque effect,—a large body of water, (such it was when I saw it,) dashing over, and gliding between dark masses of volcanic rock, skirted by wood, and emerging from native forest, backed or walled in by mountain peaks and ridges.

Not the least interesting portion of this excursion is the cultivated part of the Carib country, and that beyond, almost in a state of nature, in which a few families of the aborigines form a settlement apart. Here again, contrasts of a marked kind are observable. The cultivated portion of the country is divided into a small number of large estates, on which, capital to a large amount has been expended, and which, in consequence, are amply provided, not only with the buildings necessary for the purposes of a sugar plantation, but also for the accommodation of the planters and their agents, in comfort approaching to luxury. No where in the East Indies have I seen the cane growing more luxuriantly, or cane fields, as far as appearances are concerned, more promising. Some were pointed out to me by the proprietor as examples of great fertility, and one especially, a piece of twelve acres, which that year had yielded thirty-four hogsheads of sugar. With such fertility, one can easily understand the great value of these estates when the price of sugar was high and labour to any amount attainable. One of the smallest of these estates, one of 300 acres, I was assured, shortly before emancipation, was sold for £80,000, which now probably would not sell for so many hundreds.

Where this costly cultivation terminates the true Carib country commences,—a country of wild forest and jungle, constituting with Morne Rhonde on the opposite coast, the last refuge of the remnant of the native people whose name it bears.* Their village, the population of which is said to be about 106, consisting of scattered dwellings in the midst of fruit trees, well apart from each other and almost hid by thick surrounding foliage—is situated on an elevated ground

---

* Some interesting geological appearances are to be met with on the coast of the Carib country, especially different forms of lava. In my note book, mention is made of one spot where a great mass of lava seems to have been suddenly arrested in its flow, incumbent on a bed of ashes and its under surface slaggy. It reminded me of the fused and slowly cooled basalt of Mr. Gregory Watt,—having a good deal the grain of basalt without the columnar structure; its exposed face was fresh and nearly steel grey.

not easy of approach near an inlet of the sea, called Sandy Bay. Each hut, or cottage, (the appearance of comfort and neatness, suggesting rather the latter word as most appropriate,) is formed by its owner chiefly of a kind of reed called 'rousseau,' skilfully arranged in double rows, supported by upright timbers, and thatched, the roof overhanging so as to serve as a verandah. Those we entered consisted of two rooms, the inner used as a sleeping room, both provided with windows and shutters, but unglazed, and both well floored, the floor being made of 'terras' and earth beaten, forming a hard smooth surface impervious to moisture.* They might serve as models for labourers' cottages within the tropics, being from the nature of their flooring, and the kind of materials of which they are made, so well adapted to the climate, affording protection equally from the sun's rays, and from rain and wind, and from ground-damp and exhalations from the ground, without excluding ventilation. The art of constructing them may be said to be hereditary, having descended from father to son from the olden times, when they were a free people, for if I recollect rightly, such as these cottages are, were those which Pere Labat more than 150 years ago described as then peculiar to them. Another art they still exercise, that of making

---

* This terras, or pozzolana we are informed may be procured at a cost of about £3 a ton. Might it not be used advantageously for the flooring of barracks and prisons in the West Indies wherever malaria is suspected? It has been vaunted of as a valuable manure little inferior to guano! only the very ignorant and credulous can believe this.

canoes. Most of the boats of St. Vincent are made by them, formed of single trees scooped out. They are skilled also in basket work, and in the construction of 'fish pots,' wicker-work traps for taking fish.

The people themselves, are much in the condition that seems inevitable to all rude races when brought suddenly under control, and subjected to laws alien from their feelings and usages. In their appearance, judging from the few I saw, there is little that is distinctive. Their dress is much the same as that of the peasantry of the country, whether male or female. In complexion they reminded me of the Malay, for though they are called "Yellow Caribs," to distinguish them from the black, a race of African descent,—the color of their skin is brown, nearly that of the ripe chestnut. In features they might pass for Europeans. One woman of pure Carib blood, possessed considerable beauty,—a fine form, pleasing delicate features, and large dark lustrous eyes. She was the mother, she said, of eight children, all but three of whom were dead. In their subdued state they are described as mild and inoffensive. They are useful not only in making canoes, but also in the management of them, and in shipping produce, a thing of difficulty and of some danger on that wild coast. They grow, on the land apportioned to them, vegetables sufficient for their own use, on which, and on the fish they catch, they chiefly subsist. Most of them have received some instruction, and can read, and they all, I believe, profess themselves christians, are married as such, and have their children baptized. It is said that they still retain the ancient usage of their people, in burying their dead in the sitting posture. According to the census of 1844, their total number then, was 273, of whom 178 resided on the windward, and 95 on the leeward coast. According to the last census, that of 1851, they then were only 167,—no distinction of coast is made. So great a difference in so short a time, lessens confidence in the accuracy of the returns, and recalls the opinion expressed by the Lt. Governor, that they are not worthy of confidence,—a fault surely, which, arising from carelessness, ought not to be tolerated officially.

# CHAPTER VI.

## THE GRENADINES.

OF the Grenadines, which form a curved chain between St. Vincent and Grenada, and which from their picturesque forms and distribution tend to make the voyage from one island to the other—a distance of about sixty miles, peculiarly agreeable, reminding one of the Cyclades of the Grecian Archipelago,—I shall notice only a few particulars.

They are said to be three hundred in number. Most of them however are mere rocks. The most considerable of the islets, those which are inhabited and cultivated, are Bequia, Mustique, Union, Canouan, dependencies of St. Vincent; Carriacou, Petit Martinique* and Isle de Rhonde, dependencies of Grenada. Of the former, Bequia is the largest, comprising 3,700 acres; of the latter, Carriacou, which is about twice the size

---

* According to Pere Labat, the same kind of poisonous snake as infests St. Lucia and Martinique, was, in his time common in this Island; whence, he says it was sometimes called Petit Martinique. I have not heard of its being known there at present.

of Bequia, having an extent of 7,881 acres; the smallest of the others, Isle de Rhonde, is estimated at only 600 acres. Their relative situation and position is shown in the map prefixed. Their geological structure is but little known. From the specimens which I have obtained from Carriacou, for which, and a good deal of general information respecting that island, I am indebted to George Mitchell, Esq., of Carriacou, it would appear to be formed in part of igneous rocks, resembling those of St. Vincent, and in part of marl and of shell and coralline limestone, having a close resemblance to those similarly designated in Barbados. Its most remarkable feature, and exhibiting, I believe, these two kinds of rock-formations, is a ridge which passes through it, varying in height from five to eight hundred feet, terminating at each extremity in a bold headland, about 1000 feet above the level of the sea. This dividing ridge is described as composed of coralline limestone, resting in some places on clay, and in others on marl; whilst the terminal headlands, it is said, are formed chiefly of "hard iron stone." A portion of the former sent me, was a coarse grained crystalline limestone, in which were traces of shells, but not very distinct as to form, as if some metamorphic change had been effected in it, from the action of fire, either direct, or through the medium of some igneous eruptive rock. A specimen of the latter, which accompanied it, the so called iron stone, had the character of such a rock, and might have been the modifying cause; it was a decomposing felspathic trachyte.

Of the other islets and rocks, probably some are volcanic; others like the last described, or composed entirely of coralline and shell-limestone. They are all said to afford coral, that is, their shores. In Canouan, on its eastern side, a reef of basaltic rock occurs, forming a little harbour or carenage, in which between the reef and the land, the water is said to be only about twelve feet deep, whilst, outside, it is worthy of note, it is so deep as to be unfathomable.

The climate of the Grenadines seems to differ but little from that of St. Vincent or Grenada. It is commonly supposed that they are more subject to drought and that very much less rain falls in them, attributable to the cutting down of the wood, with which the larger ones were once covered. Both the asserted fact and its explanation seem questionable. Were rain gauges kept in them it is probable that they would be found to have their due proportion of rain. The quantity that falls on the windward side of St. Vincent, even greater as it is believed than the leeward, is in favor of this, as is also the fertility of these islands. Moreover it is difficult to imagine that spots so small, rising out of a deep sea can have any marked influence on the passing winds and currents of air. The circumstance of their situation, rising thus out of a deep tropical sea, may indeed contribute to the equability of temperature for which they are remarkable.

Though without rivers, as might be expected, they are not altogether without springs: the inhabitants however are chiefly dependent on rain for their supply of water, which is collected and preserved in tanks. Though situated within what is called the hurricane region, these islands are very little subject to hurricanes. Carriacou, for instance, has escaped on several occasions, when islands within sixty or a hundred miles, as St. Vincent, Tobago, and Barbados have suffered severely; thus showing how limited is the sphere of action of these whirlwinds.

The Grenadines have all of them the reputation of being very healthy, yet Carriacou is not exempt from visitations of sickness; they have been there of periodical occurrence, and of considerable duration, in the form of a bilious intermittent fever peculiarly fatal to children under ten years of age, and to persons turned of fifty. The severest visitations were in 1802—3, 1833—4 and recently in 1847—8. The fever is described as often ending in ague, and having as its sequel an organic disease of the abdominal viscera, especially the liver and spleen.

The population of the Grenadines is necessarily a very limited one. In 1844, the amount of it altogether, according to the census of that year, was 5,713; according to the last census, it was 6,394, denoting an increase of 681: of the total, 4,461, belonged to Carriacou and the dependencies of Grenada, and 1,933 to Bequia and the dependencies of St. Vincent. People of color constitute the great majority of the inhabitants, about 99 per cent. Whether there has been an increase or diminution of their numbers since the time of slavery appears to be doubtful, two opposite causes being in operation,—one, the natural tendency to increase, from the salubrity of climate and facility of living,—the other the withdrawing of that increase, from the inducement to emigrate, tempted by a higher rate of wages in the adjoining larger islands.

Agriculture is the main occupation of all the inhabitants. The system followed, the kind of properties, their condition, and that of the labourers, are very much the same as in St. Vincent and Grenada, as was also the manner in which they were originally settled. The soil generally, is said to be good,—in the larger islands well fitted for the sugar cane and cotton, and in the smaller for pasturage. In the time of slavery, during the years 1827—8—9, the maximum yield of sugar from nine estates in Bequia, amounted to 442,823lbs., from the labour of 364 negroes ; and in Canouan and Union Island, where cotton was exclusively grown, the yield of this article from two estates employing 624 negroes, was 46,089 lbs. Since emancipation, less sugar has been grown, and less cotton. More land belonging to the larger properties, has been applied to the feeding of stock ; and from the smaller to the growing of provisions. Of both, considerable supplies are exported,—such as sheep, pigs, poultry, vegetables,—especially to St. Vincent and Grenada.

The condition of the labouring class in all these islands is peculiarly favorable. The majority of them have small portions of land, of which they have become possessors by purchase; and their spare labour is in demand on the larger properties for money wages, or for a certain portion of the produce, on the metairie system which has recently been attempted. Many of them, the more enterprising, moreover, have small vessels or shares of them, especially in Carriacou, which are gainfully employed in trade between the islands. To this island alone, from twenty to thirty belong, of a tonnage varying from one to thirty tons. The vessels are built of the white cedar, the growth of the island, a valuable wood for ship building on account of its durability, and, from its containing a bitter principle, not being liable to be attacked by the worm; such at least are its reputed virtues.

Considered as communities, their state appears also to be favorable, having within them the elements requisite for improvement. Both in Bequia and Carriacou there is a small town,* in which is a church and an appointed minister—a resident rector, schools, a small body of police, a stipendiary magistrate, and one or more medical men. Most of the people can read and write; moreover they have their representatives, the

---

* That in Bequia is at the bottom of Admiralty Bay, which is described as an excellent and spacious harbour, and excepting want of water, every way fitted for a great naval station.

That of Carriacou is called Hillsbrough. It contains about 400 inhabitants. The Rector who resides in it, is provided with a house, a glebe of sixteen acres, with a salary of £264 a year. Service is performed on Sundays both in the morning and afternoon. The morning attendance, I was informed, was about 300 persons—the afternoon about 80. There is also in the town, a Roman Catholic Priest, and a Wesleyan Minister. According to the last census, of the total population of the Island—4,461, the Protestants amounted to 3,154; the Roman Catholics to 1,307. Of the former, 3,129 were of the Church of England, 3 of the Church of Scotland, and 22 Wesleyans.

dependencies of St. Vincent, returning two members to the house of assembly, and those of Grenada the same number. They have also in common with the larger islands the benefit of periodical visitations from the Lord Bishop of the Diocese. In 1846, on an occasion of this kind, sixty-six young persons were confirmed in the parish church of Bequia, and eighteen in that of Carriacou. Nor are they without a printing press and newspaper; the latter published weekly bears the title of "The Carriacou Observer and Grenadines Journal." It consists of a single sheet, and judging from the specimen now before me, seems well adapted as a miscellany to be useful, in accordance with its excellent motto, that, "True patriotism consists not merely in that love and zeal for one's country, which will cause the citizen to arm in defence of her rights and interests, but in an earnest and zealous endeavour to promote the well-being and social improvements of those who from adventitious circumstances are debarred the enjoyment of advantages possessed by their more favored brethren. He who does this, actuated by the principles of truth and justice, is the true benefactor of his country."

The future, in regard to these islands, is not unpromising. Well off, as are the inhabitants at present, unquestionably they might be better did they avail themselves of all their resources. The sea surrounding them is reported to abound in fish, yet no fishery has been established,—none being taken but to supply the immediate wants of the inhabitants. Coral and shell-sand are in plenty on their coast, and with marl of an excellent quality in Carriacou, might probably be exported with advantage to St. Vincent or Grenada, especially the former, in which there is such a scarcity of calcareous matter. Further, it is not unlikely that were these islands examined by a competent person, qualities of rock might be discovered of value as building material, and even ores might be found, which might be profitably worked; both in Carriacou and Canouan strong indications of iron, it is said, have been observed.

# CHAPTER VII.

## GRENADA.

THIS island greatly resembles St. Vincent. It is of nearly the same size, not less mountainous, and hardly less beautiful and picturesque. In its political history it is also similar. Well watered, almost every valley having a stream; well wooded; having a coast abounding in harbours and landing places, it was one of the last resorts as it had long been a favorite abode of the Caribs. It was, and at a late period, first colonized by the French. Their earliest settlement was made in 1650. From them, as in the instance of St. Vincent, it came into our possession by right of conquest, viz. in 1762. Nor does its after history much differ from that of its beauteous sister island; like it under similar circumstances, it had its prosperous period; and like it, its disastrous period, which still continues, and even aggravated, owing to causes, some of which are obvious, others less so and more obscure.

Grenada in length is estimated to be about 25 miles, in its greatest breadth about 12; and is supposed to include within its area about 85,000 acres, or 133 square miles.

No satisfactory account has yet been given of its geology. In a popular work,* its rocks are said to be principally varieties of schist, greywache, and sandstone. From a partial examination which I made in three several visits, I am disposed to infer that these names have been incorrectly applied, the observer having been misled by appearances, and that the greater portion, if not the whole of the island is volcanic; that is, like St. Vincent composed either of loose matter, such as dust, cinders, and lapilli, forming tufas resembling sandstone; or of lavas or liquid ejecta; or of igneous, uplifted rocks, constituting the compact or crystalline formations more or less resembling those which not long ago were called primitive, such as granite and gneiss.

From what I myself saw, and from the specimens sent me, I am disposed to infer that tufas predominate in the southern portion of the island, and crystalline rocks undergoing decomposition in the central and mountainous parts. Relative to the eastern and northern districts, I can only offer conjectures. From such information as I was able to obtain, I think it possible that the latter, on examination will be found to resemble Carriacou, and to consist not only of rocks of igneous origin, but also of those of the tertiary aqueous formation. I have been assured that marl,

* The British Colonial Library, by R. M. Martin, Vol. iv. p. 259.

containing coral and sea shells, is not uncommon there.* The windward, the eastern side of the island, judging from the few specimens I have seen from thence, I apprehend will be found to have a very mixed character, and to consist in great part of alluvium, in which is a large proportion of clay, with gravel and fragments of rock intermixed, derived from the decomposition and disintegration of the crystalline rocks of the higher hills, brought down by torrents, and more or less spread out on the lower grounds of the district.

Of the crystalline igneous rock, there is no great variety. Fine columnar basalt, containing augite and olivine is not uncommon. A good example is to be seen on the west coast, near Goyave, about ten miles from the town of St. George, the principal town, some of the columns of which are curved. In the neighbourhood of the town, there are some striking examples of cliffs, formed of successive layers of volcanic ashes, in some of which large fragments of ejected trap rocks—their angles and edges sharp—are intermixed, whilst others of the same kind of formation are overlaid by a compact rock, closely resembling lava. In one spot, by the road side, close to the river St. Jean,

* A specimen from Mount Alexander Estate, near the river Duquesne, where marl is said to abound, which I have examined, I found very similar in composition to the marl of Barbados, being composed principally of carbonate of lime, with a little carbonate of magnesia, and a trace of phosphate of lime; and also, like that of Barbados, free from infusorial remains. Included in it, was a fragment of coral, having the character of a cast. I am informed that, though not accurately ascertained, the marl district may extend at least five miles, and that limestone has been found contiguous to it.

there is a remarkable appearance,—a bed of trap rock or lava overlying clay, which, near the overlying rock, bears marks of igneous action, being red, indurated, and fissured in the form of small columns.

Still more worthy of attention, as likely to be valuable, is a metalliferous vein or bed of a large size ; a hill is said to be composed of it. It occurs on the north west part of the island, at the Sauteurs. The substance of this vein is principally magnetic iron stone, more or less crystalline. In some specimens which were sent me, I detected, intermixed and scattered through the ore of iron, granules and small nodules of carbonate of copper, (malachite), and in others, an admixture of black oxide of manganese. Rarely have I seen a richer ore, and as it is so abundant and rises to the surface, and there is no want of wood in the neighbourhood, it might probably be smelted on the spot with profit ; or should this be thought too hazardous, it might be worth while to send a certain quantity of it to England, if it could be conveyed as ballast, to be reduced on trial at some of the great iron works. But preparatory to either attempt it might be well to have the bed of ore carefully examined and reported on by a competent person.*

Mineral springs are noticed by all writers on this island. In a map of Grenada, the only one that I have ever seen, that engraved by Thomas Jeffrys in 1810

* In some loose earth sent me, taken from the surface of this vein or bed, I found a portion of vesicular scoria ; seeming to intimate, that the great metallic mass had been ejected in a fluid state, at a later period than its including walls.

from a survey made by order of Governor Scott, it is stated that "in several places, you find hot springs of sulphureous and mineral waters which might prove of great service in many disorders if their qualities and virtues were sufficiently known, but no person as yet has taken the trouble of making their analysis." This omission is still to be regretted. I endeavoured to obtain samples of them for examination, and some exact information respecting them as regards their sites and ordinary qualities, but in vain, except in one instance, that called 'the mephitic spring,'—the same I apprehend as is described by Mr. Montgomery Martin, * as "a hot spring in St. Andrew's parish, emitting considerable quantities of carbonic acid gas, possessing analogous properties to the famous 'grotto del cane,' and containing iron and lime and possessing a strong petrifactive quality." From information afforded me by assistant surgeon Sanders, who visited it at my request, and made some experiments which I suggested, it would appear that this spring, situated in the parish above mentioned on the Hermitage Estate, is a dry spring of gas, and of carbonic acid gas, issuing in a strong cool current from a hole about a foot square beneath a rock overhung with creepers. A fowl exposed to it, by placing its head in the current, was rendered insensible in ten seconds. Suspended animation was restored by dashing cold water on it, and by artificial respiration. The same fowl left at liberty,

* Colonial Lib. I. p. 254.

running into the track of the current, again dropt down apparently dead, and was restored by the same means. Why this evolution of gas should have received the name of 'mephitic spring' I do not know, unless occasionally it has a bad smell from an admixture of sulphuretted hydrogen. When examined by Mr. Sanders he found it, as he specially mentions, without odour of any kind, and also cold; 'on placing the hand on the hole' he mentions, 'a strong current of cold air was perceptible.' That the air contains carbonic acid gas may be inferred both from its effect on the fowl, its heaviness, flowing in a current close to the ground, (the hole from which it issued was about three feet above the surface of the ground,) from its having no smell, and from its extinguishing flame. The last mentioned property was ascertained by putting a piece of lighted paper into the stream of air, or into a bottle filled with the air, when the flame was immediately extinguished. It is worthy of remark, that amidst surrounding luxuriant vegetation a track perfectly bare or barren was made by the current of gas, between its exit and a stream of water below, some feet distant, the ground sloping in that direction;—demonstrating, I may remark, that carbonic acid though favorable to vegetation, when diluted and applied through the medium of water, is in its concentrated state and free, as fatal to plants as to animals. In the same track Mr. Sanders observed a dead snake, which from its position, he had no doubt owed its death to this gas, and on another visit he saw in the same place a dead bird.*

The soils of the island appear from the very partial examination I made, and from such information as I could collect respecting them, even more varied than those of St. Vincent, and hardly less so than those of Barbados. Over a large extent, especially in the central portion, and in the north east and north west districts, clay appears to predominate, mixed more or less with gravel. This soil derived from the decomposition of crystalline igneous rocks, is one of considerable fertility, is often deep, and is commonly undergoing renovation and deepening from the continued process of decomposition to which these rocks are subject. Excepting where the marl, as already mentioned, occurs, calcareous soils, or soils containing a notable portion of calcareous earth, are I believe nowhere to be met with. Occasionally a white matter found in layers, consisting chiefly of siliceous earth, has been mistaken for calcareous marl. Towards the southern extremity of the island where the soil is shallow, resting on tufa, it appears to be composed principally of volcanic ashes or sand resisting change, probably siliceous sand, and consequently of little fertility.

The climate of Grenada resembles a good deal that of St. Vincent. Though as mountainous as St. Vin-

---

* I am informed that there is another spring of similar gas, on the lake Antoine estate. The lake or pond so called, is situated on the eastern coast; it has been described to me as about 18 feet above the level of the sea, and about 50 feet in depth. It is supposed to be the crater of an extinct volcano.

cent, and its highest mountains, little, if at all inferior,
it seems to be rather more subject to drought,—
tending to shew, that mere wood, however well wooded
an island may be, is not sufficient to secure it against
this evil. The following table, limited and imperfect
as it is, of the annual fall of rain, comparing it with the
like table for St. Vincent's for the same years will
afford some proof in confirmation.*

| | 1846. | 1847. | 1848. | 1849. | 1850. | 1851. |
|---|---|---|---|---|---|---|
| January .... | | 8·20 | 1·91 | No record. | 2·84 | 6·32 |
| February.... | | 90 | 2·20 | | 2·90 | 2·60 |
| March ...... | | 1·40 | 2·38 | | 1·61 | 1·57 |
| April ...... | | 4·12 | 1·91 | 1·73 | 42 | 48 |
| May ...... | | 1·98 | 6·44 | 9·61 | 3·96 | 32 |
| June ...... | | 3·78 | 4·58 | 10·66 | 3·23 | 71 |
| July........ | 7·63 | 5·86 | 14·38 | 9·67 | 6·60 | 5·76 |
| August .... | 6·73 | 5·41 | 10·16 | 10·48 | 9·20 | 5·25 |
| September .. | 17·76 | 9·85 | 4·23 | 3·26 | 3·10 | 5·81 |
| October .... | 7·71 | 3·85 | 6·24 | 13·36 | 8·17 | No record. |
| November .. | 8·59 | 9·12 | 4·14 | 4·00 | 6·53 | |
| December .. | 16·82 | 8·27 | 15·60 | 6·10 | 6·97 | |
| | | 62·74 | 74·17 | | 55·53 | |

Grenada is also reputed to be less subject to hurricanes than either Barbados or St. Vincent. In a note

* The situation of the rain gauge was the Military Hospital, a little below
Richmond Hill barracks, about 700 feet above the level of the sea. The
crest of the hill on which the barracks stand, is said to be 750 feet. It will
be seen that the observations of the rain gauge do not reach beyond Sep.,
1851, owing, I regret to learn, to neglect on the part of the medical officer
in charge; a neglect, which if tolerated by the head of the department at
home, will probably spread to the other stations, and ere long the rain gauges
provided at the public expense, will be thrust aside or forgotten. It should
be remembered, that it is only a long series and uninterrupted of such
observations that can be essentially useful.

to the map already referred to, it is stated that "the hurricanes which make such frequent and dreadful devastations in the Antilles are not so common in this island, for the French, from their first settlement in 1650, never experienced but one single hurricane."* Nor am I aware that since this note was made in 1810, it has been less fortunate. In the minute chronological list of hurricanes appended to Sir Robert Schomburgk's history of Barbados, the name of Grenada, (a solitary exception) does not even once appear, and it brings down their history almost to the present time.

In relation to health, the climate of this island may be considered as about on a par with that of St. Vincent. Fevers are the prevailing diseases as in St. Vincent, and nearly of the same type, chiefly of the remittent kind, attacking most frequently the whites, and most fatal to them, and in a less degree, amounting often to exemption, the colored and African race.

The population of Grenada according to the last census, that of 1851 (including Carriacou) amounted to 32,671, or 246 to the square mile, shewing an increase of 3,748 compared with the aggregate of the preceding census of 1844 which was 28,923. This

---

* The hurricane, it may be inferred, of 1780, by which happily the sugar ants (as they were called) so ruinous for years to the sugar plantations of this island, were destroyed after baffling all the various methods tried, stimulated by a reward of £20,000 offered by the colonial government. For particulars, many of them very extraordinary, respecting this ant, and the hurricane, see an interesting account by John Castles, Esq., *Phil. Trans.* 1790 p. 346. It was the same hurricane, viz. that of the 10th October, already noticed, which desolated Barbados. See for details *Chronological History of the West Indies*, vol. 11, p. 473.

increase, it would appear, is partly in the native population, from the births exceeding the deaths to the extent of 3,049, or about 1½ per cent. per annum; and partly owing to an accession by immigration—Portuguese and Africans,—460 of the former, and 1,052 of the latter: these liberated Africans, have been a great boon to the planters.

The returns of the last census drawn up by Mr. S. Cockburn, commissioner general of population, are ample and minute, and are worth consulting. They are published in the reports accompanying the blue books for 1851. The four following tables, shewing the classification of the inhabitants as to places of abode, native places, occupations and religion, constitute only a small part of them.

FIRST.

| | |
|---|---:|
| Town of St. George, .. .. .. .. .. | 4,567. |
| Parish of St. George, .. .. .. .. .. | 5,413. |
| " St. David, .. .. .. .. .. .. | 2,581. |
| " St. Andrew, .. .. .. .. .. | 5,635. |
| " St. Patrick, .. .. .. .. .. | 5,160. |
| " St. Mark, .. .. .. .. .. .. | 1,738. |
| " St. John, .. .. : .. .. .. | 3,116. |
| Island of Carriacou, .. .. .. .. .. .. | 4,461. |
| | 32,671. |

SECOND.

| | |
|---|---:|
| Creoles, natives of the Colony, .. .. .. | 28,082. |
| Natives of the other British colonies, .. .. | 1,168. |
| British Europeans, including Maltese, .. .. | 410. |
| Africans, .. .. .. .. .. .. .. .. | 2,425. |
| Natives of Madeira and foreign Europeans, | 408. |
| Americans, foreign West Indians, &c. .. .. | 178. |
| | 32,671. |

THIRD.

| | |
|---|---|
| Clerical, professional, official and academical, .. | 133. |
| Agricultural, .. .. .. .. .. .. .. .. | 13,502. |
| Commercial, .. .. .. .. .. .. .. .. | 477. |
| Architectural and mechanical, .. .. .. | 1,854. |
| Maritime, .. .. .. .. .. .. .. .. | 467. |
| Domestic, (sempstresses, house servants, and washers,) .. .. .. .. .. .. .. | 3,434. |
| Porters and others, variously employed, .. | 1,777. |
| Children and others, unemployed, .. .. .. | 9,474. |
| Sick, infirm and soldiers, .. .. .. .. .. | 1.159. |
| | 32,671. |

FOURTH.

| | | |
|---|---|---|
| Of the Church | of England, .. .. .. .. | 10,025. |
| " | Scotland, .. .. .. .. | 264. |
| " | Wesleyans, .. .. .. .. | 1,657. |
| " | Moravians, .. .. .. .. | 34. |
| " | Roman Catholics, .. .. | 20,675. |
| " | Mahomedans .. .. .. | 16. |
| | | 32,671. |

The great majority of the native population, as in the instance of St. Vincent, consists of colored people of the African race. Of the foreign accessions, as immigrants for the purpose of labour, not Africans, the principal have been Maltese and Portuguese. The first were introduced at the cost of a few enterprising proprietors in 1838—39; the latter by means of a bounty, a heavy charge, allowed by the local government. Neither have answered the expectations formed. Most of the Maltese have quitted the estates for which they were engaged, there being no contract act, and have become hucksters, or have left the island.

Many of the Portuguese have died from fever and dysentery. To a return of the latter made in 1848, under the head of "general remarks," it is stated that "the majority of the reports appear to concur in the opinion that the Portuguese immigrants can never be employed to advantage for agricultural purposes, or in the cultivation of the cane, except perhaps on some of the more elevated portions of the island; but that they would be more suitably engaged as domestic servants, carters and stock keepers, or occupied in the manufacture of the sugar, attending to the mills or boiling houses." The result of the trial of them in most of the other colonies into which they have been introduced, appears to be similar, especially in Trinidad and British Guiana.

The native labouring class differ but little from those of Barbados and St. Vincent. They speak a French patois, in which, a large proportion of them being Roman catholics, they are said to be encouraged by the French and Italian missionary clergy.* Many of them are not free from the superstition of 'Obeah,' which, in the report of one of the stipendiary magistrates, is said to retard their moral improvement, and even to unfit them for the common occupations of life.† The accounts given of them generally, however, are favorable, and as improving since the time of slavery and apprenticeship, especially as regards industry, frugality,

---

* The influence of these missionaries is further shewn as well as the religious feeling of the people in the circumstance that a Roman Catholic Chapel has been raised by a voluntary contribution of the labourers.

† Blue Book 1844, p. 200.

and thrift. Most of them are described as having laid by money ; a large number as being purchasers of land, and indefatigable when working on their own account. Poverty and pauperism are almost unknown amongst them. One of the stipendiary magistrates, alluding to this happy circumstance, states, " during a residence of more than four years in this island I do not recollect an instance of being asked for alms, or stopped by a beggar on the road." Here is a later picture of them, given by the Commissioner General of population. " The peasantry appear joyful and happy in their little homesteads, many of them possessing comfortable cottages on their patches of land, upon which they grow the sugar cane, and grind it on the neighbouring estates for half the produce, by which they obtain a considerable sum, besides the provisions they send to market, and their daily money earnings whenever they choose to work on sugar plantations. Thus they are in comparatively easy circumstances, and are fast approaching an important position in the community." Their main blemishes are said to be, a disregard of truth, and a tendency to petty pilfering. That their morals should be low is not surprising, considering that they have either been slaves, or are the offspring of those who had been slaves, and that they have not had the benefit of good teaching or training. A late Governor General, Sir William Reid, who paid much attention to the school system in the West Indies, specially remarking on the schools in Grenada, observes "most of the schools are of a miserable description," adding "I have endeavoured to inculcate the principle that when schools are made practically useful, they will be popular. Instruction will be sought for when it teaches how human labour may be abridged;" *—he, I believe, being of opinion that schools for the children of the working class should be industrial schools, and the teaching made as practical as possible, even in regard to morals. To recur to their material condition; besides growing provisions as in St. Vincent, most of them cultivate either a patch of canes which gives not less than a barrel of sugar, or have a portion of land growing coffee or cocoa. The price of land sold in small portions is from five to ten pounds an acre, and the rent of such land about one pound six shillings yearly per acre. Their great ambition being to have land of their own, it is not surprising that of late years the number of small landed proprietors should have increased, that villages should have sprung up where there was not a house before, or that detached dwellings should have become widely spread and an independence acquired.

Whilst this class has so prospered since emancipation, the proprietors of the larger estates have been very unsuccessful. The Bishop of Barbados in a visitation report made in 1846 states, "that Grenada has hitherto suffered in her agriculture more than any other island in his diocese from the abolition of slavery." And from a return made to Sir William Reid, in 1848, it would appear that of the total number of estates, viz. one hundred and twenty,—twenty-one

* Blue Book, 1848, p. 26.

had been abandoned. In a letter now before me of a proprietor, it is stated "that on the 31st December, 1828, there were about one hundred and fifty estates of all kinds, of which eighty-eight at the outside are now in cultivation to about one half the extent of what they were formerly, and unless something is done speedily to sustain them, at least one half of the present number will go out of cultivation in next year, unable to meet the fatal measure of equalization of duties which is to take place in 1854, in competition with the foreign slave holders." This gentleman is the proprietor of two estates in the island. In a letter preceding that just quoted, he mentions. "I have just abandoned a fine estate which I bought in 1837 on the faith of English honor, and the guarantee of a British act of parliament, for £6,000, which now stands me in £14,000 in round numbers. I am half inclined to abandon the other which I have, and I fear that eventually I shall be obliged to do so from the want of labour to carry it on: I fear that any remedial measures now will be too late and that the proprietary body are too far gone to recover." Alluding to the final issue, the abandonment of estates, unless there be aid afforded in granting large loans and affording large immigration, he finishes with this remark. "Such of us as have youth on our side must begin the world afresh, and those who are aged, find an asylum in a workhouse, or in the grave." The diminution of agricultural produce tends to the same conclusion. In what was considered the most prosperous time of the island,—in the time of slavery, the produce of sugar, it is said, had been known to amount in one year to 39,000 hogsheads, whilst recently it has reached only about 9,000 hogsheads. Even in 1825 it had diminished more than one half; in that year the amount in pounds was returned as 31,609,587, and in 1846 as 9,462,872.\* Nor is there any compensation in an increased production of any other article; on the contrary, owing mainly as I am informed to the disappearance, by emigration or death, of the small White proprietors, by whom these crops were principally cultivated, less coffee and cocoa now is grown and exported than in the previous period of slavery.

The causes of this disastrous condition as regards the larger properties,—the sugar estates,—do not appear to be obscure; they are the same, aggravated, which have been so injurious to the interests of the planters in St. Vincent. They may be referred in part to the proprietary body, and in part to the labouring class.

A majority of the former are absentees; in 1848 out of 120, the then total number of proprietors of the large estates, 73 were such. Moreover, most of their properties are mortgaged at a high rate of interest, from 5, to 6 or 8 per cent., and in the hands of restricted agents. It is unnecessary to dwell on other minor circumstances regarding this body, tending in their influence to the same injurious effect, such as want of science, of intelligent enterprise &c., the common

\* Blue Book 1848, p. 35.

accompaniments of the two graver conditions,—absenteeism, and encumbering, paralysing debt.

As regards the labourers, though so diligent when working on their own land, they are described when employed by others,—by the planters, as the reverse of diligent; 'their attendance irregular, their work careless and of little value;' owing, it can hardly be doubted to the circumstances in which they are placed, and the temptations to which they are exposed, especially in the manner of hiring, after that followed in St. Vincent often carried to extremity, viz., in allowing them, partly in lieu of payment of wages not only a house and a portion of land to grow provisions, but almost an unlimited quantity of such land.*

The evils of the system are well described in a report of one of the stipendiary magistrates made in 1844; referring to it he says, "I cannot avoid considering the system (of giving houses and grounds as payment in part for wages) most injurious to the planter, and one which has led to the ruin of numerous properties in this island, it being so difficult to lay down any thing like a correct or practicable rule for ensuring the labour required, while such free opportunity is given to the labourer to place himself upon a different foot-

---

* The money wages in Grenada are much the same, or but little under those paid in St. Vincent, varying (in 1846) from 5s. to 6s. per week (the labourers week of five days.)  In some instances when a dollar a day, or a dollar a task was paid, an allowance of salt fish was made ; 2 *lbs.* per week, and in some an allowance of rum ; and very generally medical attendance was provided at the cost of the estates,—a plan, a modification of that in use in slavery,—resorted to from real or supposed necessity or thoughtlessly entered on without a foresight of consequences.

ing to his own advantage ; as an example ; on many
estates in this district where the mountainous character
of the land is an obstacle to its being put into cane
cultivation, the labourer through the carelessness or
apathy of the planter is allowed to cultivate land in
such parts without any restriction as to quantity, and
the result is that he gradually furnishes himself with
the means of living, without caring to work for wages,
and after slowly lessening the number of days in which
he labours for the estate, he drops it altogether, and
depends on his own resources : the proprietor of course
complains, and a rent is charged for the house and
ground : the labourer frequently refuses to pay, and
the stipendiary magistrate is called upon to eject him.
Upon an investigation of such cases it generally appears that the proprietors or managers are in error ;
and when eviction is ordered to be carried into effect
as the only means of remedy, they turn round and
state that they do not wish it, as they will then lose
the labourer altogether."* Besides this bad system
there are other causes in operation conducive to the
same end,—the withdrawal of the labourers from the
estates, and the separating the interests of the labourers
from those of the proprietors. One of these is a neglect
on the part of some, in exactness of payment of wages,
which, according to usage is expected fortnightly or
monthly ; any breach on this point, it has been officially declared, ' soon leads to the extinction of cultivation of an estate.' Another is the want of a labour

* Blue Book, 1846, p. 197.

rate; some labourers giving only seven hours labour, while others perform nine hours a day, both expecting to receive the same pay and allowances. A gentleman deeply interested in the agriculture of Grenada, the present chief Justice of St. Vincent, in a printed letter on some of the difficulties which the proprietary body have to struggle with in this island, adverting to this want, and the benefits likely to result from the correction of the evil after the example of Barbados, remarks "That step was taken there in 1835 by order of Sir Lionel Smith, who referred it to three experienced planters to ascertain and settle what ought to be considered a fair day's work in the agricultural and manufacturing labour on an estate, and to make their report to him. This was done, and their scale of work was approved by the executive council," adding, "This judicious measure has contributed greatly to promote the singularly good feeling which exists between the master and the labourer throughout that island, and complaints between them for non-performance of work, and non-payment of wages are seldom if ever heard before the magistrate."

That agriculture is not in an advanced state in Grenada under the circumstances described is no more than could be expected. Little has been done to improve it, or even to economise human labour so much needed by substituting the plough for the hoe, and by introducing other implemental aids. The only ones of which I have seen notice taken in the reports have been the hoe-harrow for weeding canes; and that seems to have

been only very partially used; and a railway from the
cane mill for conveying away the megass (the crushed
canes). In some instances,—they are few; the metairie system has been tried in consequence of the difficulty of securing labourers. The conditions are stated
to have been as follow, "if the contractor digs the cane-
holes and plants the cane, performing the other parts
of the required labour, he receives one half the produce; but if he takes a part before planted at the estates expense, and called ratoons, he gets one third
when the sugar is made; the estate finds stock to carry
the cane to the mill, manufacturing utensils &c." How
this plan has answered I am not certain: by some it is
spoken of favorably, by others the reverse.

In illustration of the depressed, and distressed state
of the planters, I shall notice briefly a few things
which I myself witnessed, or which I learnt from unquestionable authority, in my visits to the island.
And I shall commence with a case, not an extreme
one, for I believe the individual is free from debt; it is
that of a French gentleman, a man of polished manners,
well advanced in years, resident on his own property,
now merely a cattle farm, who had once been in opulent
circumstances. His house, in which he resided with
two daughters, was plain and clean, very much patched,
wood not unfrequently being the substitute for panes
of glass in the windows. The furniture was old and
worn, but decent. The dress of the gentleman and
his youngest daughter,—she a pretty, graceful, unaffected, lady-like girl just passing into womanhood,—

was coarse but clean, and yet becoming; the elder daughter, a widow, we did not see. All was in keeping and in character, denoting change from better days, but without degradation. Turned of seventy, as we were informed, he seemed hale, active and cheerful, and fond of his retirement; he spoke of going into the town with dislike. He maintained himself, we were told, chiefly by his cattle. None of his land—it was an extensive wild tract—was in cultivation excepting his garden in which he worked himself, and his hands were almost black from exposure to the sun in this occupation. He was we were assured generally respected for his worth, and amongst the officers of the garrison he was a special favorite; great indeed was the charm of his simple gentlemanly manners, his cheerfulness, and total absence of affectation. May I add, that strangers as all of the party were, but one who introduced us, and he was but a slight acquaintance, we were received kindly and courteously and entertained hospitably, I need not say not luxuriously, unless turtle's eggs,* common in that part of the island, which were offered to us, be considered so, and eau sucré, presented to us by the fair hands, and they were fair, of the young lady, and prepared by her, using brown sugar, for which her father thought proper to apologize, there being no white sugar in the house.

* There was an ample supply of them obtained from an adjoining beach, a breeding place of turtle: they were kept in salt water, which, we were assured by our experienced host, was the proper mode of preserving them.

Owing to the depressed state of agriculture in this island, and the facility of renting estates at a low rate, some enterprising planters from Barbados have been tempted to come here with sanguine expectations of success commonly ending in disappointment. As an example, a large property in the neighbourhood of the town, very favorably situated, was pointed out to me, which had been taken by a Barbadian when nearly out of cultivation ; and, though he paid a rent of only about £50 a year for it, and sublet portions of land to the same amount, I was assured that he could barely earn a subsistence, and that it was his intention to return to Barbados, where he expected to do better.

As a contrast to the state of the island last named, I may mention an incident I learnt from a planter, also a native of Barbados, renting a large estate near the town, how he found, 'discovered,' a negro family who had established themselves on a portion of it overgrown with wood, where they had cleared a piece of ground and had remained there cultivating it, quite unknown to him, and without even the pretence of any leave, for two years. This is called "squatting," which according to my informant is not unfrequently practised in Grenada where there is a large surface in forest, and capable of great fertility, the latter circumstance tempting, the former favoring the enterprise, and more than counteracting the disposition on the part of the labourers to 'associate together.'*

* In a report of a stipendiary magistrate (Blue Book 1848, p. 30) in reply to a query relative to "Squatting," it is stated "there is none, and

Considering what has been stated before,—the properties thrown out of cultivation, the majority of the estates encumbered with debt and heavily mortgaged, —these few instances may suffice; it would be easy to add to them and draw more painful pictures of distress, which I willingly avoid.

What the future fortunes of this island will be must depend on circumstances, and probably more on the decisions of the Home Government affecting these colonies than on the inhabitants themselves and the Local Government, at least in the first instance. Unless some measures be taken by the former in aid of the West India planters, it is likely, judging from the past, that either all the sugar estates in this island will be given up,—be thrown out of cultivation, or be sold for what they will bring, and pass into the possession of unincumbered hands and be more and more broken up and divided: and that, in conformity with this, the White portion of the population will more and more decrease and lose influence, and the colored increase and acquire power. What its future lot is to be, can be conjectured only: even supposing the worst, the latter, eventualities, keeping in mind the natural advantages of Grenada, and believing that the African

---

for this reason,—labourers are too fond of associating together and are too well off, to think of squatting." That labourers in the West Indies are addicted to this practice the government query denotes; it is notorious; and also that it is never attempted but where there is an excess of land, and where in consequence labour is scarce and labourers are well off; the love of perfect independence, may with some be more powerful than that of 'associating together.'

has a capacity for improvement, I cannot but think there is tolerable ground for auguring favorably of the future. I shall mention some of these advantages in addition to those already described.

The mountainous nature of a considerable part of the island, and its sheltered and well watered vallies are peculiarly suitable for small farms and for garden cultivation. The climate too, and soil, as well as the abundance of water, are well fitted for fruit-bearing trees. Even now Grenada abounds more in fruit, especially the most prized,—the orange and its varieties, and the pine-apple, first introduced by the French, than any of the British Antilles. Here they have escaped the blights which have proved fatal to them, including the cocoa nut palm, in our older colonies, viz. Barbados, St. Kitt's and Antigua, where, except in some isolated favored spot, an orange tree is never to be seen, and a cocoa nut rarely, unless in a withered, wretched state,* and where, most of the oranges in use are imported and sold at as high a price, or even higher, than in England.

Moreover, from the convenient position of the island about midway between St. Vincent and Tobago, about 80 miles from the one, and 90 miles from the other, and between Barbados and Trinidad, with excellent

---

* I well remember the impression of a painful kind made on my mind shortly after my first landing in Barbados, at the sight of the cocoa nut trees of that island, most of them dead or dying from the effects of blight; mainly, it is believed, owing to the invasion of a minute insect, contrasting this wretched appearance with the beautiful healthy groves of the same palm that I used to admire in Ceylon, and in witnessing in my visits to the other islands, the same diseased state of this useful and valuable tree, threatening its entire loss.

ports, especially that of St. George, adjoining its principal town of the same name, already the coaling station of the West India steam packets. This island is peculiarly well adapted for trade, and possesses more than ordinary facilities for exporting its excess of produce, especially fruits ; of which and of vegetables, even now large quantities are sent weekly to Trinidad and Barbados, and for obtaining supplies of all articles needed from abroad.

Even the circumstances tending to the deterioration of the larger properties and the absenteeism of their proprietors,—such as the nature of the roads, commonly little more than bridle paths,* the difficulties, owing to this state of the roads and the distance of estates, in the way of social intercourse, and moreover the almost total want of amusements of all kinds, whether ordinary or refined,—those which constitute the recreations and enjoyments of the wealthy and easy class in Europe, residing either in town or country,—would be little

* I have only heard of one wheel carriage in the island, and that a gig ; and in connexion with an accident, an upset attended nearly with the loss of life of a worthy commander of one of the packets. The extent of carriage road is almost limited to that connecting the garrison on Richmond heights with the town and port, a distance of little more than a mile. Such is my own impression of the state of the country as to roads. The following is a better account of them, at least of their extent,—with which I have been favored by a gentleman well acquainted with the island. " The roads in Grenada are not at all times passable for carriages ; in the wet season they are very bad. The extent of road round the island is about 50 miles ;—across 14 miles, and two others of 6 and 9 miles in other directions, making an aggregate of road repaired at the public expense of 79 miles."

felt or objected to, if at all, by a population such as the people of color in their present transition state, who have never known any better condition, but a worse one than their present;—who as yet can well dispense with carriage and cart roads, having neither carts nor carriages, and who are not without their own amusements, which, rude as they may be, are to their taste ; and with their daily occupations fully secure them from that tedium and weariness of spirit to which the European, especially the educated woman, is subject in these regions.

By many, the idea of an improving, advancing community of the colored race, a chief proprietary and influential body, will probably be held to be Utopian and impracticable. I cannot so consider it. I would fain believe that, were exertions made, such as are required in the way of education, in the manner proposed by Sir William Reid, that, what is visionary in the idea, would disappear, and that a well regulated and instructed community of Africans might here be formed ; affording proof that the African is not by nature degraded, labouring under a curse, and that as he has been called by philanthropists, a fellow creature and a brother, he is so, apart from sentiment, in reality.

I am glad to find a confirmation, or at least some support to these hopeful views, in the opinion expressed by the Commissioner General of Population, in his remarks on his special subject, which I subjoin in a

note.* " It appears," he continues, " that the deficiency
of increase not being traceable to physical or natural
causes, must rather be attributed to the social and
moral habits of the people. It is true, that the outward decencies of civilized life are more highly respected than formerly ; but that an unreserved pro-
miscuous intercourse of the sexes still prevails to an
undue extent, is frequently the theme of regret to all
who take an interest in the moral and religious state
of the country."

" A general system of education, accompanied with
moral and religious training, must form the basis of

---

* Concluding his report on the results of the last census, adverting to the
small annual increase of the natives, he says ; " How far their occupations,
habits, and diet are calculated to invigorate the constitution or improve the
physical and mental powers, or otherwise of the labouring population is a
question that would require a closer investigation than can at present be
instituted ; but it is remarkable that notwithstanding the advantages of a
climate proverbially healthy (to the colored race) with abundance of food,
and a description of labour not requiring an over exertion of the physical
or mental power, and the general comfortable condition of the peasantry,
the increase has yet been so far below the ratio which might have been
reasonably expected. An opinion pretty generally prevails, that although
infants and children are exempt from many of the diseases fatal to them in
European countries, the greatest mortality is to be found here amongst this
class, supposed, and with much apparent truth, to arise either from a disin-
clination on the part of parents to incur the expense, or a habit of never considering any ailment of sufficient importance to require medical assistance
until it assumes a character which renders the medical advice then sought,
totally useless."

This supposed cause of the slowly increasing population of this and most
of the other West India colonies, will, I believe, be more and more confirmed by inquiry. In Trinidad, in the town of Port of Spain, the mortality
amongst children as estimated by Dr. Gavin, is as high as 1 in 17 or 5·8 per
cent. *See Dr. Gavin's Report on the health concerns of this colony,*—an interesting and valuable document.

any future improvement. But here, as in the mother country, there are many obstacles to be overcome; and not among the least is, that education is by many supposed to imply an immunity from manual labour. The parents not having received the benefits of education, are not able to impress on the minds of their children a due appreciation of its advantage, nor prepared to make any sacrifice to obtain it. From old associations, agricultural labour has been considered degrading; and the facilities afforded of a competency from the sale of the produce of provision grounds, wood, and charcoal, render all agricultural labour uncertain and precarious. It is somewhat singular," (is it not rather the rule amongst a people emerging from barbarism?) " that the blessings of abundance should, it may be said, form the principal barrier to the moral advancement of a people;—still upon the whole, there remains sufficient evidence to warrant the lively hope that the natural resources and moral elements of the island, both latent and active, judiciously fostered and cherished by measures tending to their full development, and powerfully assisted by a total absence of all class prejudices, are sufficient to cause at no distant period, a great improvement in the general tone of society, and produce a more rapid increase in the population of the country."

Though to the naturalist, and especially the geologist, Grenada offers an ample field for enquiry, hitherto little explored, to the passing traveller it presents but few special attractions. Owing to the state of the roads and there being no places of public refreshment of any kind out of the town, excursions beyond a few miles, cannot be extemporaneously made; much arrangement and friendly aid is needed to make them to any of the more distant parts of the island, which in consequence are little known, beautiful as we are assured many of them are, except to the natives. Two spots however are within reach, not more than six or seven miles from the town, one or other of which may be very well visited, even during the few hours that the steamer stops. These are Pointe des Salines, and the Grand Etang, in going to and coming from which, a tolerable idea may be formed of the character of the country generally.

The former, the Pointe des Salines, is the extremity of the S. W. Coast, one of the many promontories of the island, and almost the chief of them. Near its termination, on one side, is a sandy beach, the resort of turtle that come there to lay their eggs; and on the opposite side, a like beach, within which is a small salt lake, the " Salines," of perhaps a hundred acres of surface, where occasionally in the dry season, pretty much salt is made by the natural process of evaporation. The year preceding my visit in 1848, thirty or forty bushels were obtained. The salt is derived from sea water, which percolates through the beach, the bottom of the lake being a little lower than the sea level, or that part of the shore on which the waves break. I was told by one of my companions that sea water is admitted by a passage cut through the sand; but this I was assured

H h

by the proprietor whom I afterwards met, is never done The country intervening between this headland and the town, is either almost level, or only slightly hilly, having a shallow poor soil, resting on tufa, mostly in horizontal beds, occasionally interrupted and broken by basalt or crystalline trap rocks. Where the path passes near the shore, some fine examples of volcanic ejecta, forming a coarse conglomerate, are to be seen; and the same on the shore itself, associated with sandstone, now in the act of formation, the cementing principle of which is carbonate of lime separated from water of the sea by the escape of its solvent carbonic acid. The quality of the vegetation of this district accords with its soil. The plants most common and conspicuous are dwarf prickly mimosæ and euphorbiæ, with long erect cylindrical leaves. None of it was in cultivation;—it was entirely given up to pasturage. We saw as we passed along, a good many cattle, some sheep, and a large flock of goats,—the latter, always an indication of poverty, which the wretched appearance of the people who took care of them visibly confirmed. In returning from the Salines we followed a road which led into the interior, and we had an opportunity of seeing a sugar estate, the natural fertility of which, judging from the vegetation, was even more remarkable than the sterility we had previously witnessed. This property was ornamented with many stately cabbage palms, with which were intermixed areka palms, of a height and size exceeding any I had ever seen before, even in Ceylon, the native soil and climate of this elegant tree.

The Grand Etang is a small mountain lake, in a ba-sin-like hollow, almost circular in its form, once pro-bably the crater of a volcano, situated nearly in the middle of the island, at a height of about 1742 feet above the level of the sea. Surrounded by native forests gently sloping from the heights above, it forms a very beautiful scene. The views from one or two of these heights, to which a path has been made, are very extensive and fine, especially over the comparatively open country, towards the windward coast, and Gran-velle Bay, which was distinguishable. A tolerable road passes close by the lake leading to the other side of the island, so that one may ride the whole distance ; and a very delightful mountain ride it is ;—the air cool, the path shaded the greater part of the way, but not so much so as to shut out the varied views, changing constantly with every turn in the ascent. The lower part of the road is through land in cultivation, chiefly the property of labourers, whose cottages, amidst fruit trees on the sides of the hills, with patches of cane ad-joining, or of cocoa plantation, realized all I had heard of their thriving condition. Higher up, all cultivation ceases, but luxuriant vegetation continues, comprised chiefly in forest trees of no mean growth, in clumps of bamboo of extreme luxuriancy, and in the beautiful tree fern. The Thermometer in the shade, at ten o'clock, stood at 74° This was on a patch of table land near the lake, on which is a police station. It is curious to notice, how accounts of the same object differ. We were told that this lake is unfathomable. According to a statement in the British Colonial Library, it is only 14 feet deep. According to the same authority, it is two miles and a half in circumference. To me, it appeared no more than a quarter of a mile. According to common report, it has no outlet, but we saw one, in a narrow piece of water, communicating with it, which, though then stagnant, ( it was a season of drought,) was evidently running water after rain, judging from the bent rushes on its sides, and their direction, and in the map of the island, the "Grande Riviere" is marked as having its source in the " Grand Etang."

# CHAPTER VIII.

## TOBAGO.

Important relative position.—Character of the country.—Its geology, soils, and climate.— Hurricanes.— Population.— Social condition.— Agriculture.— The parts of the island best worth seeing.

THIS island is the most southern of the Antilles with one exception, and next to Barbados the most windward,*—a position to which some importance has been attached, both commercially and in a military point of view, especially in relation to the adjoining island and more southern,—the great island, as De Foe calls it, of Trinidad†—and the gulf of Pera, and from the circumstance of its having at least one excellent harbour,—" Man-of-War Bay," capable of sheltering the largest fleet.‡

In point of size, though very different in form, it is

* Its principal town, Scarborough, is in latitude 11, and longitude 59·48.

† Where nearest, the distance between the two islands is said to be only six miles; but incorrectly. It is eighteen geographical miles, namely, from Brown's point, (the south west extremity of Tobago,) to Punta de Galera, (the north east point of Trinidad.)

‡ See some excellent remarks on this subject, in Sir William Young's West India Common Place Book, p. 191.

about equal to Barbados.* Its extreme length is
stated to be about 32 miles; its extreme width about
9 miles. Its character throughout is hilly rather than
mountainous, its scenery pleasing and gentle rather
than bold, its highest hills hardly reaching to an elevation of 2000 feet, and their declivities though often
steep, seldom rugged or broken into rocky precipices;
indeed the absence of naked rock excepting along its
coast, especially its abrupt northern one, has been
pointed out as one of its remarkable features. Another
of its remarkable features is the abundance of wood,
and the luxuriance of its vegetation, and also the
abundance of water. Seen at a distance it has the
appearance of a forest, no naked summits being to be
seen and not a single naked spot, except patches here
and there in the vallies, and on the sides of the hills
under cultivation and cleared in consequence; and in
travelling through it, every valley is found to have its
running stream, the largest of which when swollen
by heavy rains, are either impassable or fordable with
difficulty and risk.

The geology of the island is not without interest
and is in some respects peculiar. Most generally considered, it may be divided into three parts; one the
small and low district of Sandy Point at the south
west extremity of the island, composed of coral rock,

---

* Its area, according to authorities in the island, exceeds that of Barbados, by three or four square miles. According to another statement, it
comprises only 120 square miles, or 66 less than Barbados. As it has not
been accurately surveyed, any estimates of this kind can be considered little
more than conjectural.

calcareous marl and freestone, closely resembling those of which the larger portion of Barbados is constituted;*—another, the north east extremity of the island, more hilly and bold, commencing at King's Bay, formed of mica-slate, containing scattered through it, isolated masses or veins of colorless quartz; the third and the greater portion, lying between the two extremities, which it joins by a mountain ridge, as it were the axis or spine of the island, formed of a crystalline rock of variable appearance and composition, in some places, having a basaltic character, in others a granitic and in some a porphyritic. Felspar, hornblende, mica and quartz, and more rarely calcspar may be mentioned as its mineral ingredients, variously intermixed, and in consequence producing so many different appearances.† Its specific character I may

* Not only is the lower portion of this district formed as above described, but also the ordinary hills. At a height of at least 200 feet above the level of the sea, marl and shell limestone occur. At one spot of this elevation, I saw shells in the marl, (barnacles,) the original colour of which was still distinctly retained. Like that of Barbados, this marl consists principally of carbonate of lime in a very comminuted state, abounding in shells, mostly in fragments; traces were detected in it, both of phosphate of sulphate of lime, and also of silica.

† At Indian Walk, on the north side of the island, a highly crystalline limestone has been found, about 100 feet above the level of the sea, and a few feet under the soil. Where purest, I have detected in it only a slight trace of silica and magnesia. Some specimens from the same place have consisted of greenish colored quartz and calcspar intermixed with particles of iron pyrites.

Crossing the island from north east to south west, a rock occurs, which, judging from the specimens of it I have seen, has a considerable resemblance to basalt, being of the same dark color and fine grain. Minute fragments of it are attracted by the magnet, probably in consequence of the presence of iron in its metallic state. In some places it has a porphyritic character,

remark is oftener ambiguous than distinct, suggesting the idea, on the supposition of igneous origin, of not having been slowly cooled, and of rapid transition in consequence from the liquid to the solid form, interfering thereby with a more regular arrangement of the elements, such as we witness in our common granites and trap rocks. A common property of the crystalline rock in most of its varieties, from a predominance of felspar, is a readiness to decompose, owing to which, no doubt, there is so little naked rocky surface to be seen, and that general tameness of scenery already alluded to. The same circumstance will account for the beds of clay which are of frequent occurrence. Associated with the crystalline rock, and incumbent on it, beds or strata of shale are in some places observable; an example of the kind presents itself not far from the town of Scarborough in the neighbourhood of the government house. What is more remarkable, and worthy of note, coral rock is to be seen at one spot on the sea shore near Plymouth, resting on the crystalline rock, and this many feet above high water mark. The adjoining cliff to a considerable extent, viz. from Courland Bay to Englishman's Bay, consists of the same rock,—coral, often of considerable height, and here

from containing crystals of felspar. Its width, considering it as a vein or dike, I could not learn. It forms the cliff and shore-rocks at Bacolet in the neighbourhood of Scarborough.

In many places the rock is almost entirely felspar finely granular, and not unfrequently a mixture of felspar and horneblende,—more rarely of felspar and mica, and of these and of quartz, (seldom crystallized.) I have one specimen of resplendent horneblende, a fragment apparently of a rock of this quality, but its locality is not noted.

too, probably resting on the crystalline rock. On the same N. W. coast instances occur of a rock formation in actual progress,—the sand becoming converted into sandstone by a precipitation of cementing carbonate of lime from sea water.

Though earthquakes are reported to be not uncommon, of a slight kind, I could not hear of any traces having been observed of volcanic action, nor of any appearance of an extinct crater. The island is without any hot springs or mineral waters, at least none have yet been discovered; nor have any infusoria or mineral pitch or coal been yet found such as occur in the smaller hilly district of Barbados. There is also a remarkable absence of lakes or ponds. These are negative circumstances, but not on that account undeserving of note.

The soils of the island vary more or less with the formations on which they are incumbent. In Sandy Point district a calcareous loam is common, and a light calcareous sand. In the N. E. district a clay soil is prevalent; as it is also over the larger portion of the island, being derived from the decomposition of the crystalline rock, and is more or less mixed with gravel or minute fragments of the parent rock least disposed to disintegrate. Generally the soils of Tobago appear to be very good and deep, and well supplied with the inorganic elements most essential to fertility, as may be inferred from the luxuriancy of the wild vegetation, even without considering their composition, and the manner in which the most of them are formed. In the

district of Sandy Point it is said that the soil once excellent and fertile, is to a great extent exhausted, which
is as little to be credited, as that that of Barbados was
exhausted a hundred years ago, according to the then
popular belief.

This brief notice of the rock formations and soils of
the island is given almost entirely from my own observations, which were more limited than I could have
wished: were minute enquiry made, no doubt much
useful information might be collected respecting both.

As regards climate, Tobago can hardly be considered
in any way peculiar, comparing it with the islands
already noticed, Barbados, St. Vincent, and Grenada.
As in these islands, the driest months are commonly
those of January, February, March and April. Heavy
or considerable rain is expected, it is popularly said,
about the 1st full moon in May or shortly after.
Occasionally in this month there is a deluge of rain;
eleven inches have been known to fall in 24 hours; it
amounted to this on the 18th of May, 1848. These
heavy falls of rain are not unfrequently attended with
slips of land; in some places constituting a feature of
the landscape, from the nakedness of the denuded space
on the hill side, the more remarkable in the midst of
forest the prevailing clothing of the country.

The following table contains the results of six years'
recorded observations with the rain gauge kept at Fort
King George, at the moderate elevation of 420 feet
above the sea level.

| | 1847. | 1848. | 1849. | 1850. | 1851. | 1852. |
|---|---|---|---|---|---|---|
| January .... | 6·54 | 3·18 | 4·39 | 3·34 | 2·92 | 3·52 |
| February.... | ·69 | 3·67 | 0·78 | ·76 | 2·11 | 2·24 |
| March ...... | 1·41 | 3·63 | ·65 | 1·50 | 1·26 | 3·30 |
| April ...... | 1·80 | 2·49 | ·65 | 2·00 | 4.40 | 1·99 |
| May ...... | 1·58 | 15·13 | 12·23 | 5·50 | 9·00 | 7·81 |
| June ...... | 4·14 | 9·37 | 11·61 | 17·00 | 11·87 | 6·50 |
| July........ | 10·61 | 8·50 | 8·26 | 9·72 | 6·90 | 11·13 |
| August .... | 5·91 | 6·52 | 14·08 | 11·37 | 9·26 | 5·18 |
| September .. | 10·21 | 9·21 | 9·82 | 5·54 | 9·86 | 9·84 |
| October .... | 7·51 | 6·71 | 18·39 | 5·68 | 8·17 | 15·39 |
| November .. | 15·20 | 6·77 | 8·55 | 7·83 | 4·99 | 10.24 |
| December .. | 11·93 | 10·16 | 8·09 | 7·59 | 8·65 | 7·59 |
| | 77·53 | 85·14 | 97·50 | 77·83 | 79·39 | 84·73 |

As regards atmospheric temperature, its vicissitudes, from what I could learn, are somewhat greater than in the other islands already mentioned, the heat by day being often greater, and the cold by night,—both owing to calms, and the former, in addition, it may be, from the prevailing wind being mostly south of east. The island has had the reputation of being exempt from hurricanes, and that entirely. It would be more correct to say that they are here of comparatively rare occurrence. One took place in 1831,—another so recently as the 11th November, 1847, which, in its destructive effects, was but little inferior to some of the most severe that have laid waste the islands more especially subject to them.*

* Eighteen persons were killed by buildings falling on them; a large number were wounded, of whom eighteen died. The greater number of the houses throughout the island were more or less damaged. Many of them, and some of the most substantial, were levelled with the ground. The de-

As regards salubrity, the climate of this island is inferior to that of St. Vincent and Barbados. It is nowise favorable to the constitution of the European, and judging from its limited population, hardly more fa-

vastation effected in the native forests was great; stripped of their leaves, they presented, as described by an eye witness, a most unnatural wintry appearance,—and great was the damage done to the cocoa nut trees, and the cane crop, at the time approaching maturity. When I visited the island, in the following July, eight months after, marks of its devastations were everywhere to be seen, especially in the forests. What was most remarkable, was the variety of effect that these exhibited, and that within a very limited space ; conveying the idea that the air in motion constituting the hurricane acted in narrow currents. Thus it was not unusual to see one line of wood completely overwhelmed, a perfect wreck, levelled with the ground; while the adjoining trees had either suffered but little, being deprived only of a portion of their branches or, altogether unhurt. The same appearance of partial action, I was told, was witnessed in the cane fields, in which I was assured narrow strips were destroyed, in straight lines two or three feet wide, the adjoining canes not suffering. And instances of the same, and even more remarkable, I heard related, as witnessed by individuals. The Lieutenant Governor told me, that when the storm was at its height, and he was obliged to leave his house, to seek shelter in a cellar below, he carried a lighted candle the short distance he had to go, in the open air, and it was not blown out. This he mentioned, when another not less singular anecdote was related;— how in a house in which the books were blown from their shelves, and scattered about the room, one of the candles that was burning on a table, in the middle of the room, was not extinguished. Many other examples of such partial action were spoken of,—instances of frail structures escaping, when strong buildings adjoining were blown down,—of windows strongly barred, forced open, when others in the same house, the shutters of which were fastened only by a weak bit of cord, remained closed and uninjured. When the tempest was raging at the N. E. extremity of the island, the Barometer at Fort King George was little affected, though shortly afterwards, when it was witnessed in its violence there, the fall of the mercury was considerable—a little more than half an inch. This I learnt from the Medical Officer of the station, who watched the instrument at the time. During the hurricane there was a veering of the wind, almost in a circle, and the vortical tendency of the air in motion, I was told, was well exemplified in the gyration of the Reindeer steam packet, in the gulph of Pera, off Port of Spain, Trinidad, over part of which island the hurricane passed ; and though with diminished, not without destructive effect.

vorable to that of the colored race. This seems to be indicated by the returns of the population made at successive periods. An opinion is common that since 1839 there has been an increased salubrity, that the rain since then has been less heavy, with diminished danger from quicksands at the mouths of rivers, —a danger often experienced here,—less thunder and lightening, and as regards febrile diseases, the greatly predominant class,—a larger number of the common continued kind, and a smaller one, of the more dangerous and fatal, the remittent.

The population of Tobago, in accordance with the preceding remark, for many years past has been almost stationary,—the colored races, if at all, but very slowly increasing, and the white, but in a more marked manner, diminishing in numbers, as shewn by the following table formed from official returns.

|      | WHITES. | FREE-COLORED. | SLAVES. | TOTAL. |
|------|---------|---------------|---------|--------|
| 1776 | 2,800   | 1,050         | 10,300  | 14,150 |
| 1787 | 1,397   | 1,050         | 10,539  | 12,986 |
| 1805 | 900     | 700           | 14,883  | 16,483 |
| 1833 | 304     | 1,266         | 11,628  | 13,198 |
| 1844 | 212     | 12,996        |         | 13,208 |
| 1851 | 122 ?   | 14,256        |         | 14,378 |

In the last census the distinction of color is not made, and in consequence the number given as whites is somewhat uncertain, and may be below the truth, being derived from the distinctions of nations and countries, these being as many as thirty-three altogether, but comprising (exclusive of the colored Creoles,) only 1634 individuals, of whom 38 are English, 66 Scotch, and 6 Irish.

The increase of nearly 1000 since 1844 is chiefly owing to immigration, imported labourers, most of them from Barbados.

The island, territorially, is divided into districts and parishes; the former three, the leeward with a population of 5876, the windward of 2349, the middle of 6153: the latter seven. The following table shews the number of acres and inhabitants in each parish, from the census return of 1844: in the last census, districts have taken the place of parishes.

| PARISH. | ACRES. | MALES. | FEMALES. |
|---|---|---|---|
| St. John ...... | 10,520 | .... 479 | .... 461 |
| Paul ...... | 7,558 | .... 488 | .... 453 |
| Mary...... | 10,447 | .... 410 | .... 382 |
| George .... | 11,192 | .... 771 | .... 794 |
| Andrew.... | 3,170 | .... 930 | .... 1,079 |
| Patrick .... | 5,801 | .... 1,100 | .... 1,170 |
| David .... | 8,720 | .... 1,358 | .... 1,354 |
| TOWN OF | | | |
| Scarborough | | .... 579 | .... 764 |
| Plymouth .. | | .... 229 | .... 235 |
| | 57,408 | 6,334 | 6,692 |

An analysis of the census might afford some singular and curious results, and is deserving the attention of those interested in statistical enquiries. I shall notice only a very few points. Of the white portion as given in 1844, of which the individuals are specified, it is remarkable that of the whole 212, the women were only 39, and the children only 39! Of occupations,—extending to 53 different ones,—5644 in the last census are returned as labourers, 103 as planters, and 5 only as proprietors and freeholders, and only the same number of "doctors,"—whilst there are 9 clergymen and ministers, 25 schoolmasters, and 11 printers. It is satisfactory to find, though in a place where one would least expect to find it, (a return of occupations) that the number of lunatics is only 4.

As to society in Tobago, except in a very limited way, and very much confined to the neighbourhood of the town of Scarborough, and the garrison, it can hardly be said to exist; the white population being so small and scattered; so few of the proprietors of estates being resident, and not many of their managers educated men. And the same remark applies to the men of color, even to those who, by their industry and intelligence, have rendered themselves independent of bodily labour for a subsistence. Another inpediment to agreeable social intercourse, is the disparity of the sexes ; the proportionately few white women, and the very many misalliances in consequence, with colored women. A clergyman told me, that his wife, during four years, had only once seen a white woman cross her threshold. It is right to mention, that they resided in a remote parish ; but that the largest of the seven.

Of the working class, the peasantry of Tobago, a favorable account may be given. Their condition is prosperous. They have the character of being sober ; and this, notwithstanding that, on many of the estates, an allowance of rum is made to them as a boon. In appearance, they seemed to me superior to the colored race in Barbados. By the stipendiary magistrates, who ought to be well acquainted with them, they are thus favorably spoken of. Further they are described as well clad and well housed; most of them having dwellings consisting of two rooms, made of wood, and provided with glass windows. As in the other islands, their taste in dress is not formed,—propriety and keeping, in dress, is not considered. On Sundays and holidays, the labourer and the planter are dressed so much alike, that by dress alone they can hardly be distinguished. A late Lieutenant Governor, (Major Græme,) writing in 1848, remarks, —" The peasantry of the West Indies, unlike those of Europe, have no distinctive dress, as a class; and upon Sundays and feast days, the employer and labourer are habited much alike." He adds,—"This is carried, not unfrequently, to an extravagant length; and until the labourer can be satisfied to adopt a cheaper holiday costume, and one more suited to his station and pursuits, I see little chance of the planter establishing a decreased rate of wages." " This propensity," he states, " was at one time encouraged; as an opinion very generally prevailed, after emancipation, that in order to secure the services of the field negro, it was expedient to foster a taste for articles of luxury; and unfortunately encouragement was given to the wearing of fine apparel, rather than to the acquiring the more profitable comforts and conveniences of civilized life." This opinion perhaps may be a little too strongly ex-

pressed, many of them, most of them having many comforts ; most of them have a sheep, a goat, or a pig, and
provision and garden ground ample for the labour they
can bestow on it.    Those who are industrious easily
better their condition.    An example of the advantages
they have, and that a common one, is given in the note
below*. Their ordinary wages, with a house and ground
more than needful, are 8*d*. a day, or 1*s*. without these.†
Attempts have been made to effect a reduction which
have commonly failed.

* "One of the Barbados labourers who arrived here in 1843, has since he
has been in the island, worked at Prospect Estate, at 1*s*. a day, and 20*s*. for
extra labour as a fireman at the coppers.    He has a good house, rent free, together with fire, provision-grounds, and medical attendance and medicine
found.    Besides which, he has worked land without rent or tax.    The following is the result of his industry this year :—

|  | £ | s. | d. |
|---|---|---|---|
| Two barrels of sugar, supposed weight 450lbs., at 25*s*.,  ......... | 5 | 12 | 6 |
| Four barrels of sugar, in canes, weight (estimated) 900lbs., at } 25*s*., less 25*s*., for labour, ..................................... | 10 | 0 | 0 |
|  | £ 15 | 12 | 6 |
| Deduct half of the above, for making the sugar. | 7 | 16 | 3 |
|  | £7 | 16 | 3 |

                *West Indies*, P. P., *August*, 1845, p. 136.

† In 1845, the rate of wages was returned, as follows in the middle district ; it varied but little in the other two, the leeward and windward.

"Labourers having a house and ground, rent free, together with medical
aid, and medicine, (received per diem.) 1st class, 8*d*. to 1*s*. ;—2nd class,
6*d*. to 8*d*. ; 3rd class, 3*d*. to 6*d*.

Labourers not enjoying a house and grounds, &c. free ; 1st class, 1*s*. to
1*s*. 4*d*. ; 2nd class, 8*d*. ; 3rd class, 3*d*. to 6*d*.

Tradesmen domiciled on plantations, from 8*d*. to 1*s*. 6*d*.

Journeymen tradesmen, from 1*s*. 6*d*. to 3*s*."

                        P. P., *August 8th* 1845, p. 134.

"On some estates, in the leeward district, a house and an acre of ground,
are let at 40*s*. a year, or thereabout; and by mutual agreement, the rent
is compromised for 40 days labour, in cane cultivation."

                        *Paper cited*, p. 136.

                    K k

On emancipation, about one third of the labourers withdrew from the estates, many of them becoming small freeholders. Since then, the latter have much increased; the number in 1847 amounted to 786.* It is expected, that ere long, their numbers continuing to augment, and their condition to improve, they will form a middle class. This opinion has been expressed by the late Lieutenant Governor,—a man not over sanguine in his views; and it is to be hoped that the expectation will be realized, if due attention be given to the rising generation; and that there is a fair prospect of this, may be inferred from the increasing number of schools and the increase of places of worship.† And, the terms on which freehold land

* The following return applies to 1845 :—

| | Freeholders. | Labourers. | Persons living in villages, built since emancipation, including scattered hamlets. |
|---|---|---|---|
| Windward D. | 49 | 1274 | „ |
| Middle „ | 151 | 1371 | 1263 |
| Leeward „ | 195 | 2298 | 235 |
| Scarborough | 263 | „ | „ |
| | 658 | 4943 | 1498 |

The last mentioned persons, in 1847 had increased to 2311.

† In 1845, there were eighteen public schools; in 1833, only one. In a report of the Stipendiary Magistrate of the middle District, for the same year, (1845,) it is stated:—" the churches and chapels are crowded on Sundays; except perhaps the church of Scarborough, where, nevertheless, there is not an unlet sitting to be procured. The pews are rented from 12s. to 20s. each sitting room, and in the church of Scarborough, about one-third are reserved as free seats." The Stipendiary Magistrate of the Leeward District, reporting in the same year, mentions, that a new church is being built at Plymouth, the negro population and others contributing towards its erection. Blue Book, August, 1845.

According to the later reports, the schools are rather fewer in number, and their efficiency not increased. Only £240 in 1849 was allowed from

can be obtained, and the moderate rent of land, and the profit resulting from judicious management, must greatly favor the formation of such a class.*

The favorable opinion thus expressed of the improving condition of the labouring class in Tobago I regret to see is not that entertained by a late resident proprietor, the author of *The Planters' Guide*, who not only denies that those of African descent have improved, but affirms that they have deteriorated, doubting even their capacity to improve, asserting that "all the churches, chapels, mission houses, and school houses that have been built, and the most praiseworthy labour that has been bestowed by the clergy and missionaries on the improvement of our black population, both as regards education, and religion is unquestionably so much labour lost, and money thrown away; which," he adds, "I fearlessly assert in spite of all the returns, or documents that can be produced."

the Colonial Treasury, in aid of them; and their masters were so poorly paid, that a labourer working on a sugar estate was said to earn as much as the stipend of a teacher.

* The price of land, in small portions, in 1848, was about £ 20 an acre; its rent from 24s. to 32s. an acre. Some land was let on the metairie system, which has been reported on favorably. When I was in Tobago, and it was in the same year, a property was pointed out to me, that had nearly ruined its proprietor, attempting the cultivation of it himself, which was then let to a colored man, at £ 150 a year,—a rent regularly paid; and another property, that was returning to its proprietor, (an absentee,) on an average three or four per cent. of invested capital, under the direction of an intelligent colored manager. These people manage to great advantage. They are not liable to be imposed on;—the attempt is not made. They know how to deal with the labourers, and are economical in all things.

I regret also to see that he refers in support of his extreme notion to an authority in *The Dictionary of Sciences, Literature, and Art,* inculcating the same views both of the inferior mental and moral qualities of the African, and of his incapacity to improve.

It would, I apprehend, be a waste of time and a trial of patience to enter into the details brought forward in support of the argument. That the improvement since emancipation,—so short a period, not yet 20 years, not more than 14 comprising the apprenticeship term,—has not been very great is surely not surprising if we consider the history of any people, the difficulty there is in changing habits, and in effecting civilization. Considering how these people or their parents were so short a time ago slaves without any moral or religious training or next to none, ought we not rather to be surprised that they are not worse, and that they have borne emancipation so well. The West India regiments, composed entirely of Africans excepting their officers, are a proof how tractable they are as men, how amenable to rule and discipline, and under good training how capable of making good soldiers, i. e. men having a sense of duties, with firmness and resolution to perform them.

The writer referred to by the author of *The Planters' Guide* ends his argument with the line, "Cœlum non animum mutant trans mare currunt,"—an aphorism no doubt true in part but only partially true, as we know too well that conduct changes with circumstances, and that high morality was not the characteristic of the

West India planters during slavery, on the contrary in relation to the sex a license, a licentiousness prevailed, even now too much indulged in, especially in this island. In one respect, if we may give credence to an account of the social habits of former times, as described by a writer in the island almanack, the saying is true; it is introduced after a good caution against exposure to the night air,—the "night's dark noon," after rising from the festive board, lightly clad, as is too frequently the case; adding, "Tobago has been always eminently distinguished for the hospitality of its inhabitants, and in former times, he would have been considered unsocial and unneighbourly who would not (whatever the nature of his constitution) "sit it out,'—In fact "who last beside his chair" was too commonly the order of the evening, and many instances are fresh on record of the *spree* having been kept up much in the spirit of Tam O' Shanter and Souter Johnny as so inimitably described by the immortal Burns, "they had been fou for weeks thegither." The most surprising matter is, that several who had figured so conspicuously on such occasions are "*still to the fore;*"—may we not add as "*devils' decoys.*" The writer continues,—"but these days have passed away. The reduced condition of the island of late years has tended to mar such *reunions*, though the rites of hospitality are still duly observed in general;"— to which I can bear testimony from my own experience during little more than a week that I had the pleasure of spending in the island in 1848, but it was a hospitality of the moderate and better kind, not carried to any excess, or tasking the powers of the constitution.

The agriculture of this island and its condition depending on its agriculture, has not been so prosperous as at one time was expected; it has been remarkable indeed for reverses similar to those of St. Vincent and Grenada, and even in a higher degree. After having been ceded to England by France in 1763, great exertions were made in colonizing it, with a large expenditure of capital. Between 1765 and 1771, land to the extent of 57,408 acres was sold, the property of the crown, leaving only a few hundred acres to be disposed of on the summit of the main ridge, of little value. Between 6000 and 7000 acres of these were cleared and brought into cultivation,* in addition to what had been previously in culture. Notwithstanding the capture of the island by the French in 1781, its cession to that power in 1783 by whom it was retained till retaken by a British force in 1793,—notwithstanding these adverse circumstances so great was the advancement of its agriculture that in 1805 it produced 15,327 hogsheads of sugar, and Sir William Young, (a competent judge being well acquainted with the island*) looked forward to the time when it would yield nearly double that quantity besides other valuable produce, such as long cotton-wool, fustic and hard

* See Blue Book, 1848.

† He was the governor of the island, from 1807 to 1815, the year of his death; and had visited it previously. So much was he esteemed and respected, that during his lifetime the sum of £2,000 was twice presented to him by the votes of the legislature, "for his unremitting exertions for the public good;" and on his decease, a tombstone of marble, with a suitable inscription, was placed to his memory, at the public expense. Good and able governors have been rare in these colonies; no where, judging from the manner in which they have been received, have they been more prized and honored.

woods the growth of its hills.* For the realizing of this expectation a continuance of the state of things then prevailing was required, viz.—the slave trade, or an ample supply of labour, a highly remunerating sugar price, and liberal advances of capital. These ceasing, the tide of affairs changed, a decline followed advance, increasing as regards the proprietary body up to the last few years if not up to the present time.

Of late years, the quantity of sugar produced, has not, on an average, exceeded from 3000 to 4000 hogsheads. No cotton has been grown for many years ; in 1780, it amounted to 2,619,000 lbs. ; and little wood has been exported, and that, merely fire wood, to the adjoining islands, chiefly Barbados. Most of the timber now used in the islands, has not been of native growth, but imported from the United States. The extent of land under cultivation at the present time, in canes and provisions, has been estimated at about 5462 acres.†

This diminution of produce, is commonly attributed to want of sufficient labour, and of continuous labour. This is the most obvious cause, but perhaps there are others, and not less potential,—such as defect of skill

* See West Indian Common Place Book, p. 24.

† The acres in sugar and provisions in the several parishes have been stated as follows :—St. John's, 556.—St. Paul's, 597.—St. Mary's, 431.—St. David's, 1174¾.—St. George's, 865.—St. Andrew's, 645.—St. Patrick's, 1093½.—Towns,—Scarborough and Plymouth, 100.——Total, 5462¼. Blue Book, 1843. The number of sugar estates specified in the map published in 1832, is 74,—the number of slaves then on them varying from 43 to 300 ; more above than below 100. Notwithstanding the distressed state of the island, at present, I can find notice of one estate only abandoned.

and science, and of good and economical management, most of the proprietors of estates being absentees, and the persons engaged to supply their place being generally unenlightened. Some efforts have been made to improve both the agriculture and the process of making sugar. In 1844, an institution was established on a proprietary basis, on a small scale, to promote the knowledge of agricultural chemistry, the analysis of soils, and the natural history of the colony. About the same time, an agricultural society was formed, which had its periodical meetings, and shows, and ploughing matches, offering prizes. The first, I am informed, like the Reid school of practical chemistry in Barbados, owing to want of zeal, has dwindled to nothing; and the second, after having been conducted with spirit for a while, and been, as it is believed, of service, has come to an end; no meetings having been held since 1850. Moreover, exertions have been made to increase the number of labourers by bounties on immigrants, at the rate of £3 10s. per head. In 1844, 600 labourers were obtained chiefly from Barbados, at a cost to the colony of £2000; but with less advantage then was expected, many of them, as is commonly the case with immigrants tempted by a bounty, being either men of worthless character, or indifferent workmen. Another, and a more successful measure, for securing estate-labour has been the settling of labourers on, or in the immediate vicinity of the properties, by letting them have by purchase, at a moderate price, small portions of land, on which to build a cottage, and form a garden.

The planters who have adopted this plan, have thus secured to themselves a certain amount of labour, if not a sufficiency, being sure (cæteris paribus,) that in the competition for labour, they will have the preference, and may be able to turn to account, even the labours of the old and very young, for light work, such as weeding.

As regards the system of agriculture in Tobago, it appears to be nowise much advanced. The sugar cane solely has the attention of the planter,—the growing of provisions, as in St. Vincent's, being left to the labourers. The breeding of cattle is neglected, and the fattening of them for the market. Even those required for supplying the meat rations of the troops, as indeed in most of the other islands, the contractor imports from America. No stock are stall-fed, and little manure is obtained from them, and no liquid manure.* Instead of exercising a just economy in this respect, guano and other expensive manures have been imported. The manner of planting the canes, and the management of them appear to be defective, close planting being em-

---

* The following is a return of stock, made in 1843 ; which, if critically examined, may be considered a good index of the flourishing condition of the labourers :—

| PARISH. | HORSES. | MULES. | ASSES. | CATTLE. | SHEEP. | GOATS. | SWINE. |
|---|---|---|---|---|---|---|---|
| St. John's, ...... | 9 ... | 64 ... | 11 ... | 214 ... | 190 ·... | 146 ... | 137 |
| St. Paul's, ...... | 25 ... | 97 ... | 16 ... | 447 ... | 332 ... | 179 ... | 227 |
| St. Mary's, ...... | 18 ... | 74 ... | 14 ... | 268 ... | 204 ... | 215 ... | 211 |
| St. George's .... | 23 ... | 117 ... | 40 ... | 425 ... | 361 ... | 315 ... | 464 |
| St. Andrews', ... | 60 ... | 105 ... | 29 ... | 405 ... | 358 ... | 638 ... | 653 |
| St. Patrick's, ... | 33 ... | 154 ... | 31 ... | 671 ... | 420 ... | 427 ... | 787 |
| St. David's, .... | 59 ... | 225 ... | 70 ... | 669 ... | 525 ... | 687 ... | 1143 |
| | 227 | 836 | 211 | 3099 | 2390 | 2607 | 3622 |

ployed, as close as 4 feet by 4, each cane-plant being restricted to a square of 4 feet,—a hole with raised banks, in the form of a flower pot, the middle portion of which only has been broken, and that to the depth only of about 6 inches. The trashing is often effected with guinea grass, owing to which, probably, this grass has become to be considered a troublesome weed. The weeding of the cane fields is imperfectly effected. Lastly, it may be mentioned, that the practice which has been justly deprecated of clearing the trash by burning, is often practised. These, and some other defects in the agriculture of Tobago, as commonly followed, have been well pointed out in the " Planters' Guide," published in 1846, the work of the gentleman already alluded to, as holding unfavorable views of the African character,—nevertheless, an experienced planter and resident. According to him, owing to the bad methods in use, the produce of sugar was only half what it might be under better culture of the cane. Exhorting the planters, he says, after offering certain suggestions relative to deep ploughing, manuring, the introduction of tram roads, &c., &c. ;—" Commence in right earnest, and keep pace with the advancement of the age we live in ; and you will see a cultivation of fifty acres yielding you a return of one hundred hogsheads of sugar ; instead of as is at present the case, one hundred acres scarcely yielding fifty hogsheads, provided the seasons of former years are vouchsafed to us."

Though some improvements have taken place since this gentleman wrote, as may be inferred from the official reports, such as the use of the plough, the hoe-harrow, and other implements, and tile draining,—they are of a very limited extent; in part owing to the straitened, and indeed distressed condition of the greater number of the proprietary body, almost para-lysing exertion.

Though in the manufacturing process the planters have some advantages, as in the water power largely available, and in abundance of wood on most estates where the steam engine is employed, I am not aware that any improvements have been made under this head.*

The revenue of the island and the public expenditure is each inconsiderable, yet the one is hardly adequate to the other; and the island finances are embarassed. The revenue is raised in a manner bearing particularly heavy on the labouring class, and in one respect, perhaps if not heavily, injudiciously, on the planters; —on the latter class inasmuch as the staple is taxed when exported, 7s. viz., on every hogshead;—on the former owing to the direct taxes in a high ratio bearing on them, such as the house tax, the tax on guns, mules, and dogs, and on retail shops,† and further, in the circumstance that all imports, whether

---

* According to the almanack published in 1849, the number of water wheels then in use for crushing the canes, was 22,— of steam engines, 23,— of windmills, 13,—and of mills worked by mules, 5.

† All houses but those on plantations are taxed five per cent. on rent ;— every small trader keeping a shop, 5s. ;—the merchant, wholesale and retail, only 10s. ;—every dog, 6s. ;—every mule, not employed on an estate, 1s. ;— having a gun, 1s.

of the first necessity or mere luxuries are indiscriminately charged a 5 per cent. duty.

I have adverted to a few particulars of a negative kind in the physical history of Tobago; I may mention some in its civil,—deficiences chiefly owing to the limited and scattered population of the island. There is no civil hospital, no lunatic asylum, no public library, or public reading room, or bookseller's shop. There is however a printing press, employed chiefly in the service of a weekly newspaper, the Tobago Chronicle. The roads, excepting for a short space in the neighbourhood of the town of Scarborough, are little more than bridle paths,—the best of them passable only for carts; the numerous streams to be crossed are without bridges, and there are no resting places, or places of refreshment, with one or two poor exceptions, not deserving the name of inns, so that the traveller is dependent, in a great measure, on the hospitality of the planters, which, rarely tried by strangers, is seldom, if ever, refused or stinted.

The traveller visiting the West Indies can hardly err in passing a few days in this island. He might become tolerably acquainted with it, making the most of his time, even in a few days, coming in the mail steam packet, and rejoining it, on its return from Demerara.

Landing in Courland bay and proceeding thence to the principal town, he would see an interesting little district, called Les Côteaux, a region of rounded little hills and hummocks, several hundred feet above the level of the sea, which brought vividly to my recollection the district of Uva in the interior of Ceylon, but better cultivated and more peopled. From the town, he might make two or three excursions with pleasure and profit even in the limited time mentioned. One to Manawawa bay, commonly misnamed "Man-of-War Bay," near the north east extremity of the island on its windward coast; another to Sandy Point in the opposite direction and at the other extremity of Tobago. The first named may be made with ease, in three or four days on horseback, the only mode, except walking, of travelling there, and provided with letters of introduction to the few planters in the way, on whose hospitality it is necessary to be dependent, if not for refreshment, at least for shelter. The other, a much shorter distance may be accomplished in a day, between an early breakfast and a late dinner.

In the first excursion, an opportunity is afforded of seeing a large portion of the leeward coast, a good deal of picturesque and wild country, and of gaining an insight into the state of agriculture, and the manner of life, far from luxurious, of a Tobago planter. The road, from one end of the island to the other, as already remarked, is little more than a bridle path, and in some places, where carried along the face of a cliff, with a precipice above and a precipice below, somewhat trying to the nerves, and not less so in others, where rivers are to be forded, and where one may have to swim one's horse. Such rough travelling however is not without its charm, that of enterprise and exertion, especially for the first time, when that of novelty is superadded.

Generally the character of the country is such as to
convey the idea of a newly settled one, the cultivation
is so partial, the native forest so extensive, and wild
animals, especially birds, so numerous.*

During the whole distance only four or five sugar
estates are passed. This character of scenery, and of
tropical scenery, with its luxuriant and varied vegetation is by itself sufficiently interesting, and on this
excursion the interest is heightened by the distant
views ; in fine weather when the atmosphere is clear,
not only a considerable portion of the bold northern
shore of Trinidad being conspicuous, and its mountains,
but also the more distant ones,—those of Venezuela
on the continent of America, with their triple summits.

---

* The natural history of the island, opens a large field for research. 123
different species of birds, most of them breeding in the island, are already
known as its inmates, and more probably will be discovered.—Amongst them
are five different species of humming bird, –two or three kinds of hawk,—
as many different species of swallow,—a large horned owl,—a fine bird, peculiar to Tobago, called the Tobago pheasant, or, more commonly, the Ka-
traka, of the genus penelope. Snakes too, are numerous ; but of the many
species met with, (some of them of great beauty,) not one is poisonous. Of
the batrachians, at least, three species are common,—a toad, like that of
Barbados,—a piping frog, whose shrill note is almost incessant at night,—
and a tree frog, a singularly inactive and apparently apathetic being. Of
scorpions, there are two kinds,—the black and brown,—the sting of each of
which, is said to be dangerous, occasionally even fatal ; and when so, preceded by "numbness and sinking."—Such were the words used by my
medical informant. Of the mammalia, the wild species best known, are the
agouti and pecora,—the hog and ox,—the two latter, the domesticated
species, run wild, and in their wild state, especially the boar and bull are formidable to encounter. Of the winged mammalia, the bats, there are said to
be several species ; but fortunately no blood sucker. The sportsman in
Tobago, if his zeal carry him into the forest, need never want excitement;
in no other of the Antilles, of which I have any knowledge, can he find
such variety and abundance of game.

If fortunate in a companion or guide, moreover, interest of another kind may be added, from the relation of local anecdotes and incidents. When I travelled through this part of the island so soon after the hurricane, the effects of which were everywhere visible, there was no lack of either. One of my informants was the doctor of the district who had himself had an almost miraculous escape amidst falling roofs and walls. Another was the clergyman of the parish, one thirty miles in circumference, whose family we found inhabiting, and contented with, a rude hut of extempore construction, consisting of one room, the sleeping part merely separated by a curtain, a substitute for their own once comfortable dwelling then in ruins. Of the information which they gave, not the least instructive was that relating to their own professional duties in such a country, and in such a climate with a population so scanty, broken up and scattered, entailing difficulties, and requiring exertions compared with which the life of a country practitioner at home or of an English clergyman, however hard worked, may almost be considered one of ease.

From a part of the road leading to Manawawa bay, Little Tobago, situated in front of Tyrrel's bay, at the eastern end of the island, comes into view. It is a bold rocky wooded islet about two miles in length and one in width, in a state of nature, and uninhabited excepting by an individual who leads there the life of a hermit, self-banished in consequence of unconquerably intemperate habits, and consciousness of his inability to resist the temptation of drink when it comes in his way. He is a white man. His small house, which was built for him by his former neighbours, is conspicuous at the bottom of a little cove on the shore. He subsists, I am told, by growing provisions, and by "turning turtle," each shell, which he sells, bringing him in about 10*s.* He has no boat; when he wishes to cross and mix with his fellow men, which is very seldom, the occurrence always ending in a fit of drunkenness, he makes a signal, and some old acquaintance ferries him over. Not unlikely, De Foe's romance suggested to him the kind of life he has adopted,—Tobago, according in situation with the island on which the hero of the story was shipwrecked, and the tale being a specially popular one there,—not less so than if the fiction were a reality,—a cave even having acquired the name of Robinson Crusoe's.*

The excursion to the Sandy Point district will enable the stranger to see that part of the island which has been longest, and is best cultivated; a Moravian school, nearly self-supporting, highly creditable to that sect in which, when I was there about 300 native children were usefully educated;†

* The cave so called, is situate on the shore, in the Sandy Point district.
† " The Moravians commence by a system of discipline, which, without fatiguing either the mind or the body, is well suited to children of the most tender age. The more advanced are, (as a part of their education,) required to sweep the school-rooms, trim the walks, and ornament the grounds within the confines of the mission establishment ;—plant and weed guinea grass ;— raise indian corn, bananas, and other fruits and vegetables. To the elder

—and in addition, if his taste incline that way, some interesting geological appearances, especially at a spot on the shore, a small headland called Jago Hall near Plymouth, where a crystalline rock, somewhat of the nature of a fine grained granite, already mentioned, containing contemporaneous veins, is capped with coral, and this at least 30 feet above the sea level.*

pupils are assigned small plots of land, for their own exclusive benefit; and I am informed, that in many cases the boys are not unfrequently at work in their gardens early and late. By such means as these, industrial habits are implanted, and obedience and regularity inculcated imperceptibly on the minds of the rising generation;—elements of instruction almost as necessary to the future comfort and happiness of the labouring classes, as the intellectual acquirements of reading and writing."—*Extract of a report, from late Lt. Governor Græme, dated Tobago, March,* 1847.

* This crystalline rock I have described in my note book, as consisting chiefly of felspar, in which a few scales of mica are visible, and grains of quartz, but more frequently hornblende is mixed with the felspar. The contemporaneous veins, it is mentioned, are darker than the including rock, and contain more hornblende. Apparently irregular in their course, in one or two places they are seen broken or interrupted, as if from dislocation. The crystalline rock rises from the sea abruptly, precipitously shelving, so as to be rather difficult of access. The coral rock, that rests on it, may be about two or three feet thick. It is crystalline in structure, and compact,—leading to the inference that it is of the nature of a cast, or truly mineralized or petrified coral.

# CHAPTER IX.

## ST. LUCIA.

Importance of its position, general character and aspect.—Geological structure.—Soufriere.—Mineral and hot springs.—Snakes.—Climate.—Population, circumstances unfavorable to its advancement.—State of education.—Cultivation and agriculture.—Its past state and future prospects.—Condition of the labourers and of the small proprietors.—Taxation and expenditure.—Objects worthy of notice to the passing traveller.

St. Lucia, the last of the five islands belonging to the windward government, is on many accounts, not the least important of them, or least interesting, especially as regards its position, its size, and capabilities, and its instructive and melancholy history.

Its importance, in relation to situation, is owing to the circumstance, that whilst it is so near to St. Vincent and Martinique, standing between them, it is to the windward of the latter, and of all the other French islands also; so that in time of war, it may be of the greatest service, both for defensive and offensive operations; and the more so, from its possessing excellent harbours.* Till accurately surveyed, its size was ex-

* See Mr. Breen's valuable work on St. Lucia.—("*St. Lucia, Historical, Statistical and Descriptive,*" by F. H. Breen, Esq., London, 1844,) for detailed information on these points. Lord Rodney estimated its value so

aggerated. From accurate admeasurement, it has been
found to be not 42 miles in length as previously es-

highly, that he concludes a letter to the minister of that time, on the necessity of securing possession of St. Lucia, in the following impressive words:—
" Pardon, my Lord, the trouble I give you, in perusing this letter ; but the
observations I made, when I commanded in those seas, and my frequent
reflections since, on the infinite importance of St. Lucia, or Martinique, (he
gave the preference to the former,) to a maritime power, have convinced me,
that either of these islands in the hands of Great Britain, must, whilst she
remains a great maritime power, make her sovereign of the West Indies."
    Mr. Breen, who quotes this letter, justly remarks,—" It is no insignificant
honor to St. Lucia, that by the position which Admiral Rodney occupied in
one of its harbours, he was enabled to watch the motions of the enemy's
fleet, to pursue them at an hour's notice ;—and, finally, on the 12th of April,
1782, to achieve that glorious and decisive victory over Count de Grasse
whereby the mighty projects of the coalesced powers were annihilated, and
Britain's dominion on the ocean secured." The same author, adverting to its
fine harbours, states, that "at its northern extremity, and within the short
distance of three miles from each other, it possesses four of the largest and
safest bays in the West Indies, namely :—Cul-de-Sac, Castries, Choc, and
Gros-ilet, all situated within view of Port Royal, the chief naval depôt of
the French, and all admirably protected, by the batteries of Morne, Fortunè,
the Vigie, and Pigeon Island ; "— well remarking, that "deprived of this post,
the British commanders must have been secluded within the remote and inconvenient stations of Barbados, on the one hand, and Antigua on the other;
whereas, with St. Lucia in their possession, they were enabled to command
the Archipelago. There they concentrated their forces during the war;
and thence, as from an impregnable fortress, the British seaman,—guided
by the dauntless spirit of a Rodney or a Hood, bore down upon the enemy
in every part of the Antilles, French, Spanish, and Dutch; pursuing their
fleets—capturing their convoys —storming their fortresses—blockading their
ports."
    The port of Castries, the safest of these ports, it is supposed was given up
as a naval depôt, on account of the marked insalubrity of its air. This objection is considered now, by a very competent authority, Lt. Col. Torrens,
who ably administered the government of the island for about three years,
as no longer of weight. His words are,—" I believe that it, (the port of
Castries,) was abandoned as a naval station since the war, in consequence of
its insalubrity. Since that time however, the pestilential swamps, which
caused the miasma, then fatal, have been drained, and filled in ; and among
the crews of the merchant ships in harbour, fever and other diseases are now

timated, only 27; and in greatest width, not 42 miles, only 13½; and to include in its area not 158,620 acres, only 114,400 acres, or 178¾ square miles. A map formed from this survey is published in the Blue Book reports of 1848.

Of volcanic origin, like St. Vincent and Grenada, it is of the same mountainous and picturesque character, and is little inferior to either of these islands in the wild and luxuriant beauty of its scenery. Like them too, it is well watered and wooded, even to excess, every valley having its stream, and every mountain its forest, indeed, with the exception of the cultivated portions of the island, its whole surface may be said to be covered with forest or rank vegetation. Even as regards the irregularities of its surface, the resemblance to St. Vincent and Grenada does not fail. As in them, its central portion is most elevated, a mountain chain extending from north to south, dividing it into a windward and leeward district; the shores with few exceptions being comparatively low, the most remarkable being those two grand and majestic headlands the great and little Piton,—inaccessible mountains which rise like vast columns abruptly from the sea to an elevation exceeding even the highest of the central region of which

almost unknown : " adding,—"it would be easy to verify the correctness of this statement, by enquiry." *P. P.*, 8*th August*, 1845, p. 166.

One great advantage of this secure port is, that from the narrowness of its entrance, it is more than ordinarily safe from being taken by surprise by an enemy. Even by day, a pilot is required to enter it; and by night, without a signal light, it can hardly be made,—much less entered, even with a pilot.

any accurate measurement has been taken, the great Piton being no less than 2710 feet in height, and the little Piton on the opposite side of the bay 2680.*

Of the geology of St. Lucia, owing to limited research, but little is known; very much remains to be ascertained. From what I observed when I visited the island and from all the information I was able to collect, its rocks with one exception, are all igneous; either crystalline, of the nature of trap rocks, exhibiting much variety of structure; or uncrystalline, composed of volcanic ashes, constituting tufas. The exception alluded to is a limestone, of which the great Piton appears in part to be composed. A specimen of it which I obtained broken from a rock in the side of the mountain, about 400 feet above the level of the sea, was found to consist chiefly of carbonate of lime with a little carbonate of magnesia, and a trace of phosphate of lime. It is worthy of remark that it exhibited unmistakeably a coralline structure, leading to the inference that it was coral, and uplifted by volcanic action. Another example of the same kind occurs in the adjoining quarter of Soufriere on the Morne Courbaril estate in a quarry two hundred feet above the level of the sea, and about as many yards from the sea shore. It is spoken of as the only spot where limestone is found in the island. A specimen of it which was sent

---

* In the best map of the island—that by Detaille and Jamet, published in 1847, besides the Pitons, the only other mountains, the elevations of which are given, are the Pitons des Canaries, ( 2585 feet,) Grand Magazin, 1467, and Piton Flore, (1530 feet.)

me resembled closely the coral limestone of Barbados. It exhibited as regards stucture, the transition from coral limestone into compact limestone rock, without other traces of organic remains.

Of the crystalline and trap rocks, many resemble basalt and greenstone, whilst some are an approach to granite or sienite. The only spot where I saw columnar basalt was Pigeon island.

Though no volcanic eruption has taken place at St. Lucia since it has been known to Europeans, it has a Solfoterra and hot mineral springs. These are situated in the quarter, deriving its name from them, of the Soufriere. Mr. Breen, in his history of the island, remarks. "The greatest natural curiosity in the island is the Soufriere or sulphureous mountain situated in the parish to which it has given its name. It is about half an hour's ride from the town of Soufriere, and two miles to the east of the Pitons. The crater appears at an elevation of 1000 feet above the level of the sea, between two small hills, totally denuded of vegetation. It occupies a space of three acres, and is crusted over with sulphur,* alum, cinders, and other volcanic matters,

* As a source of sulphur the Soufriere of this island is deserving of attention. Considerable quantities have already been obtained from it. In 1836, the quantity exported, was 540 tons; in 1838, 60 tons; in 1840, 160. A duty of 16s. on every ton of purified sulphur, exported, imposed by the island authorities, put a stop to the process of collecting it,—a work undertaken by two enterprising gentlemen of Antigua, Messrs. Bennet and Wood. (See Mr. Breen's St. Lucia, p.296.) In case of war, and any difficulty occurring, in obtaining sulphur from Sicily, a supply of the article, almost to any amount, might be obtained from this and the other like Solfoterras of the British Antilles, they being, so long as the natural distillation of the substance continues active as at present, inexhaustible.

in the midst of which are to be seen several cauldrons in a perpetual state of ebullition. In some the water is remarkably clear, but in the larger ones it is quite black, and boils up to the height of two or three feet, constantly emitting dense clouds of sulphureous steam accompanied by the most offensive and suffocating stench."

Of the hot springs and mineral waters of the Soufriere, the same author remarks that they "were celebrated in former days for their medicinal properties, and continue to be advantageously used by convalescents to this day." He describes how the baths were established upon an extensive scale in the time of the French, under the superintendence of the Baron de Laborié in 1785 between the crater and the village of Soufriere,—how they continued for many years to be the resort of invalids from the neighbouring islands ; and how after going out of repair, attempts to restore them in 1836 were frustrated in consequence of the ground on which the works stood, having been laid claim to, and with effect, before the royal court as private property.*

* These mineral waters are very similar in composition, to some that have acquired much celebrity in Martinique ; and where too, it would appear from an official report made on them, and published by order of the governor in 1846, the baths for their use have not had due attention paid to them. A specimen of the water taken from one of the baths still in use, on the Diamond Estate, about three quarters of a mile from the town of Soufriere, and about the same distance from the Solfoterra, sent to me, at Barbados, I found slightly impregnated with the following substances, viz : carbonate of magnesia, with a trace of phosphate of lime, and some carbonic acid. So dilute was it, that its specific gravity did not exceed 10,006 ; and 44 cubic inches, or about 11,000 grains, yielded, on evaporation,

St. Lucia is without lakes, if two small 'Etangs' as they are called in the Soufriere district, be excepted, one communicating with the Soufriere river, about half a mile from the Solfoterra, the other a little more distant contributing a branch to the Riviére Anse l'Ivrogne. The basins holding these small collections of waters may be considered as extinct volcanic craters. They are supposed by some to communicate with the

only about 4 grains of solid matter; of which, 1·7 grains were principally carbonate of lime and magnesia; 1·3 sesqui carbonate of soda, and 1 grain chiefly silica. The bath from which this water was taken, I am informed, was of the temperature 106, Faht.

Of the springs in violent ebullition in the crater, two specimens were sent me, both from near its centre. One was turbid and white from siliceous, or other insoluble earthy matter, with a little sulphur, in a finely divided state, suspended in it. Its chief constituents in solution were, free sulphuric acid, and sulphate of lime, with a small proportion of sulphate of magnesia, a trace of phosphate of lime, and of sulphuretted hydrogen. The other was turbid and black, from sulphuret of iron in fine particles suspended in it, and sulphur and volcanic dust. It differed chiefly from the former in composition in holding in solution a little sulphate of iron, sulphate of soda, and common salt. Both waters were very dilute, being of the specific gravity 10,009.

Adjoining the baths, and in hot springs of water of the same quality, in the neighbourhood, stony deposits, and incrustations are found : they consist chiefly of silica and carbonate of lime, exhibiting what are called petrified leaves and cane stalks,—these having served as moulds, and the matter deposited representing their casts. The manner in which these petrifactions have been formed, may serve to explain, how others, and more perfect, have been produced ;—such for instance, as siliceous petrified wood, of which a fine example has recently been found in the neighbourhood of Castries ; a trunk of a tree, of about a foot in diameter, from which all the vegetable matter has been removed, its place supplied by siliceous matter, in part with a loss, in part with a perfect retention of structure, exhibiting, not only the woody fibre, but also its tubes. It is curious to observe, in the mineralized portions, those in which the vegetable structure is entirely lost, the many varieties of appearance the siliceous matter has taken, such as of jasper flint, chalcedony, &c., &c.

Solfoterra and to be sources of the water that is inces-
santly ejected in its boiling springs; but this conjecture seems to be very doubtful.

From the little that is known of the natural history
of the island, in its fauna and flora, it appears to
resemble less its nearest neighbours, Martinique, St.
Vincent, and Grenada, than the more distant Tobago and
Trinidad, especially in the number of species of birds.
In common with Martinique, however, it possesses besides several kinds of harmless snakes*, one that is
poisonous, the rat-tailed snake† (*trigonocephalus lanceo-
latus*) and scorpions (*scorpio Americanus*) of large size,

---

* Of the harmless snakes, one is a Boa, that often attains a large size,
called by the natives, Tête chien. Another is a coluber, the natural enemy
of the poisonous species, known by the name of Cribo, which is said, and I
believe from the enquiry I made, truly, to attack the rat-tailed snake, whenever they meet, and commonly with success, swallowing it after having killed
it. I was assured by an accurate observer, that he knew an instance of a
Cribo having been wounded in the palate, in the act of swallowing its vic-
tim; and it was supposed by the poison fang, and yet without experiencing
any bad effect. Could it be proved, that the wound was not inflicted by an
ordinary tooth, the inference would be almost unavoidable, that the Cribo
is proof against the poison of the other; and if so, (which can only be determined by further experiments,) it need not have recourse to the plant
(pied poule, as it is called,) by rolling itself in the leaves of which, it is supposed to effect a cure of the poisoned wound, when bitten.

† This snake is found chiefly in damp places. It is said to swim, and
prey upon fish; birds and reptiles are probably its more common food. It
is not found in Pigeon Island, a very dry situation;—in some places it
abounds; I was told by the Lt. Governor, that in clearing a piece of land,
of no great extent, near government house, as many as thirty rat-tailed snakes
were killed. Several instances have occurred, of soldiers of the garrison having
been bitten by this snake. The symptoms have been severe; but no case
has been fatal. In a late report of the Lt. Governor, it is stated, that not
less than 19 persons have been killed in one small parish, by the bite of this
snake; and, from the manner in which the statement is made, it may be
inferred, that all these casualties occurred in one year—that of 1849.

the sting of which is believed by the natives to be not less virulent than the bite of the snake, if not more so.*

In climate,—that is as regards weather and the sensible qualities of atmosphere, not those properties affecting health,—St. Lucia does not appear to differ much from St. Vincent; nor in the other qualities,—those less manifest, affecting health,—from Tobago.

The following table, shewing the result of observations on the rain gauge kept at Morne Fortuné, about 800 feet above the level of the sea, is instructive in relation to the seasons. It shows that whilst the first six months of the year are commonly the driest as in the other islands, the fall of rain as to quantity in any one of these months, is uncertain and more or less variable.

| | 1846. | 1847. | 1848. | 1849. | 1850. | 1851. | 1852. |
|---|---|---|---|---|---|---|---|
| January .. | | 6·60 | 2·30 | 3·87 | 3·52 | 5·40 | 3·38 |
| February.. | | 2·51 | 2·04 | 2·10 | 1·62 | 6·27 | 6·99 |
| March .... | | 4·41 | 3·15 | 4·71 | 6·94 | 1·79 | 2·62 |
| April .... | | 4·89 | 4·48 | 11·53 | 2·90 | No record. | 5·94 |
| May...... | | 3·14 | 8·71 | 8·70 | 6·50 | | 26.06 |
| June .... | | 2·24 | 5·19 | 6·00 | 44·60 | | 10·69 |
| July...... | 11·47 | 8·51 | 11·49 | 8·18 | 16·19 | 5·16 | 7·76 |
| August .. | 11·20 | 10·27 | 7·12 | 10·59 | 11·93 | 10·26 | 15·43 |
| September | 14·91 | 10·33 | 3·71 | 2·17 | 10·59 | 7·23 | 5·80 |
| October .. | 5·63 | 8·87 | 14·95 | 16·52 | 5·90 | 13·12 | 10·38 |
| November | 6·90 | 7·88 | 4·97 | 2·22 | 9·94 | 7·50 | 13·91 |
| December | 21·80 | 8·81 | 10·80 | 9·26 | 18·80 | 4·94 | 14·84 |
| | | 78·46 | 78·91 | 85·85 | 139·43 | | 123·80 |

* A respectable planter stated confidently, that the sting of the scorpion of St. Lucia is more fatal to the negroes, than that of the rat-tailed snake; and that he had known many die of it. This information I had from the principal medical officer of the station, who believed it to be correct.

It is to be regretted that rain gauges have not been kept in other parts of the island for the sake of comparison. It is believed that the proportion of rain that falls in its different regions would exhibit, if recorded, a wide range,—large as we have seen in the hilly and mountainous district of the central portion, and small probably in the comparatively low tract of country at its northern and southern extremity. Mr. Breen mentions "a blighting drought during nine months of the year," as commonly occurring at Vieux Fort, situated at the extreme southern point of the island.

In an official report,—that of the officer administering the government in 1846, it is stated that "St. Lucia presents a remarkable variety of climate. In certain localities, and at certain elevations,—the heights of Soufriere for instance, and in other parts of the island, the climate is cool and healthy and in an eminent degree favorable to the European constitution; but in the marshy vallies,—insalubrious in proportion to their extreme fertility,—it is deadly, the very negro of the island, unless a native of such localities, suffering equally with the European."*

According to the same authority, the climate of the island is subject to certain irregular changes, especially as regards humidity, and fall of rain, on the degree of which the success of the planter so much depends. Referring to a low average of crops since

* B. B., July, 1846, p. 85.

the abolition of slavery, Lieutenant Colonel Torrens makes the following remark, bearing on the point in question. "The latter average would doubtless have been higher,* had it not been for the constant succession of dry seasons, which had for the last eight years been so regular as to constitute an entire change in the climate and seasons,—formerly during nine months in the year subject to continual rains, which were eminently favorable to the growth of the sugar cane."† The continual rains alluded to, I cannot but think doubtful. It is probable, if a rain gauge had been kept before the presumed dry period set in, it would have denoted much the same monthly fall of rain as after, viz. from 1847 to 1851, years which have been considered seasonable and productive. In all reports on the climate of a country, generally made, that is, not founded on precise instrumental observations, it seems necessary to make allowance for expressions more or less exaggerated.

As might be expected from its situation, St. Lucia is within the range of hurricanes, and is subject to earthquakes. Hitherto the latter have never occasioned much damage; even those of the 11th of January 1839, of the 7th of May 1842, and of the 8th of February, 1843, which were so fatally destructive in Martinique, St. Domingo, and Guadaloupe, were not

---

* Before the abolition of slavery, from 1831 to 1837, the average yearly produce of sugar, was 5,556,972,—after the abolition, from 1838 to 1845, it was 5,467,925.

† Loc cit., p. 85.

the occasion of any loss of life in St. Lucia and of partial injury only to houses and cane fields.* The effects of the hurricanes which the island has experienced have been more severe, second only in point of severity to those of Barbados, and in more than one instance almost ruinous, arresting at least for a time a prosperous progress. The author of the history of St. Lucia adverting to the devastation and loss of life occasioned, remarks,—"so intense is the feeling of awe with which the public mind is impressed by these phenomena, that the ' Miserere mei Deus ' and other prayers are offered up in the churches during the continuance of the hurricane months, and at the conclusion, the ' Te Deum' is sung as a public thanksgiving." †

In relation to its size St. Lucia is one of the most thinly peopled of the West India Islands.‡ According to the last census, that of 1851, in the proportion of

* In Mr. Breen's work is to be found some account of these catastrophes and also of the hurricanes, from which the island has suffered.

† Since 1756, the island has suffered from six hurricanes ; the most severe were the following, that of 1780, on the 10th of October;—of 1817, on the 21st of the same month ;—and of 1831, on the 11th of August. That of 1780, is described as tremendous in its effects, as having destroyed the estates' works, and as having thrown a damp upon industry and enterprise, from which, followed as it was soon after by the still more discouraging effects of the revolution, (the island was then in possession of the French,) the inhabitants were unable to recover for a series of years." *History of St. Lucia*, p. 277.

‡ In 1772 the total population amounted to 15,476, of whom 2,198 were whites ; in 1789 it was 21,778 ; in 1844 it was nearly the same, viz. 21,000, of whom 1,039 only were whites : (see *History St. Lucia*, p. 165 and *census return*,) in 1851 it had risen according to the census of that year to 24,318, of which number 11,763 were males, and 12,527 were females.

137 to the square mile. Many causes have contributed to this remarkably sparse population. Three may be specially mentioned, viz. the general unhealthiness of its climate, the late period at which it was taken possession of by Europeans,* and its unsettled distracted state for many years during the successive wars waged between England and France.† Others may be briefly adverted to,—circumstances tending to render the island unattractive, such as the want of agreeable society, the difficulty of communicating from the badness of the roads,‡ want of books, the almost total abnegation of intellectual life;§ the form of local government,‖ and the forms of judicial

* The first attempt made to colonize it, it would appear, was by some English settlers in 1637, who the following year were massacred by the native Caribs. The next and more successful one was by the French in 1651.

† It was first captured by the English in 1762; restored to France in 1763 by the treaty of Paris;—was next captured in 1778, and restored in 1783 at the peace of Versailles; its next capture was in 1794,—its recapture by the French occurring in the following year. The year after, viz. in 1796, it was retaken by the English; it was again restored to the French in 1802 at the peace of Amiens. The following year it was once more taken from them, since which time it has formed a part of the British Empire, as a crown colony, having been obtained by conquest.

‡ Of the 114 miles of road in the island, only about 9 miles are tolerable, (the road from the principal town to the garrison of Morne Fortuné); generally the roads are, as in Tobago, little more than bridle paths : I am not aware that there is any carriage in the island.

§ In the town, even of Castries, there is no reading room or public library; there is however a printing press, and two newspapers are published weekly.

‖ By an officer administering the government, the Lieutenant Governor has been called the government. He must initiate all the measures for discussion in council. Without a house of assembly, as in the older and chartered colonies, the people are nowise represented; public spirit is not fostered, nor self-reliance.

proceedings for the most part foreign, alien to British feelings and usages.*

Scanty as the population is, it is very mixed. The remark applies both to the white and colored races. Of the former, the French constitute the larger portion; theirs is the language which is most spoken, and a patois, formed from it, is that of the mass of the people. Of the British settlers or residents the majority are Scotch. On the subject of the several classes of inhabitants, and the state of society in St. Lucia, ample details are to be found in Mr. Breen's history of the island, and are well worthy of being consulted by those interested in its affairs. He gives a favorable account both of the people of color, the half-castes, and of the negroes; the former as rapidly advancing in numbers, wealth, and respectability; and the latter as improved in physical appearance with the bettering of their condition. Of their moral and religious state he is no-wise laudatory. In both however that there is improvement may be inferred from what we know of the character and disposition of these people as men, and from the efforts made since emancipation to instruct them. During slavery Lieutenant Colonel Torrens states there was only one priest in the island, and now, (that was in 1846) there are nine parish churches and as many curés, and two protestant clergymen; and he adds that

* The laws are in part the old laws of France anterior to the Code Napoleon, in part English, and in part local. A minute respecting them drawn up by a late chief justice of the island is appended to the annual report of the officer administering the government in 1846, published in *Blue Book*, July, 1846, p. 89.

most of the churches have been built or enlarged at the sole expense of the enfranchised negroes; the cost of schools at the same time exceeding £1000 yearly, and these chiefly for the use of the children of the labouring class.* Notwithstanding, further and greater exertions appear to be required, for in a report of the Lieutenant Governor in 1848, it is stated, that more than one half of the people are not in the habit of attending any place of public worship, and that only about one twelfth of the children are receiving daily instruction in schools.

As in Grenada the people of color and especially the negroes, are said to adhere to their African superstitions and the practice of sorcery—'Obeah,' or 'Kembois' as it is called in St. Lucia. Such an addiction can hardly be considered surprising amongst them, generally ignorant as they are, when we reflect on the influence that mesmerism, clairvoyance, and other delusions have exercised on the minds of even educated persons in Europe. It is satisfactory to think, as an anonymous writer has well said, that "the systems of African superstition have no deep foundations, possess little which imposes upon the imagination or affects the heart," and that "the only reason offered by them, the Africans, for their indefinite and obscure notions of a future state, and for their absurd religious customs, is, thus our fathers believed, and such was their practice," adding, "their natural disposition renders them peculiarly susceptible of religious impressions, and that they are mild, docile, and strong in their attachments, and acted upon with-

* Blue Book, July, 1846.

out difficulty by superior intelligence ; * a statement, the truth of which is borne out by all we know of the character of these people, and of the little hold that the superstitious practices alluded to, such as those of Obeah, have had over their minds, where they have been met by sound instruction.

In accordance with the sparse population of St. Lucia, is the limited cultivation of its surface. Of the 158,620 acres,† its estimated extent, about 12,182, or only about one thirteenth, are supposed to be under any kind of culture. Of these about 7,547 are in canes,‡ about 460 in coffee, 275 in cocoa, and 3,920 in provisions. The land in pasture is not included in the above total, not being under culture, but, as in the West Indies generally, left, even more neglected than our fells, in a state of nature. The extent of these natural pastures is stated as about 3,000 acres. The number of sugar estates are stated to be 81 ; the number of coffee plantations about 20 ; and the number of small freeholds about 100.§ Agriculture though far from being in an

* Colonization Herald, July, 1852.

† Of these 45,000 are stated by Mr. Breen to belong to private persons ; the remaining, be it remembered, to the crown including the cinquante pas du Rois,—that is the breadth of 50 paces inwards round the island, measuring from the beach.

‡ All these figures can only be considered approximate. When Mr. Breen wrote, the land in cane cultivation was supposed not to exceed 5,245 acres. The higher estimate given above is that of Lieutenant Colonel Torrens in 1845, by whom it is considered more correct.

§ The number of estates is given on the authority of Mr. Breen. In the statistical summary for 1847 the number of freeholders is returned as 1,333,—the majority of whom, it must be inferred, are small proprietors of a few acres.

advanced state in St. Lucia is believed to be improving. It has the advantage of a very productive soil,* the disadvantages of a want of capital,—of few of the principal proprietors being resident or independent,†

* The soil is formed chiefly by the decomposition and disintegration of volcanic rocks. It is, as is generally the case when so formed, very productive, especially in its virgin state. M. Beauce ("notice sur l'île Sainte Lucie," quoted by Mr. Breen) affirms that in some places it is twelve times more productive than that of Europe, and that half an acre of land is sufficient to supply the wants of a man. Mr. Darling, the able late Lieutenant Governor in a letter to Sir Wm. Reid, written in 1848, makes the following statement, "according to the information I have received when visiting various estates, I find that an average production of a ton per acre is not unusual, and in many cases the land yields more the second and third year in ratoon than in the first year when in plant. I notice this more particularly because in the fine sugar colony of Jamaica, an average yielding of an ordinary hogshead (18 cwt.) was considered in its most prosperous days an excellent result; and it is of some importance to this colony (St. Lucia) to show that the productive powers of the soil are still so little impaired." *Blue Book*, 5 September, 1848, p. 65.

† Almost every estate in the island has been disposed of by judicial sale on account of debt; those sold between 1833 (when the encumbered estates act for this island first came into operation) and 1844, was 76. It is mentioned by Mr. Breen, who gives an interesting account of the proceedings, that during the first eighteen months that an office had been opened for registering debts, liabilities were recorded to the enormous amount of 1,089,965, sterling. These debts, be it remembered, were contracted during the period of slavery. Great good, such as is expected to occur in Ireland from the late act for that country, has resulted from it in St. Lucia. On this point Mr. Breen who witnessed its effects, makes the following remark, "above all it, (the mortgage office) by facilitating the sale of real property, enabled the colony to assume an attitude of independence on the advent of final emancipation. This portentous change, a change fraught with embarrassment under the most favorable aspect, found the greater portion of the estates in the hands of persons in easy and even affluent circumstances, instead of finding them in the possession of a generation of bankrupts, planters of the old school whose passions and prejudices fostered rather by the folly of the system than the fault of the men, had long since unfitted them for the management of willing slaves, and could have awakened but little sympathy for the

besides others arising out of these circumstances, and the existing backward state of its society.

As in all the West India Colonies, the sugar cane, that which was last brought into cultivation, (it was first planted in 1765,) is now the principal crop, comparatively little coffee, and less cocoa being grown; and as in the other islands, excepting Barbados, of which we have hitherto treated, its culture, to the exclusion of provisions, is attended to on the larger estates;—provisions being grown only for supplying the market by the labourers on their own small portions of land, or on the estates' land, which, to secure their labour, they are privileged to till.

The St. Lucia planters have been described by a competent judge, a late Lieut. Governor, as imperfectly acquainted with their business, either as agriculturists or manufacturers. They have been noticed also as careless in their dwelling houses and establishments, not keeping them in good or neat order; regardless of all ornamental planting and gardening, and even the culture of fruit trees; a neglect too widely prevailing in all our West India Colonies. Notwithstanding their exclusiveness, restricted to one species of culture, implemental husbandry, for which the land generally is well adapted, is little in use, even the plough is only beginning to be brought into common use, and the process of making sugar is generally conducted in the

intercourse of free cultivators." For this act so highly beneficial, the island is indebted to an old public servant, the late Sir John Jereme, when filling the office of first president of the court of appeal. How greatly it would have been for the advantage of the other West India Colonies, had the law been extended to them.

old way, without any aid from modern science. There are however exceptions. One energetic proprietor has had a tram-way made on his estates connecting four and terminating at an embarking place on the shore, and on more than one estate along with the plough, the hoe-harrow and sub-soil plough are reported as beginning to be employed, and also an improved machinery for the manufacture of sugar.

The metairie system, that " primæval plan of division between landlord and tenant, which dates from the very origin of civilization," as it has been described by one writer, has had a trial and pretty extensively, and seems to have had tolerable success.*

* It has been a subject of much discussion and difference of opinion. It has been considered a make shift system, incompatible with a good system of agriculture,—as fatal to progressive improvement, and threatening to render the labourer independent of the proprietor. In a " Return of the sugar-growing colonies for 1848," some valuable observations are to be found respecting it, both in a letter from the Lieutenant Governor to the Governor in chief, and in one from a planter to the former. The Lieutenant Governor points out one advantage in the system, viz. the possibility of reducing the cost of superintendence. He adds, although concurring in the opinion which is so generally received amongst agriculturists that the metairie system is at best but an expedient to maintain the cultivation of estates deficient in pecuniary resources, and that the most profitable results are produced by the liberal and judicious application of capital as well in the field as in the factory, it is yet worthy of remark that in cases now referred to (instances of estates cultivated on this plan and in the ordinary way) there is strong proof that the metairie system does not necessarily involve slovenly farming or an inferior manner of dealing with the soil." How these evils are prevented, is described in the letter from the planter to the Lieutenant Governor, appended to that of the latter. His method of prevention consists mainly in his preparing the land and putting in the plants at his own expense,—the labourers doing all the rest of the work. The details he gives, both of the bad effects of the metairie plan as conducted in the ordinary way, and of their avoidance by his method, are

Limited as is the supply of labour and also the capital that is available,—the two great checks where there is such an excess of fertile land to a more extended cultivation,—it is satisfactory to find that in the produce of the island as regards its great staple, sugar, there has, on an average, been no falling off either since St. Lucia became a British colony, or since emancipation, and this even though the cane fields are cultivated with fewer hands, it is believed, now than in the time of slavery, either in its most palmy time under French rule or later under British. 1789 is considered the year (that preceding the revolution) when the island was most prosperous under the French,—then there were only 43 sugar estates; now there are 84; then the *matériel* of the estates, consisted of thirty-two water mills, eighteen cattle mills, and three windmills; in 1843 it consisted of fifty-one water mills, twenty-six cattle mills, six

well deserving of the attention of the practical agriculturist, as are also his views and suggestions as regards estate economy, and the making available the island-resources.

"In this island," he says, "we can make our staves, hoops, headings, boards, shingles, temper lime, building lime, &c., &c., (hitherto imported monthly) at one half at least of the price those articles will cost in the market. We are well provided with all sorts of hard wood. Upon many estates we can raise a sufficient quantity of stock for the work of the estate, our breed of cattle being very good ; in fact situated as we are, if we could only put our head to it, we should succeed in making cheap sugar,"—adding, "If we cannot get a good day's work at the reduced wages, we must then work our estates on the metairie system, not as it is done here by the generality of planters who allow the labourers to cultivate a small portion of land where they please and almost without any control from the proprietor, but it should be done on a more judicious plan, in order to keep up the crops on the estate." His details respecting the two plans follow.

windmills, and fourteen steam engines; in the former year the population amounted to 21,778 including 17,992 blacks, and 1,588 people of color; in the latter it hardly reached this amount, the total being 20,694, of whom 14,368 were blacks, and 5,287 were colored people.* The sugar produced before the abolition of slavery, during the last seven years, averaged 5,556,972 lbs, since that event, taking the average of eight years, the quantity was nearly the same, viz. 5,467,925. Lieutenant Colonel Torrens who makes the comparison remarks,—" In few islands, perhaps has the experiment of free labour been more successful in spite of an insufficient population, a lack of capital, much waste land, and plentiful and cheap food."†

This success it must be remembered was, when there was a differential duty in favor of British colonial sugar; whether it will be maintained when the duties on slave and free grown sugar are equalized, is doubtful and precarious, depending as the result necessarily must, on so many contingencies.

All the accounts we have of the labourers in St. Lucia, especially the official reports, leave no room to doubt, that their condition at least is prosperous. One of the special justices in his report for 1845 remarks, "many of them have purchased small lots of land in the neighbourhood of the large estates, where they establish the nicest cottages, and most comfortable abodes, keeping sheep, cows, horses, and

* History of St. Lucia, p.p. 165, 291.　　† P. P. July, 1826, p. 84.

every description of live stock. Several are now creating sugar plantations. A labourer of the Beau séjour estate named Victor, has an establishment of this kind upon which he made last year 18 hogsheads of sugar, in halves with the owner of the estate; and this man assured me that none but his wife and two children helped in the culture of the canes from which this quantity of sugar was obtained."*

By another special justice in his report for the same year, speaking of the dwellings of the labourers on coffee and cocoa estates cultivated on the metairie plan, he describes them situated at equal distances from each other, presenting "the delightful aspect of a hamlet surrounded by a luxuriant vegetation," and having "abundance and variety of provisions of every description," and this on the easy condition of two days labour weekly from one member of each family, so located, without any payment in money for the privileges enjoyed, viz. the rearing of pigs and poultry within their enclosures, and cultivating vegetables and provisions on almost as much of the estate land as they please."

Even the common labourers who have money wages possess the like advantages, viz. a cottage on the estate, without payment of rent, and land almost without limit to cultivate on their own account. Recently a charge has been made of 5*d.* a day for house and land as rent, when the labourer is not working on the estate on which he resides; but this is rather an exception than the rule, and is mentioned as only

* P. P. August, 1845, p. 146.

having been begun to be tried in one district. Wages for day labour previous to the disastrous year of 1847—8 were from 1s. 3d. to 2s. Since then they have fallen from 10d. to 1s. 3d., the higher rate to the employed in the boiling house.

Another advantage which the labouring man has in St. Lucia is the cheapness of land owing to its abundance. Its rent price is stated to be from £2 to £3 the carré, of about three acres: what the sale price of small portions of land is I cannot find officially noticed; but judging from the low price at which the larger properties have been disposed of under the encumbered estates act, it cannot be high.*

Besides growing provisions, the labourers and small farmers of St. Lucia are becoming sugar planters on their own land, and with this peculiarity, that of being in many instances also manufacturers, after the manner of the same class in India,—the makers of jaggery, the coarse sugar of that country. The extract in the subjoined note from a report of one of the special justices for 1845, referring to this practice, cannot, I think, be read without interest.†

---

* Of the 76 estates disposed of by judicial sale, between the 1st. January, 1833, and the 1st. January, 1844, varying in extent each from 1740 acres to 37, no one sold for so much as £7000, most of them were disposed of for less than £2000. See appendix to chapter viii. of Mr. Breen's work for the particulars of all these sales.

† " I have already had occasion to notice the fondness of home, causing many of the negroes to establish themselves on small lots of land in the vicinity of the estates on which they had formerly been slaves. These settlements augment every day, and many of them are assuming the character of small sugar estates. There are not less than 40 small establishments of this kind whereon canes are grown by the negroes, to be afterwards manufactured in halves on the nearest estate, and more than ten of the

Another element of their prosperous state is the light manner in which they are taxed,* the local government being conducted in an economical manner. Lieutenant Governor Darling adverting to this subject in 1848, remarks.—"When I find that the average taxation for this year need not exceed 12s. a head upon the whole population (the expenditure being estimated at £13,750, and the population fairly assumed at 23,000, and when I remember that this includes, besides the ordinary provision for the executive government, and the administration of justice, effectual relief both in-doors and out to the really indigent and sick, stipends for 11 ministers of religion, assistance to schools, and repairs of wharves, together with interest and sinking fund of the capital borrowed to carry out the great public object of bringing water into the town of Castries; and when moreover I look to the extent of the island, forty miles in length by twenty in its greatest breadth, I cannot think there is much ground of complaint on this head or that hopes may be reasonably entertained of such a reduction in the public expenditure, as would afford material relief to individual tax payers."†

forty, have erected small wooden hand mills, and manufacture their own produce, while others by means of an ingenious contrivance, forming a sort of lever attached to some strong tree, squeeze out the juice of their canes, which they boil down to a coarse syrup, and for which they find a ready sale in the town."

* "A tax of 4s. per acre on cultivated lands and a money payment of 18s. sterling; a personal service of sixty hours in repairing the public roads, the only taxes paid by the labouring class."

† P. P. September 5th, 1848, p. 70.

These remarks were made by him in forwarding to the Governor-in-chief a petition to the Queen from all the principal landholders, they believing that they were unduly taxed and beyond their means. Some other remarks which he makes in refutation are very judicious, and well worthy of attention. Mr. Darling came, it may be mentioned, from the highly taxed island of Jamaica with its costly local government, and House of Assembly regulating the taxation, to St. Lucia without such an assembly, and in a manner absolutely governed. He continues—

"I rather indeed believe that considerations not less of moral responsibility than of policy and prudence call for a considerable increase of the amount now appropriated to the religious instruction of the people, and the education of the rising generation; nor do I perceive that the taxes press unduly upon any particular class. The estimate of ways and means for this year, shows that the duty on rum consumed in the island, and on imports generally has been calculated to produce £9,500, or more than two thirds of the whole expenditure. In so far as the petitioners imply, that excessive taxation is the result of the present constitution of this government, and that a reduction of burdens would be the effect of a change to a representative system, I fear they are too sanguine in their expectations.

Among the undoubted advantages and good effects resulting from the representative form of government, to a society qualified for the exercise of the rights

which it confers, I confess I have not observed that
consistent economy is one ; on the contrary, headlong
expenditure in seasons of prosperity, and inconsiderate
retrenchment when less favorable times arrive frequently mark the proceedings of elective bodies in
limited communities, where private and class interests
exercise a more than ordinary influence, and where
present relief from the burden of taxation is too often
the paramount consideration, without reference to the
permanent injury inflicted on the best interests of
society, which that relief may involve."

Relative to the objects most worthy of the notice of
the passing traveller, the Solfoterra, and the Pitons may
be mentioned as the principal, and the garrison of
Morne Fortuné and Pigeon Island.   The two former
are easily accessible by sea, passage boats going daily
from Castries the principal town, to Port Soufriere, from
which both the crater and the mountains are, if he be
actively disposed, within walking distances.   Pigeon
Island is also best approached in a boat starting from
Castries.   Morne Fortuné is a short and pleasant walk
or ride from the town.   In making these excursions
he will have an opportunity of seeing a large portion
of the island, at least of its coast and mountainous
district, which rises nobly in view, as seen in coasting
the leeward shore, exhibiting almost an Alpine aspect.
If however he should be intent on research, so much
needed in St. Lucia, if undertaken by a competent
person, and the thorough exploring of the country, he
must travel mostly on horseback and should be well

supplied with notes of introduction to the principal planters, on whose hospitality be must be dependent for shelter and entertainment,—a hospitality little tried and burdened, and consequently when required generally, if I may speak from report, freely and kindly exercised.

## CHAPTER X.

### TRINIDAD

THIS, next to Jamaica the largest of the British West India islands, and considering its position and great capabilities, hardly second to Jamaica in importance, is about ninety miles in length, fifty in width, and contains not less than 2,400 square miles, or 1,536,000 acres.

Its interesting position in relation to the continent of America and the Gulph of Paria, is best understood by consulting the map, shewing how it is separated from the former only by two narrow straits, the celebrated Dragon's and Serpent's mouths, and how it acts towards the latter, intervening between this inland sea and the ocean, as a great barrier or natural breakwater.

In its geological structure, Trinidad displays much variety, in accordance with which are its surface and

scenery. Great is the contrast between the northern
and southern portions of the island, and more especially
between its extreme parts, the north western and the
south western, by which on one hand it approaches the
bold lofty coast of Venezuela, and on the other the low
delta of the Orinoco; the former, the northern and
north western, completely mountainous in character,
not unlike Wales or the Lake district of England, composed of clay, siliceous and mica slate, rising to the
height of two and three thousand feet above the level
of the sea; the latter, the southern and south western,
almost level, or rising only into low hills, and formed
chiefly of alluvial matter, sands and clays, and even
mud. Intermediately, the country exhibits more or less
of gradation, a hilly region succeeding the mountainous,
and succeeding that an undulating one varied with
hills and plains in which limestone* and sandstone are
the prevailing rocks.

It has been said that no organic remains have been
discovered in the rock-formations of this island. This
is true, I believe, as regards the slate formation and the
compact limestone approaching to marble in its grain,

---

* Good examples of the compact limestone, commonly grey or bluish,
are to be seen in the neighbourhood of Port of Spain, where it is quarried
as a building stone and for making lime, and also in the Five Islands, and in
the Bocas in junction with the schistose rocks, and singularly intermixed,
and where acted on by the waters of the gulf singularly water worn and
deeply grooved. Specimens of shell limestone I have obtained from
Point à Pierre, of infusorial chalk from the Naparima district, and of coral
limestone from the Tamana hill, situated nearly in the centre of the island.
This last was slightly crystalline, and the coral form, was almost ob-
literated.

adjoining the slate, but not so as regards the formations remote from these in the lower hills, for in these I know that shell-limestone occurs and chalk, or a chalk-like aggregate very like that of Barbados, abounding in infusorial remains.

Of mineral productions the list is not large; but, without doubt, it will be increased when a more careful and more extended inquiry has been made. Quartz in veins and included masses is common in the slate; and calcspar in the limestone. Gypsum has been found, of the granular kind, white and massive, in one spot, and has already been quarried and used for making plaster of Paris. Mineral pitch and asphaltum, as is well known, are yielded in abundance, indeed in inexhaustible quantities by the great Pitch Lake.* No bituminous coal has yet been dis-

* This remarkable lake, at no great distance of time, will probably be a source of wealth to Trinidad, when the population of the island is increased and the price of labour diminished, and also when the properties of the mineral pitch and asphaltum are better understood. That both will be applicable to many useful purposes there can be no doubt; even the limited trials hitherto made afford sufficient proof of this. For flooring, mixed with sand, the pitch is found to answer admirably well, under cover, where not exposed to the sun. As fuel, it and the asphaltum (the pitch indurated by exposure to the air) are likely, with management, to prove a good substitute for coal. Admiral lord Dundonald, when commander in chief in the West Indies, gave it a trial in a war steamer, and was satisfied with the result obtained, though made under very disadvantageous circumstances. The details of his experiment are to be found in his "Notes on the West Indies," published by Ridgway in 1851. For the production of gas, the pitch is not without promise. Should it be used for this purpose, a fine coke will be obtained and naptha and other products of more or less value. In one experiment that I have made, I found the proportion of coke yielded to be about 10·8 per cent. exclusive of ash. The ash or earthy incombustible part, in all the specimens I have examined, has been considerable in

covered; but, were it sought after by boring in the neighbourhood of the Pitch Lake, probably it would be found, as the rocks of that quarter, chiefly siliceous sandstone, are similar to those of Barbados where coal of this kind occurs as well as mineral pitch. Lignite is not uncommon and amber has been met with;—the former in the alluvium, the latter in the mud formation.*

Schistose rocks and the transition limestones are often metalliferous and probably those of this island are not an exception. I have a specimen of galena of excellent quality, said to have been found in the northern district: could a vein of it be discovered it might be worth working, and form a source of wealth; and, I have recently learnt from Dr. Mitchell, that both quick-

quantity varying from 32 to 36 per cent.;—a circumstance, which if constant, must necessarily detract from the value of the substance as fuel. The ash was found to consist chiefly of silica, with some alumine, lime, magnesia, and sulphate of lime, and a trace also of peroxide of iron, phosphate of lime and of a fixed alkali. It may be deserving of mention, and as lowering its value both as fuel and for the purpose of making gas, that in every specimen I have tried, I have detected the presence of sulphur —a distinct odour of sulphureous acid gas is perceptible even from the charred residue when burning. The quality of the ash, I may remark, accords well with the rock-formation of siliceous sandstone in which the lake is situated, as does also the quality of the sand on the adjoining sea shore, which I found to be almost purely siliceous. Neither in the ash, the sandstone or the sand, was I able to detect with the microscope any infusoria or other organic remains.

* Since the above was written I have learned from Dr. Mitchell, of Trinidad, that bituminous coal has been found on the eastern coast, and that "there is every indication that this seam extends quite across to the western shore, near the Pitch Lake." He adds,—"nearly the whole southern coast below high water mark is composed of strata of sandstone with vegetable imprints,—strata about three feet apart—the intermediate space filled with lignite, containing 63 per cent. of combustible matter,— their position, unfortunately vertical."

silver and cinnabar have been met with to the east of the town of Port of Spain, and also,—but without its locality having been ascertained—an ore of antimony.

Of the varied scenery of the island I shall not attempt any laboured description. In harmony with the geological features of the country, it may suffice to remark that boldness is the characteristic of the northern and mountainous district, and tameness and monotony of the greater portion of the remainder. The monotony here alluded to, in part arises from the little diversity of surface, and in part, and perhaps chiefly, from the vast extent of forest which covers its face, interrupted only here and there by patches of cultivated land, or by tracts of grass land—native Savannas. Excess of wood even prevails in the mountainous district, but there, covering the declivities of the hills confining the cultivated vallies, the effect is commonly pleasing and often beautiful,—especially when the vallies open, as they often do, on the Gulf of Paria studded with its charming wooded islets, and the view is bounded, as it often is, by the mountain peaks of the continent rising in the far distance above the clouds.

The soils of the island also are very much in accordance with the geological formations on which they rest. Judging from the many specimens from different parts of the country that I have examined, the majority of them may be referred to three kinds. First, a gravelly loam, abounding in siliceous matter containing but little clay, and destitute of carbonate of lime or nearly so, and consequently not effervescing with an acid.—Secondly, a calcareous marl, containing a large proportion of carbonate of lime, with a portion of clay and siliceous sand and more or less of the remains of infusoria, effervescing strongly with an acid.—Thirdly, a stiff clay, either altogether without carbonate of lime or containing only a small portion.

The first description of soil belongs to the mountainous district, and I believe, to the Savannas; the second occurs in the Naparima district; the third, in the southern part of the island: the extent of both these remains to be ascertained.

It is worthy of remark, that notwithstanding the rich vegetation almost every-where prevailing and the excess of forest, none of the soils, not even the forest soils, abound in vegetable matter,—tending to prove that the decomposing influence of moisture and of a high temperature is sufficient to prevent such accumulation, and more than adequate, at least at the surface, to prevent decaying vegetable matter from passing into the state of peat. It is also worthy of remark, that these soils, in most places, are of great depth, and in consequence the more favorable for productive vegetation. The first mentioned seems to be peculiarly fitted for the growth of timber and for perennial plants such as the spice-bearing trees and shrubs, requiring not a rich soil but a deep one with moisture.* And, in accor-

* The soils of Ceylon in which the cinnamon tree comes to the greatest perfection, are of the same kind,—deep, poor and moist, (moist from situation) chiefly formed of siliceous matter; as are also, I believe, those of the islands of the eastern Archipelago where the nutmeg flourishes most.

dance, nowhere in the West Indies that an attempt has been made to cultivate spices, has the result been so successful and encouraging as here. It seems also very suitable, and for the same reasons to the coffee and cocoa plant, the latter of which has been and still is largely grown, affording fruit of excellent quality. The second kind of soil and also the third is of a richer quality; and both seem better adapted for crops of more luxuriant growth, and not perennial,—such as the sugar cane, Indian corn, Guinea corn, the yam, and sweet potatoe, and others commonly designated as " ground provisions."

In climate, Trinidad does not differ much from the adjoining islands; the seasons are much the same, and the prevailing winds, and as far as can be inferred from limited experience, the quantity of rain that falls. The following table contains the results of six years' observation with the rain gauge at St. James's barracks in the neighbourhood of Port of Spain, only a few feet above the level of the sea.

| | 1846. | 1847. | 1848. | 1849. | 1850. | 1851. | 1852. |
|---|---|---|---|---|---|---|---|
| January | | 5·545 | 3·69 | 1·95 | 2·97 | 1·14 | 4·77 |
| February | | ·565 | 2·69 | ·70 | ·90 | 1·70 | 2·25 |
| March | | 1·545 | ·37 | 2·20 | ·82 | ·67 | 2·34 |
| April | | 1·820 | 2·22 | 2·90 | 3·97 | ·07 | 2·00 |
| May | | 1·970 | 8·16 | 9·00 | 6·35 | 5·23 | 4·97 |
| June | | 3·910 | 6·53 | 13·67 | 8·89 | 11·36 | 6·28 |
| July | 10·52 | 10·340 | 14·38 | 7·30 | 9·54 | 10·56 | 4·27 |
| August | 8·80 | 11·620 | 10·16 | 10·86 | 16·22 | 16·81 | 5·90 |
| September | 14·94 | 12·390 | 4·23 | 5·24 | 7·70 | 15·60 | 3·64 |
| October | 9·36 | 9·360 | 5·66 | 10·15 | 7·57 | 10·47 | 5·92 |
| November | 5·10 | 6·670 | 8·60 | 10·51 | 5·94 | 10·69 | 5·86 |
| December | 12·32 | 3·700 | 7·60 | 5·20 | 2·95 | 6·84 | 9·92 |
| | | 69·435 | 74·29 | 79·68 | 73·82 | 91·14 | 58·12 |

Though thus generally resembling that of the other islands, it is not without its peculiarities. Those best known are of a happy kind,—an almost total exemption from hurricanes, and in being little subject to suffer from droughts and blighting winds. And others probably would be discovered were meteorological observations carefully made and recorded in different parts of the country, especially on the windward and leeward coasts, in the central district, and at the north and south-western extremities. Port of Spain with its immediate vicinity affords an instance; it may be taken as an example of the leeward coast. Sheltered by mountains which rise behind it and a hilly country, there the nights are calmer, the temperature by night lower, and dew more abundant than is commonly experienced on the windward shore, and on the smaller islands. In relation to salubrity even greater differences appear to exist in different situations than are observable in the more marked and appreciable atmospheric qualities,—some spots having the character of extreme unhealthiness, in which fevers have more than decimated the troops stationed there, such as Orange Grove and St. George's Hill, not far distant from each other, and very differently situated; whilst others have the opposite character of healthiness, such as St. Ann's, St. Joseph, and San Fernando; differences,—in ignorance of the true causes,—that can be determined only by experience, and which, may be, and probably are, rather occasional than constant.

Where so much rain falls, where a drought of a month's duration is rare, it is hardly necessary to remark that want of water is seldom felt. Few islands are better supplied with streams and springs. Of the former some are considerable rivers, navigable even in their present state for small vessels; and of the latter, a few from their elevated temperature, and the substances which they hold in solution, may be considered deserving of the name of medicinal and mineral waters.

Of these, one occurs at Point á Pierre, called a hot spring; but the exact temperature of which I could not learn: a portion of it which was sent me for examination I found to be slightly impregnated with sulphuretted hydrogen and carbonic acid gas, and to contain minute quantities of carbonate of potash, lime, and magnesia, and of silica, with a trace of phosphate of lime: so dilute was it that its specific gravity did not exceed 10,003.

Another, also called a hot spring, occurs at a spot as it was described, on the south side of a hill about six miles north of San Fernando, the water of which on examination I found so similar to that of the preceding, that I am doubtful whether they are not identical: its temperature I was informed had been ascertained to be so high as 120° Fahrenheit, and that it was in repute amongst the natives for the cure of rheumatic and scrofulous affections.

Another which I had an opportunity of visiting occurs in the shingly bank of a small stream, a feeder of the San Joseph river, in a secluded spot not to be found without a guide. The water of the spring, which is but a feeble one, was only a few degrees above that of the rivulet; the one was 85° when the other was 78°, the air being 82° : it had a strong smell of sulphuretted hydrogen, and there was a disengagement of gas in bubbles at its surface. From an examination of a portion which I took with me, I found it contain the following ingredients, viz. carbonate and sulphate of lime, carbonate of potash, common salt, and a trace of silica, and to be impregnated with sulphuretted hydrogen, and that pretty strongly, and with carbonic acid gas: its specific gravity was 10,016. This spring when I visited it in 1847, had then been only recently discovered and had been but little used. Considering how strongly it is impregnated with sulphuretted hydrogen, it is likely to be efficacious in cutaneous diseases and other ailments, in the treatment of which waters of this kind have been found useful. To be made easily available, it requires the attention of the government. At a trifling expense, a path might be made to it and a cistern formed to collect the water to be employed as a bath. At present it can be taken only internally.

The mud volcanos may be considered as mineral springs. Of these the most remarkable and best known are at Cedros near the south west extremity of the island. Another occurs in the Savanna Grande a few miles from San Fernando. As the phenomena of their eruption have been often described, it is unnecessary to dwell on them here: I shall restrict myself to the ejecta and chiefly to the water. This water, muddy as it is in its ordinary state from clay suspended in it, when filtered becomes perfectly clear and transparent. A portion so treated procured from Cedros was of specific gravity 10,147, had a faint bituminous smell and a slight pure saline taste. Its chief ingredient I found was common salt. It contains besides a notable proportion of iodine probably in the form of hydriodate of soda with a trace of carbonate of soda, and carbonate of lime. A specimen of muddy water from the mud volcano in the Savanna Grande,—this at a considerable distance from the sea, afforded on examination results very similar to the preceding, seeming to indicate a common source and origin * The mud, from the former, a fine clay, was found to be composed chiefly of alumine and silica with a little carbonate of lime and magnesia, and a trace of protoxide of iron and sulphur and vegetable matter. It effervesced slightly with an acid and fused into a greenish slag when strongly heated. It may be useful for making bricks, pottery, and draining tubes. What is worthy of note in the waters of the mud eruptions is, the presence of iodine, a substance of high

* The circumstance that the mud ejected is cold, at least of ordinary temperature, is unfavorable to the idea of true volcanic action being concerned in the phenomena. It is more easy to reconcile these with the action of currents produced by the flow of the Orinoco into the Atlantic and these penetrating into chasms and caves underlying that part of Trinidad where the eruptions occur. The ejecta all favor this notion, the quality of the water, the quality of the clay, the nodules of iron pyrites, and amber, (that already alluded to—a solitary specimen, was found near the last mentioned mud volcano) with other and very dissimilar matters, such as bituminous limestone, sandstone in pebbles, dark siliceous pebbles, differing but little from flint—all of them decidedly water worn ; I may add that the mud from the Savanna Grande had a fishy smell.

medicinal powers. Containing it, they may be deserving of trial and may prove efficacious in all those ailments, in the treatment of which iodine has been found to be beneficial.

The population of Trinidad is small for the extent of the island, and very scattered, and is even more remarkable for the number of races of which it is composed and the different modes in which they have been brought together. According to the last census, that of 1851, the total is 68,600, which is in the ratio of only 28·4 to the square mile. How scattered this population is and irregularly distributed, is strikingly shown, in its distribution according to districts or counties which territorially do not differ much in extent. Of the eight counties into which the island is divided, four are western, four eastern. Belonging to the former the population is in

| | | | |
|---|---|---|---|
| The county of St. George, | ... | ... | 38,630. |
| " Caroni, | ... ... | ... | 7,107. |
| " Victoria, | ... ... | ... | 15,490. |
| " St. Patrick, | ... | ... | 4,574. |
| To the latter in St. David's, | ... | ... | 913. |
| " St. Andrew, | ... | ... | 257. |
| " Nariva, | ... ... | ... | 184. |
| " Mayaro | ... ... | ... | 980. |

So miscellaneous and mixed is this population, that it is not easily defined with any accuracy. The principal races are, the descendants of the Spaniards, by whom the island was first colonized; those of the

French who found an asylum here when flying from
St. Domingo; our own countrymen, whose advent
followed the conquest of the island; a remnant of the
Aborigines, once a numerous people, now reduced to a
few families of Indians; Africans of various tribes
imported during the time of slavery and since emancipation; and lastly the more recently imported Portuguese and Hindoos or (as commonly called) Coolies,
the former brought from the island of Madeira, the
latter from the coast of India, from Calcutta and
Madras. In a late despatch of the governor, Lord
Harris, he makes the following distinctions, with their
numerical proportions:—

| | |
|---|---:|
| British, ... ... ... ... ... ... | 727. |
| Other Europeans, ... ... ... ... | 767. |
| Creoles, ... ... ... ... ... ... | 39,913. |
| From the British colonies, chiefly the West Indies, ... ... ... ... ... | 10,800. |
| Coolies, ... ... ... ... ... ... | 3,993. |
| Africans, ... ... ... ... ... ... | 8,010. |

Foreign Colonies and other countries completing
the amount.*

In another despatch of his lordship's, he illustrates
in a striking manner the varied character of the
population by the fact, that at one and the same
time there were in the gaol of Port of Spain 44
prisoners, natives of different countries.†—Singular
is the effect of such an assemblage of people of

* Reports (1852.) P. P. p. 163.
† Return—Emigration, P. P. 31st July, 1850, p. 219.

so many races and complexions, of so many languages and dialects, religious creeds and superstitions, of manners and costumes, notwithstanding the natural tendency always active, conducing to a certain degree of conformity.

French is the prevailing language of the mass of the people; the Roman Catholic the prevailing form of religion; the former introduced by the St. Domingo immigrants, and a strong proof of their influence. English is becoming more and more spoken, and is used in the courts of law. Spanish is least of all used, and is rarely heard, excepting in the remote and more secluded parts, those least exposed to change.

With a population thus constituted, it may well be supposed that the state of society here is not advanced, especially when it is known, that there are few resident landed proprietors, few merchants deserving the name from the largeness of their commercial transactions,—few well educated professional men, no schools for the higher branches of education, no college, no university. The majority of those who do not earn their bread by the sweat of their brow, (if the expression may be used where there are so many idlers living it is hardly known how) are mainly intent on gaining a subsistence either in petty trade, or as clerks, overseers and managers of estates.

This admixture of races, this state of society, moreover, has no wise been favorable to the formation of a representative form of government, or of any

kind of self government or independent local rule;
—the absence of which in turn has tended to cramp
the faculties and feelings and to check patriotism and
public spirit.

Of the peasantry of the island, two descriptions
stand out prominently, the old and the new,—the
Creoles or natives, and the immigrants or newly
arrived. These are remarkably contrasted. The
former, on whom the cultivation of the estates,
at one time, entirely depended, have gradually become
in a measure independent of estate-labour. After
emancipation and the termination of apprenticeship,
receiving high wages and good allowances, a cottage
and land free from rent, they were able to lay by
money, and to purchase lots of land, when land became
cheap, as after the panic of 1847, and wages fell : then
restricting themselves to the cultivation of their own
little properties, subsisting on the produce, they are
described as leading an indolent life, smoking and
sleeping more than working, and in danger of degenerating, low even as they were before, and falling
back into the savage state. Exceptions are made and
these many,—such as the more intelligent and enterprising, but who are equally withdrawn from the
labour-market, having become petty shop-keepers
and hucksters, spirit-dealers, and more useful car-
riers. The danger alluded to of degeneracy, seems
to arise out of neglected education,—inefficient educational means, in a country and climate where the
tendency is to ease and idleness rather than to
exertion and industry. So imperfect is the ordinary

school-teaching, that we are assured, of the few of the natives of this class who have learnt to read, most have been taught by the female members of their masters' families and not in schools;—and that even in attending schools, such as they are at present, or, have recently been, the chances are, not that the children will be morally improved, but on the contrary, be morally contaminated,*—demoralised.

Of the immigrants, whether Africans or Coolies, by whom the cultivation of the estates is now principally carried on, the latest accounts we have are on the whole not unfavorable, judging from the official reports which have been published and from the M. S. reports of the superintendent, which, through his kindness in sending them, I have had the advantage of consulting. The Africans are described as intelligent, industrious, and fond of money, and as more honest and truthful than their half civilized brethren, the natives of the island;—the Coolies, as not inferior in industry and intelligence, but I infer, morally inferior, being addicted to false swearing and subornation of evi-

* What is stated above, I derive from a report of a competent judge, the superintendent of immigrants, Dr. Mitchell, made in 1851, in which he advocates domestic instruction,—a home-teaching, and that daily, however simple, rather than school teaching on Saturday or Sunday, at the distance, it may be of many miles from home.—

School statistics undoubtedly are useful,—but by themselves are inadequate to give any just idea of what education is accomplishing.—From a return made in 1846, it would appear, that at that time, of the whole population, only 1,932 children were at day schools; and only 685 were attending Sunday schools; the number of schools 54; 27 of the Church of England; 13 of the Church of Rome; 6 Wesleyan; 4 Presbyterian; 1 National school;—in the majority of instances the instruction gratuitous. i. e., the parents contributing nothing.

dence. Those from Calcutta bear a better character than those from Madras. The latter are said to be less docile, less honest, and more intemperate. All are described as at present doing well as labourers, and giving satisfaction to their employers; the climate, as agreeing with them; the Coolies as of a hardy constitution, exempt from intermittent fever, and the endemics of the West Indies, and peculiarly fitted in consequence, to labour on the lower grounds. They are stated to be improving under the judicious ordinance regulations enacted in 1850; as becoming attached to the country, some of them permanently settling; a few of them, as converted to christianity; many of them as growing provisions, rice, corn, &c. on the portions of land allowed them, and gaining prizes, offered for the maximum of produce thus raised, without interfering with the estate-labour, and also for continuous labour. So healthy are they, that when well conducted, and properly cared for, their mortality is under two per cent.,—a marvellously small one. Their greatest defect is a want of steadiness, considered merely as labourers, and a disposition to wander. This told terribly at one time, when "the Cooley Regulations," as they were called, made by the governor but not legalized in the form of an ordinance, were withdrawn by order of Lord Grey, then the colonial secretary, no doubt with good intention but with the worst result,—leading to the breaking of engagements, insecurity of labour, to vagrancy, to loss on the part of the planters, and to disease, misery, and starvation in a large number of instances on the part of the labourers. The Attorney-General of Trinidad, in an able speech on the subject of immigration and labour, delivered at the colonial board in 1848, alluding to the effect of the suspension of these regulations, said, "the Coolies broke forth at once and wandered over the country; became the victims of disease; and in spite often of all the exertions of the government, lay down by the road sides and perished miserably:"—adding, "that of about 5000 Coolies introduced since 1844 at the public expense, there are not now, it is believed, above 2,110, whose labour is available for the production of the staples of the colony."*—And, here, may I be allowed to pause and remark,—reflecting on such misery and on other and many instances of suffering endured by immigrants,† as well as on their general condition and

---

* I quote from the report of this speech (the whole deserving of being read, as well as that of Lord Harris preceding it) as given in the *Morning Post* of the 5th of December, 1848.

† Lord Harris, in a despatch of the 8th March, 1848,—noticing the neglect of immigrants in many instances, states,—"other cases of great neglect have come to my notice. One I may mention in which it was reported to me that a number of Coolies were in a very wretched condition on an estate in a distant part of the island. I immediately ordered them to be inspected, and if the report proved true, to be sent to the hospital in Port of Spain. In consequence, between thirty and forty were forwarded, and a more wretched set of beings I never beheld,—all in a state of starvation and more or less of disease; though every care was taken of them after their arrival, scarcely any survived. As far as I could learn they had received neither wages, clothing, or medical aid, and but the smallest modicum of food." "This," he adds, "is the worst instance which has come to my knowledge, but there have been numerous cases in which great neglect has been shown."— adding farther,—" It would appear so palpably the interest of the planters looking at the matter even in the lowest point of view, viz., as to pecuniary return, to take a proper care

treatment and the object for which they are imported, —that the history of immigration in this island and in the West Indies at large, almost without exception, is little less painful than that of the slave-trade ; and probably for the same reason ;—the one, as it has been commonly conducted, hitherto, being as much a mercenary transaction as the other, having in view merely one object, the profit of the planter,—not the general good of society,—not the formation of a well organized and well conducted community, united by common and worthy interests. There are some remarks applicable to the subject, but of wider import, to be found in a despatch of Lord Harris, of the 19th of June, 1848,— words of wisdom, founded on experience, deserving of being inscribed on the portals of the colonial office in letters of gold. " One of the many errors (he says) which have been committed since the granting of emancipation is the little attention paid to any legislation having for its end, the formation of a society, on true, sound, and lasting principles. That such an object could be attained at once, was and is not to be expected, but undoubtedly had proper measures been adopted, much greater progress might have been made. As the question at present stands, a race has been freed, but a society has not been formed. Liberty has been given to a

of the immigrants, that a stranger to the facts will hardly credit the negligence which has been manifested in this respect." *Correspondence*, p. 774, P. P.

I may add an instance of another kind of abuse,—well authenticated,— I heard it at his lordship's table ; how the manager of an estate had seven women returned in his account as labourers—but were doing no work, being with child all by him.

heterogeneous mass of individuals, who can only com‧ prehend license,—a partition in the rights and privileges and duties of civilized society has been granted to them ; they are only capable of enjoying its vices."

The agriculture of Trinidad, the main support of its inhabitants, exhibits in its history great changes and reverses. Until the arrival of the French immi‧ grants from St. Domingo, it was principally restricted to the cultivation of cocoa, coffee, tobacco, and cotton. Previously, we are told, there was not a single sugar estate, and that the extent of land under culture was inconsiderable, as were also the number of hands employed. These immigrants first introduced the cultivation of the cane on a large scale. In the short period of ten years, viz. from 1787 to 1797, we are assured that no less than 150 sugar estates were made and in full and productive operation. After the conquest of the island in the latter year by the English, a fresh impulse was given to the cultivation of the cane by the pouring in of British capital, applied to the purchase of slaves in greatly increased numbers, and the breaking up of virgin soil, affording ample profit at the then high price of the commodity.* This was continued with little stint, till the abolition

* We are told that the government refused to grant lands to settlers unless they provided a certain number of slaves, to insure cultivation,—the number in proportion to the number of acres ; and that so exigent were the authorities on this point, that if the slaves died and the regulated number was not maintained by births or the purchase of fresh hands, the lands were forfeited to the crown. See Mrs. Carmichael's *Five years in Trinidad*, &c., London, 1834.

of the slave trade. That was the first check. The next was the act of emancipation. The last, and perhaps the greatest, was the alteration of the sugar duties and the admission of slave-grown sugars into the British market. The effects of this and of the mercantile failure consequent on it were most severely felt, hardly less than in British Guiana. The following is a brief and graphic account of them, by the Attorney General, in a speech already alluded to, —" Previously to the passing of the act of 1846, the high price of sugar created an excessive competition, and the consequent high rate of wages commanded a certain supply of labour; but even then labour was neither regular nor continuous.* Since that measure passed, two years have elapsed. In the interval nearly all the English and Scotch houses connected with this colony have been struck down. In this island 64 petitions of insolvency have been filed; estate after estate thrown upon the market, and no purchaser found. Even where there has been no insolvency, many estates have been abandoned from the inability to raise money on the faith of the coming

---

* Sir H. Mac Leod in a despatch of the 19th of April, 1845, states, "Wages are now 1s. 8d. per task, or 2s. 1d. per diem, with the task and day's labour considerably increased in quantity ; and such is the competition, that all [the planters] give house and grounds rent free, and many employers allowances besides. In short, so great is the scarcity of labour, that I know of one instance, where from the want of hands, the proprietor is obliged to work two of his estates alternate days, as he cannot keep both at work at the same time."—In 1848, according to Lord Harris, wages per task (from four to five hours' work) were 1s. 3d., or 1s. 8d. per day of eight hours. *Correspondence*, &c. p. 767.

crop. I may mention a particular instance, which proves, were proof wanting, the extent of the depression of property. Within the last few weeks, the Jordan Hill estate, in the quarter of South Naparima, with a crop of 450 to 500 hogsheads on the ground, and on which about £1,500 were expended last year, in laying down tram roads—an estate which, previous to 1838, gave I am told, an annual average income of about £3,000,—has been sold for £4,000. This sale has taken place, not under an insolvency or bankruptcy, nor to meet the pressure of creditors, but by persons of wealth and respectability; and men here wonder, not at the sacrifice of the vendors so much as at the rashness of the purchaser."

Of the whole extent of land, 1,536,000 acres, as it has been roughly estimated, constituting the entire surface of the island, the enormous proportion of 1,336,000, is the property of the crown, only about 200,000 being claimed as private property. The former is almost entirely waste, a great part of it covered with forest, much of it yet unexplored. Of the latter, it has been estimated that about 92,000 acres belong to proprietors of sugar estates,—206 in number,* of which about 32,000 acres are in cane cultivation; and that about 35,000 belong to cocoa

---

* This was the number in 1838, when freedom was fully granted;—since then no new estates have been formed; but some of the old ones have been abandoned: up to the end of 1847, the number in cultivation was 193. See P. P. of 5th September, 1848, p. 320.

estates, of which only 9,000 are in cocoa cultivation; and 1,000 in coffee; leaving 53,000 acres unassigned, which may be supposed to belong to the smaller proprietors, at least 7,000, and to be appropriated in part to the growing of provisions.*

Of the state of agriculture in Trinidad, where nature has been so bountiful,† those most competent to pass judgment seem to be but of one opinion, viz. that it is deserving of no commendation, rather open to censure, as unskilled and unscientific, and "conducted generally on the old routine principle," with as little regard to order and method as to economy.

* See despatch of governor Lord Harris of 19th July, 1848, in P. P. of 11th May, 1849.—The above estimates must be considered at best merely proximate ones,—no survey of the island having yet been made,—and the extent of many of the estates not accurately ascertained by measurement. Well authenticated instances are given of proprietors, over estimating by more than one third the extent of their land in cultivation,—and paying for job-work in proportional excess. See Dr. De Verteuil's *Prize Essay on the cultivation of the sugar cane in Trinidad*, p. 72.

The price of land, of average quality of the several kinds, has been estimated as follows, per acre, in cane cultivation, £13; in cocoa, coffee, or provision grounds £6. 10s.; Savanna (pasture) £3. 5s.; uncultivated— uncleared £1. 10s. How low in comparison with the value of the produce!

† "In considering our position in comparison with that of the other colonies, we have many reasons to be grateful for the advantages bestowed upon us by a bountiful providence. Our lands are fertile; our seasons are regular and propitious. We are little harrassed by those various insects and animals, which at times commit such ravages to the canes in other colonies; nor do we know what it is to have our canes prostrated by a gale of wind. Heaven hath done everything for us; we little for ourselves." Thus writes an intelligent native, A. W. Anderson, Esq., in 1848. The extract is from his Essay (p. 232) on the cultivation of the cane, one of the three published by Lord Harris.

The few proprietors resident, are alluded to as ignorant,—and this by a native writer.* The managers are described as generally uninstructed and careless, and with the overseers as well as attorneys no wise qualified to take charge of the interests of the proprietors.

In none of the other colonies with the exception of Jamaica and British Guiana are the estates so large;—a circumstance, no doubt conducive to an improvident husbandry and to an exhaustive system; to which also has contributed the neglect of green crops, scarcely any ground provisions being grown on the large properties, nor, indeed sufficient in the island altogether for the use of the inhabitants, a large proportion we are assured being imported.

The situation moreover of many of these sugar estates cannot but have an injurious influence, unfavorable to improvement,—scattered over a vast area,—separated by extensive forests, and even where adjoining,—the rarer occurrence,—approachable only by roads that are for some months in the year almost impassable for horsemen.†

---

* Mr. Anderson in the essay already referred to, after noticing the system, as regards the education of the planter, followed in Barbados, asks—" But where do we find anything like this system amongst ourselves? A youth is taken from a store and placed on a sugar estate to superintend its management. A youngster arrives from England and is placed on an estate as an overseer, and as soon as he is found to possess intelligence and probity, he is invested with a management; or if he be of mature age, he is inducted at once into a management, without undergoing any previous probation. It takes years of apprenticeship to make a tailor or shoe maker, but in Trinidad a planter is made in a day." p. 225.

† Lord Harris, *Correspondence* &c., p. 770.

Another disadvantage laboured under, is that most of these estates are of recent formation. An intelligent writer, on this point remarks:—"It should be remembered that as compared with all the English and French Islands, and most other sugar growing countries, Trinidad is but of modern date. No capital invested by father or son through the successive generations of two or three centuries in permanent works,—such as bridges, or cutting hills for roads, making wells, water-tanks or aqueducts, substantial buildings capable of enduring through ages,—is to be found on any sugar estate purchased in this colony. Every thing is comparatively new and temporary. Wood is the prevailing material used in our works and buildings, which perish and are to be renewed every few years. The brick wall of the boiling-house, or the setting of a steam engine, is probably all that would mark the site of our present sugar estates in twenty years hence were the plantations to be now abandoned."*

From the same author we have a graphic account of the condition of the works of an estate, about the time he wrote, viz. in 1848,—though, perhaps *pró pudor*,—he carries the reader back two or three years. He says,—"Go two or three years back and from the gate by which you entered upon the estate (made most probably by a pole thrust through holes in the gate posts) up to the works, the negligence, filth and waste, the consequence of want of labour (?) were

* Essays quoted, p. 272.

everywhere visible. Around the works you would see stagnant pools, whilst the broken machinery, cast wheels and old boilers lying about would be sufficient to set up a dealer in old stores and old iron for life. Go inside the boiling house, and it is nine chances to one that the floor is of naked earth and covered with filth, and you are told that the boilers don't boil well, or the fuel is wet from want of hands to attend to it, or that grinding has been stopped for a week from not being able to get some engineer or other to his business; for the stint of labouring hands is by no means confined entirely to the negro cultivation." Adding, "Where nothing in the shape of labour can be commanded at the moment of being required, and where the whole amount of it to be procured is far short of what is wanted, surely nothing like order and perfection can be attained or expected."*

Respecting the agricultural processes employed and the details of farm operations, it is not necessary to be minute,—most of them, so far as they have come to my knowledge, being of a defective kind. The ratooning system is very generally adopted, and the substituting for decaying canes others from some abandoned cane piece,—that is, entire plants with their roots,—a practice open to great objection. Both the burning of the trash remaining on the land after the cutting of the canes, and of the brushwood preparatory to planting are also operations in common use, and are surely proofs of a low state of agriculture.

* Essays quoted, p. 271.

Close planting is commonly adopted—three cuttings in each hole placed obliquely. The hoe, not the plough is most in use,—in brief, hand, rather than implemental labour, is the rule. Little manure is made on the plantations; little is applied to the land; and that little is oftener an expensive foreign one, such as guano, than home made. The process of weeding has not the attention paid to it which it deserves; this, thrice repeated whilst the canes are in progress of growth, is held to be sufficient, and in consequence, the land is but for a short time free from weeds. As in St. Vincent, at a certain stage of growth, the canes are trashed,—the term here meaning the stripping off of some of the under leaves with the intention of favoring the ripening of the cane by the admission of more light and wind to the soil and the promoting an increased evaporation; a process likely in some situations to be beneficial, if conducted with moderation, but in others, where there is no excess of moisture, to be rather injurious. The canes in harvest time are commonly cut, and not closely, and with the cutlass, rather than with the handy and more efficient bill-hook, such as is used in Barbados. The age at which they are reaped varies greatly,—an extraordinary latitude being allowed, viz. from ten or twelve months to twenty-four or twenty-eight,—both, there is good reason to consider, especially the latter, injurious extremes. It is believed that plants of the first year reach maturity here, in from fifteen to eighten months, and ratoons about two months earlier.

It would appear that live stock—the cattle and mules employed on the estates in carting the canes to the mill and the sugar to the place of embarkation,—are sadly neglected, ill fed and worse housed, and that in consequence, they are feeble and sickly and subject to a great mortality. "Perhaps (says a resident) I should not make an under calculation in stating that 15 per cent. of the working animals alone die every year from these detrimental causes."* And in calculating the loss thus sustained, let it be remembered that almost all the live stock is imported, scarcely any being bred in the island.

Of the manufacturing processes now or till very lately in use, those pertaining to the expressing of the juice of the cane and the reducing it to sugar, but little I believe can be said in praise. They are commonly admitted to be inferior, or less carefully conducted than those adopted in Barbados, and to turn out a worse sugar. It is worthy of remark, however, and with commendation, that steam power is more employed in crushing the canes than in most of the other islands ; † and this probably owing to the command

* Dr. De Verteuil, Essays, p. 17.

† Twenty years ago, it would appear, that the steam engine, now most in use in Trinidad, was attempted, and the first tried considered a failure,— Mrs. Carmichael, who published her *Five years in Trinidad*, &c. in 1832, says,—"A steam engine was introduced upon the Eldorado Estate in Trinidad ; but I heard that it disappointed the expectations formed of it.

of capital, which occasionally for other purposes has been lavishly expended. In a few instances the imperfect cattle mill is still employed; in a few the water mill; and in still fewer the windmill. The greater use of the latter has been recommended with a view to economy. The propriety of this is very doubtful; it being a power so uncertain, so little under control, and labouring under so many disadvantages, especially where labour is scarce and precarious. Experience in Barbados, where the windmill has been so long and generally in use, is more and more opposed to it.

With these notices I shall take leave of the agriculture of Trinidad,—it must be confessed rather a disheartening subject, but not a hopeless one. To those who wish for more minute details and for further information, I would refer them to the Essays so often quoted "on the cultivation of the sugar cane in Trinidad," to the writings of Dr. Mitchell, residing in the island, who has meritoriously exerted himself, to improve the manufacturing processes; and lastly to the Parliamentary Papers, and to those more especially as affording proof of the depressed state of agriculture and of the wretched condition of the country in consequence.

That cane cultivation should have been for many years a losing concern generally in Trinidad, all

The mill had been a cattle mill formerly, and the steam engine was intended to reduce the expense of so many cattle. However it was found upon trial that the labour required from the cattle or mules in going for fuel to supply the engine was equal to that which the mill had demanded from them."

T t

things considered, can easily be believed. We are assured on high authority that since complete emancipation, since 1838, no less than a million sterling has been sunk unprofitably in this island in the production of sugar;—and, that there has been, notwithstanding, no falling off, either in the extent of land under cultivation, or in the amount of produce obtained; the crops during the last few years having more than equalled the largest in the time of slavery. The losses sustained are not difficult of explanation;— they are all well accounted for in the subjoined note.*
The continuance of cultivation is more remarkable

* In the despatch already quoted, Lord Harris gives his views on the subject,—views carefully formed, as we are sure they are, and founded on the best information.

"I find," he states, "that in 1838, when freedom was fully granted, there were 206 estates in cultivation ; at the end of last year there were 193, so that only 13 had been abandoned up to that period. From the increase in the quantity of land planted on some estates, the number of acres was probably the same at both periods. Previously to emancipation the slaves on the estates amounted to about,—prædial labourers 12,000; artisans 4,000; total 16,000. At present, by the return inserted in my annual despatch, it is shewn, that there are about 10,000 at the command of the estates.

The crop has however increased; notwithstanding this it can be shewn without much difficulty that since 1838 not only has there been no net profit, taking the whole term of years and all the estates together, but that, on the contrary, there has been a dead loss of British capital to the amount of at least £1,000,000 sterling.

Out of the 193 estates, about 17 may be considered as having given a profit; about the same number may have held their ground neither gaining nor losing; the rest have been kept up at the loss above mentioned.

It is also necessary to take into view some of the peculiarities of these 17 profitable estates.

On examination it is found that most of them belong to resident proprietors who have sold their sugar in the island, who have adopted few if any

and seemingly paradoxical. It is referrible, doubtless, to various causes,—bad and good; to partial and occasional successes encouraging perseverance, and with perseverance hopes of better times; to tenacity of property, even under losses, with the same hope; and to the change of hands of properties and the purchases from bankrupt owners; and not least, perhaps, to that love of speculation, amounting to gambling, which has been too much practised in all times in our colonies, and too much encouraged by the facility with which credit has been given and advances made.

Relative to the future,—and whether the cultivation of the cane is to be continued in Trinidad or altogether

improvements, and thus saved any outlay of capital, and who have exercised the strictest economy.

It may be asked, how is it that sale in the island has proved more lucrative than shipping home; is it that a foreign market offers higher prices?—Not at all. There are scarcely any demands here for foreign markets.

The American and Canadian traders prefer going for better and cheaper sugar to Santa Cruz or Porto Rico, and all shipments that have been made hence to the American continent have proved bad speculations.

The advantage of selling in the island has been occasioned solely by speculation, and that almost entirely by one mercantile house, which kept up the price in the island for some years to the detriment of the merchants at home; it failed last year for £190,000.

Thus there has been in reality no net profit; the estates have been kept up by a drain on British capital, and the public income has been chiefly defrayed from the same source.

Two serious evils have resulted from the facility with which money has been procured. The cultivation has been spread over a larger surface than was judicious, even on estates belonging to capitalists, and, together with the comparative ease of obtaining the labour of immigrants many persons without capital have been enabled to undertake sugar making, whose circumstances in no way authorised such an enterprise.

Thus far up to the end of 1847.

It does not appear probable that more than six estates are likely to make a profit this year." (1848.)

abandoned,—whether this fine island with all its natural advantages is to advance in prosperity or decline, must be determined by circumstances, most of which it may be predicated are internal—that is, within the power of the inhabitants themselves. Take the two main classes forming the island-society,—the proprietary and the labouring: let us suppose the former, resident, intelligent, prudent, with practical knowledge and science sufficient to take the management of their own estates and turn them to the most account; and at the same time competent and ready to perform the public duties belonging to their station: let us also suppose, the latter educated, as labourers might and should be, morally and religiously as well as industrially, so as to be made more honest, industrious and skilful;—can the result be doubted that success would be eminent,—and that even under existing difficulties, Trinidad might compete with any slave colony and afford proof that free labour on the whole is more efficient and cheaper, as well as more secure than slave labour? What is supposed, no doubt, will be called Utopian and unattainable. It may be so in the fullest extent; but, it is to be hoped that it is attainable at least in part; indeed, unless it be,—even successful planting would little benefit society. Let it ever be remembered, that mere wealth does not create happiness or give power.*

---

* At present the ordinary revenue of the island derived chiefly from import duties and the duty on rum made in the island exceeds considerably the ordinary expenditure and has for two or three years past been increasing; in 1850, the former was £ 88,660. 10s. 6½d.; in 1851 it had ncreased by £ 5,000; in 1850, the latter was £ 77,402. 8s. 1d.

Fortunately for the island, the nobleman presiding over its government is fully sensible of the real wants of the community, (as the extracts given from his despatches and his whole conduct, since he took office in 1846, shew,) and is exerting himself to supply them,—creating hope where there was almost despair, confidence where there was mistrust, and, it is believed, a more healthy tone and feeling generally. He has instituted model and training schools, which it is said are doing well and to be of great promise, as to an improved and efficient system of education. Under his directing influence, the roads are undergoing substantial repair, and new ones of communication are being made; bridges are being constructed; pipes laid for supplying water to the principal town; and other public works of a useful kind are in progress.* Under his encouragement, efforts are making both to improve planting and the manufacture of sugar, and it is said with marked success:† And, what is most important of all,

---

* In 1850, the sum of £3,132 was advanced for the repairs of the roads, to be repaid by the road fund ; in 1851, the sum of £2,460 ; and for public works and buildings, £4104.

† Shortly after his arrival, as Governor, Lord Harris, amongst other prizes placed £100 sterling, at the disposal of the Trinidad Agricultural Society, as "a premium for the best Essay upon the cultivation of the sugar cane in Trinidad ; its merits and defects as compared with its progress in other cane growing countries and also with the actual state of agricultural science, with remarks on the processes at present in use in the manufacture of sugar and the improvements in contemplation." Of eight essays, which this offer produced, three of the best were published, containing much valuable information.

Dr. Mitchell, in a letter with which he has favored me, speaks sanguinely

measures have been taken and are in operation likely to conduce to that better social state, the neglect of which as expressed in a former extract, has long had his lordship's attention.*

of the improvements that have taken place and are in progress, both in planting and sugar making. He estimates the crop of the preceding year (1851) at 40,000 hogsheads, then by far the largest ever obtained in the island. He makes mention of one field of 50 acres, yielding without manure 200 hogsheads, and of many yielding 3 hogsheads per acre of 40 inch truss. Results, which he appears to attribute more to an improved agriculture and the use of better machinery for manufacturing, than to the season. He adverts to estates in the fertile district of the Naparimas, on which sugar can be made for 25 dollars the hogshead,—virgin soils requiring only from one to three weedings, which, (the labourers receiving from 20 to 25 cents a task) paying in four years, the amount of purchase money, and that notwithstanding the low prices of the produce.

From the letter of the same gentleman, I regret to find that implemental husbandry is not making progress, those in charge of estates not being sufficiently versed in its application to teach the labourers.

Another subject surely of regret, and raising doubts as to real advancement in agriculture, is that the agricultural societies which were formed,—shortly after Lord Harris's arrival, I believe in consequence of his exertions, —have come to an end,—" died away."

* The more important of these are, first the sub-division of the counties, into wards, each with a warder and the other requirements necessary, supported by ward rates, to keep order, and check the formation of small disorderly settlements, becoming a serious evil in the wilder parts of the country ; secondly, the establishment of an inland post, with improved roads ; thirdly, the establishment of a public board of health. Alluding to the first measure, his lordship in a despatch of the 10th February, 1851, referring to the ordinance authorizing it, the working of which had not come into play fully till the preceding year, remarks ; " on the whole I have so far reason to be satisfied with it, as confirming my conviction, that the principle on which it is founded is correct."—"In a colony (he adds) constituted as this is, there are of course many difficulties, independently of direct opposition, to overcome in establishing a system of this nature. The confusion of races and languages, the suspicions of the population, the want of a landed gentry, the general distress, and the very irregular habits in matters of business, in which delay and postponement are the only ones which can be certainly counted on, all tended to weaken the probabilities of success."—Reports (for 1850) 1851. P. P. p. 141.

Relative to the secondary, but not unimportant articles of cultivation such as cocoa, coffee, rice, ground provisions, spices, and fruits, I must pass them by, as the consideration of them would occupy too much space. I may briefly remark, that from such information as I have been able to obtain, no country appears better adapted for these and other varied productions of the tropics than Trinidad, both as regards soil and climate; and that such experience as has been obtained, whether largely in the culture of the cocoa, cotton, or coffee, or on a smaller scale in that of rice, and some of the spices, especially the nutmeg as already pointed out, is corroborative, and encouraging in a very satisfactory manner. For the same reason—a necessary limit,—I must avoid even touching on any part of the natural history of the island, a most ample subject for research, both in its fauna and flora, and which hitherto but little explored, cannot fail to reward any enquirer who with leisure at command and opportunities can devote himself to it.

I shall conclude with noticing a few excursions, which the passing traveller, restricted as to time, might make, who may be desirous of acquiring some general idea of the island and of seeing in it what is most remarkable.

Entering the gulf of Paria by one of the Bocas del Drago in approaching the Port of Spain, the capital of the island, (one of those mouths by which Columbus passed in his last adventurous voyage, when he discovered the continent of America,) he will see the most beautiful part of the gulf, that portion of it which is studded with wooded islets crested with their pretty marine villas, the occasional cool retreats of the wealthier citizens, and also of witnessing some noble coast scenery, such as is that of the northern portion of Trinidad, where the richly clad mountains and fertile vallies abruptly terminate, or gently open on this inland sea. After arriving at Port of Spain, a day may be well, and most agreeably spent in seeing the immediate neighbourhood of the city.

First, he should visit St. Ann's, the charming residence of the govenor, about two miles distant, with its beautiful grounds, where there are tastefully planted, a remarkable variety of palms—I allude to the number of species—and of other tropical trees, many of them of elegant forms and some of them of great size, showing the wonderfully rapid growth of timber within the tropics, where favourably situated, the largest and oldest not having been planted half a century,* though judging from their appearance, it might be conjectured that they were at least two or three hundred years old. The nutmeg trees at St. Ann's, of noble growth, and in full bearing are particularly deserving of attention.

Next, he will do well to drive or ride into the beautiful valley of Diago Martin and quitting his vehicle or horse where the road terminates, ascend through native forests to the commanding height called North Post,

---

* They were planted most of them,—all the older ones by Sir R. Woodford, who became Governor in 1813; previously the ground was a sugar plantation·

about 800, or 1000 feet above the level of the sea, where there is a signal-station, and from whence, whether in the direction of the ocean or that of the coast, the views are peculiarly fine, in the latter beauty and grandeur being singularly combined,—beauty from the wild woodland reaching even to the water's edge,—grandeur from the boldness of the mountain cliffs. In the valley he will have an opportunity of seeing examples of the manner in which cane cultivation is carried on, on some of the best managed, and most productive estates in the island.*

A day,—and a day may suffice,—should be given to the seeing of the Pitch Lake, certainly a most remarkable scene of its kind, the like of which of such amplitude, (it is about a league in circumference,) is I believe nowhere else to be witnessed in the world. It is about a mile and a half inland from the low shore of La Brea which is about thirty miles distant from Port of Spain. A steam boat which leaves Port of Spain in the morning, and returns in the evening will convey him to the little village of La Brea, from whence he can either ride (a horse being commonly procurable) or walk, as he may feel inclined, to the lake, or as it may be more properly called the lagoon of pitch. The voyage going and returning he will find amusing and

* When I made this little excursion in 1847 there was pointed out to me a property belonging to a planter who had started in life as a manager, and who was then said to be in the receipt of a large income, (seven or eight thousand pounds a year was mentioned) the profits of the estates under his own skilled and prudent management. I was told he exported his own sugars. There was a very powerful water wheel for crushing the canes, amply supplied with water close to his boiling house.

interesting; amusing from the strange and mixed
assemblage of people commonly crowding the passage
boat; interesting from the extent of country he has
an opportunity of seeing in this little coasting voyage,
the boat keeping near the shore the greater part of the
way,—a shore low, with a low undulating wooded country within it, so far as the eye can reach. About two
thirds of the way, at the foot of the highest hill that
is passed, he will touch at San Fernando,—a scattered
village, the most considerable in the island, and adjoining to it many rich sugar estates, which with leisure
an agriculturist would do well to examine. On making
the landing place at La Brea, indications appear denoting what is to follow. The boat in nearing the shore,
glides over a pitchy bottom imparting a dismal hue of
gloom to the water. On stepping on the beach one has
to scramble amongst rocks of pitch or asphaltum rising
out of the white siliceous sand. The road to the lake
a broad one, rugged and black, is also of asphaltum;
seemingly, and so it is supposed, formed of a descending stream of pitch from the lagoon that has moved
down slowly after the manner of semi-fluid lava in
motion, or of a glacier. It is skirted by plants of congenial aspect, reeds, coarse grasses, cacti, low palms,
and numerous creepers, with black bare patches of
pitch intermixed; the whole country, excepting a few
patches of cassava, an uncultivated waste. The Lagoon
itself, surrounded by a wilderness of wood or jungle, is
a strange sight, it is so incongruous, such a mixture of
things rarely seen together;—pitch the prevailing sur-

face, a plain of pitch most irregular in its outline, breaking off here and there, expanding and narrowing, so that
it is difficult to have a just idea of its extent and form;
water in pools, tar water, brown and clear, or of the
same color and turbid, here and there resting on the
pitch, which on examination is found mostly to be
fresh, probably rain water which has not had time to
evaporate, or land water in part that may have flowed in
from the skirting jungle; lines of green grass on the
black surface, where a little dark soil has been collected
in cracks or crevices; and lastly, to mention the most
striking, certainly the least expected feature, insulated
little groves, scattered oases, as it were, in this pitchy
desert, exhibiting all the luxuriancy of tropical vegetation, trees, shrubs, and creepers intermixed forming
almost impenetrable thickets, rich not only in flowers,—
there being a large number of flowering shrubs,—but
abounding also in animal life, seen on closer examination, insects, lizards, and especially birds, many of
gay plumage such as humming birds, and parrots;
and this luxuriant vegetation seemingly rooted in or
springing from the pitch, and though not actually so,
what is hardly less surprising, from a bituminous soil
differing but little from pitch.* A walk on the Lagoon,

* It is of a rich brown color; burns with a flame like bitumen, not
however liquifying, and leaves a white ash, hardly more in quantity than
the pitch itself; its constituents are similar to those of the ash of the pitch:
whence, probably, it may be inferred to be derived from the pitch, and to
owe its fertilizing powers mainly to its inorganic elements. Roots and the
remains of insects were found in the specimen I carried away and
examined.

on its varied and heated surface, heated by the sun's rays, if it be in the middle of the day, cannot fail to prove interesting, though it will hardly be found agreeable, the surface in many places being soft, and yielding, and in some places to such a degree as to be hazardous. The enquirer here will have ample room for speculation, but with comparatively few data for coming to any satisfactory conclusion, relative to the origin of this vast accumulation of pitch, no attempts having been made to explore its basin or sub-strata by boring. He will see in many places portions of wood, like lignite, protruding through the pitch, as if they had been thrown up: they may have fallen in and have acquired their character from partial decay, and from becoming impregnated with pitch. In some places he may observe a whitening of the surface of the lake, owing to an incrustation, the material of which I have ascertained to be chiefly carbonate of lime derived probably from the land water. Should he taste the water collected in the shallow depressions, though he will find it fresh in most instances, in a small number it will prove to be brackish. Moreover should he use a thermometer to ascertain the temperature, he will probably satisfy himself that the heat of the surface is solely owing to the sun's rays absorbed by a black ground, and not derived from any subterraneous source. Where of highest temperature, when I visited the lake, which was on the 28th of March, it did not exceed 140°, the air at the time, the middle of the day, a few feet above the surface being 88°.

Between the hours of landing and embarking at La Brea, there is time sufficient, not only to see the Pitch Lake well, but also to visit (if there be means of conveyance, and a guide with an invitation) some one of the scattered sugar estates in this wild country, and see, what is well worth seeing, a property in cultivation thus situated, thoroughly isolated, "self-contained" (to use a Scotticism) and which could hardly have been formed except in times of slavery. In my own case I had the good fortune to have the opportunity offered me, of which I gladly availed myself, to gratify my curiosity, having met in the steamer a gentleman with whom I was acquainted who was going on business, to an estate of his uncle, one of the nearest to the lake, and who invited me to accompany him. His equipage, which was waiting his arrival on the shore was a light open wagon, seatless, drawn by four active mules. He was his own charioteer, and with a long heavy whip, the cracking of which made the forests resound, and reins in hand, standing erect in the old classical manner, he drove at rattling speed regardless of the rough road, and it was an asphaltum road, the greater part of the way, in all the energy of youthful vigour. And, let me add, to show how easily man in youth can adapt himself to circumstances, he was not a Creole, nor had he been long in the island, but was an Englishman, or Scotchman, I hardly know which, probably somewhat tenderly reared, his father being a man of wealth. What I witnessed on this estate chiefly about the works, (and I saw little else except at a distance,) though not so bad as that given as an example in a former page, appeared to me far from good, and as my friend admitted, capable of great improvement. His main object in coming was to pay the labourers,—a monthly transaction, for which he brought a good bag of dollars. The manager, I learnt, was an American, one of those enterprising and intelligent young men who seek their fortunes in the wide world. He was laid up, labouring under a chronic malady, aud as it seemed to me, in a piteous state, so solitary, having only ignorant labourers about him, and so much out of the reach of medical advice. My arrival he considered a piece of good fortune.

One other excursion I shall mention as specially worthy of being made, viz. to the old village of San Joseph, and beyond it to the fine valley and remarkable water fall of Maracca. San Joseph dignified with the name of town, contains about 1,000 inhabitants, is the abode of a stipendiary magistrate, and a military post; here the early Spanish colonists first established themselves, making it their head quarters; here history tells us Sir Walter Raleigh did a gracious act in liberating two Indian Caciques whom he found imprisoned when he took the place. It is about seven miles in the interior from Port of Spain, with which it is connected by a tolerable carriage road,—tolerable at least in dry weather. The water fall of Maracca may be about the same distance from San Joseph; a track, a bridle path requiring a guide, leads to the forest in which it is situated, and in which there is hardly a foot path. This excursion affords the opportunity of seeing some of the most picturesque and beautiful country, whether native forest or river scenery, in Trinidad and this in its most picturesque and beautiful part,—the hilly region along the banks, frequently crossed, of the San Joseph river,—a clear mountain stream of good volume exposed to the accidents to which a mountain stream is subject. It affords also an opportunity to see cocoa plantations, some of the most productive in the island being in the Maracca valley.* When I made the excursion our guide from San Joseph, was a mounted native, armed with a naked cutlass, not for the purpose of defence, but for use as a pioneer to aid us in cutting a way through the jungle. I shall not enter into any minute description of

---

* These plantations have little labour bestowed on them, little more than to keep down the brushwood which is done with the cutlass. The cocoa trees are planted between others, affording shade,—here the bois immortelle, the " Erithryna" la " Madre del cacao," as it was called by the Spaniards, a very beautiful tree especially when in flower,—then most gorgeous. When I passed in the first week in April, the fruit of the cocoa was then ripening, some purple, some yellow, the pods hanging from the trunk and larger branches. The seed, or cocoa berries are enclosed in the pod in the midst of a succulent, sweet agreeable pulp. By the magistrate at San Joseph I was informed, that the method now in use of saving the berries, is different from that formerly  employed by the Spaniards. By them the pods were collected in heaps and allowed to ferment; after the fermentation the cocoa " grains " were got in a state of purity. Of late years, he said, the grains had been dried in the pulp and cuticle, and in consequence were less delicate and more bitter,—adding, that prepared in the old way the grains were brown, in the latter reddish. He said, thus saved they are preferred for the English market, admitting more easily of adulteration and of dilution with flour &c. in the process of making chocolate.

the scenery. What impressed me most were, first on one side, beyond San Joseph, the great Coroni plain, the Savanna grande seen in the distance, stretching away interminably as it seemed, like the ocean, and in the distance, much of the same colour as the ocean;—next on entering the woodland, the vast and beautiful clusters of bamboos often arching the road, especially along the course of the river;—then the noble forest trees on the mountain sides, well apart below, meeting above with their far spreading and dense foliage, intercepting the sun's rays and producing a very agreeable cool shade, most refreshing after having been heated by rather hard riding; and last and not least, the water-fall itself in the midst of the forest,—the stream, a tributary of the San Joseph, descending, scattered by the wind, a perpendicular black rock,—a vast naked mass, perhaps two or three hundred feet high,—its source unseen, as if it came from the clouds, nothing but sky above and beyond meeting the eye.

I need hardly remark that many other excursions might be made with advantage by the traveller who has time at command and who is strong and active, able to endure fatigue, and can submit to some privations, such as are unavoidable in exploring a wild, and in a great part, unreclaimed and unopened wooded tropical country, such as Trinidad is at present. On the spot he would have no difficulty in learning which are the localities best worth seeing. The selection must depend in some measure on his tastes and pursuits. The geologist probably would be most attracted by the mud volcanoes in the district of Cedros, a beautiful district I am informed, and easily reached, as the steamer already mentioned goes to it regularly. The botanist and one intent on witnessing wild tropical forest scenery in its primæval state might gratify himself by a journey rarely made, and not easily, to the hill of Tamana, nearly in the centre of Trinidad, and from the summit of which I am informed the greater part of the island is visible. The sportsman too might have his enjoyment, and that of a very animating kind, by taking to boat and joining the whale fishers at the Boccas at the northern entrance of the gulf, where the scenery may without exaggeration be said to be of matchless beauty. And not to omit the military man,—the officer directing his attention to the defence of the colonies, will find ample subject for observation, and probably instruction in the study of the ground in the neighbourhood of Port of Spain, the situation where the barracks now are, open to so much objection, and the old works at St. George's hill now deserted and hid in jungle, which at one time promised to become another Gibraltar, and on which vast sums were expended, but in vain, owing to the then unhealthiness of the position.

# CHAPTER XI.

## BRITISH GUIANA.

Approach to the Colony.—Its coast and peculiarities.—Boundaries, and territorial division.—Geographical sketch.—Soils.—Springs.—Climate.—Population.—Agricultural details, past and present.—Fallen fortunes of Planters exemplified.—Labourers, their kinds and characters.—Schools and education.—Circumstances favorable to a hopeful future.—Objects and subjects of enquiry likely to interest the passing traveller.

THE approach to British Guiana is remarkable and characteristic. Passing from the blue waters of the ocean, you enter a shallow sea, turbid and dirty, not dangerous from rocks or coral reefs, from which indeed, it is entirely free, but from shifting shoals, and banks of sand and mud.* Even at the distance of 60 miles

---

\* The nearer you approach to land, not only is the water more turbid, but also as might be expected, more dilute. In the first week of June 1847, I found the water washing the shore of George Town, nearly fresh and of specific gravity ... ... ... ... ... ... ... ... ... 10,036.

About a quarter of a mile from shore ... ... ... ... ... 10,091.
" 4 miles ... ... ... ... ... ... ... ... ... 10,210.
" 11 " ... ... ... ... ... ... ... ... ... 10,236.
" 18 " ... ... ... ... ... ... ... ... ... 10,249.
" 25 " (a heavy shower when taken up) ... ... 10,236.
" 32 " ... ... ... ... ... ... ... ... ... 10,249.
" 39 " ... ... ... ... ... ... ... ... ... 10,258.
" 46 " (in blue water) ... ... ... ... ... ... 10,266.

From these specimens a sediment (a sediment of fine clay) appeared on rest, diminishing with the distance from shore till entering blue water.

The fresh water,—the freshes of the great rivers of Guiana,—occasionally affect the sea even to a great distance, flowing probably in superficial

from land, there are soundings and those of no great depth.* Further, as you advance and near the land, the first objects that meet the eye, are not the bold cliff and headland or mountain chain, such as fix the attention of the mariner in making any of the West Indian islands, but merely a low, and at first, a doubtful line of trees (the Courida, Avicenna nitida,) their heads emerging as it were from the waves and bounding the near horizon. On landing, moreover, everything you see is in accordance with this approach, denoting an alluvial country,—the low country of South America, formed by deposition of matters—sand, gravel, and clay, (the same as constitute the bed of the shallow sea,) brought down by rivers, and spread out into plains, and these so low as to be subject to frequent invasion of the sea, which is excluded only (and with difficulty) by embankments, constructed by a people, the first colonists, well trained to such works in their native country, Holland;—in brief, on landing you find yourself in what has often and well been called a tropical Holland.

British Guiana, of which this low country at present is the most important portion, taken in its whole ex-

currents. In July 1848, when I found the water of the Gulf of Paria about half a mile from the shore of Port of Spain, of the specific gravity 10,264, water in the open sea just outside the Boccas (the southern) was only 10,145, and in Courland Bay, Tobago, about a quarter of a mile from shore, only 10,161. The vast alluviums, the result of [the sedimentary deposit, whether emerging or still submerged,—whether constituting the low lands, or the bed of the sea as described, are surely unmistakable records both of the agency and antiquity of these mighty rivers.

* I was assured that at this distance the depth did not exceed 7 fathoms.

tent, is the largest by far of our possessions in the West Indies and the only one belonging to Great Britain on the South American continent. Its coast on the Atlantic is about 200 miles in length,—from its shallowness without a single harbour,—reaching from near the mouths of the Orinoco in the west, to the river Corantin in the east; bordered on one side by the territory of Venezuela, on the other by Surinam or Dutch Guiana. Inland, its boundaries are hardly yet determined; the greater portion of its interior being still unexplored,—a vast region of far-stretching forests, mountain and table land, thinly inhabited, where not entirely desert, by rude native tribes.*

Territorially considered,—that is the low land apart from its little known interior, British Guiana, consists of three districts or counties which were united and formed into one colony in 1831, each bordering on, and deriving its name from one of the three principal rivers by which the country is intersected, viz., the Essequibo, the Demerara, and the Berbice. Of these Demerara is the largest; in it is the capital and seat of government. Berbice is next in point of importance; New Amsterdam is its county town. The district of Essequibo is without a town; and though at present least notable, it is perhaps from the noble river belonging to it, and its many advantages of soil and climate, likely to become the most valuable, and most in request of the whole.

---

* Their numbers have been roughly estimated at 10,000, divided into five tribes, under the general name of Arawak Indians.

The geology of British Guiana, so far as it is yet known, is little varied and little interesting. Such of the mountains of the interior as have been examined, have been found to consist of primary rocks,—chiefly granite. Between them and the coast, no secondary or tertiary rocks, none containing organic remains, have been observed; and indeed over a wide extent, from the foot of the mountains to the shore of the Atlantic and even beyond, as already mentioned, no rock of any kind is to be seen, not even a detached stone or pebble. The intervening country is entirely alluvial;—where rising into low hills, formed of a fine siliceous sand,—where spread out into plains, composed of clay, or of clay, sand, and gravel alternating, —an alternation made known by boring.* Near the

* The following is a note of what was found in boring, on Plantation Woodlands, one mile from the mouth of the Mahaica river;—" 1 to 3 feet, surface soil; 6 feet, layer of caddy (siliceous sand more or less colored by peroxide of iron); 7 to 9 feet, blue clay; 9 to 39 feet, soft mud mixed with caddy, in which the auger went down by its own weight; 39 to 53 feet, rotten wood and pegass or decayed vegetable matter; 53 to 55 feet, bluish grey clay, stiff; 55 to 57 feet, clay a little red and grey; 57 to 70 feet, reddish clay; 70 to 82 feet 10 inches, yellowish grey clay with a little sand and ochre, very stiff; 82 feet 10 inches to 86 feet 8 inches, bluish grey clay, streaked; 86 feet 8 inches to 92 feet, bluish grey clay, streaked, more yellow. The bed of sand from which the water is obtained was reached at a depth of 118 feet and the same stratum was found at the depth of 125 feet." *Catalogue of articles sent to the Great Exhibition in 1851.*

In conducting a similar operation in the neighbourhood of George Town, it was said, that a portion of iron, and as it was believed, wrought iron, had been brought up from a considerable depth, giving rise to the speculation, that it might have been derived from some shipwrecked vessel stranded at a remote period long before the time of Columbus,—some Carthaginian or Tyrian galley. With some difficulty I obtained the specimen; it was indeed of iron, but it was the black oxide, the streak of which was rather more lustrous than usual!

shore, on the higher inequalities, lying loose, shells
are often abundant, and these little altered, denoting
a recent formation of the banks and ridges, on which
they are scattered.

At an early period the interior of British Guiana
was considered an Ophir, a land of promise, rich in the
precious metals. Here the gallant and unfortunate
Raleigh sought his El Dorado, probably deluded by
the glittering particles of mica derived from the
granite. Hitherto, the only metallic ores found, have
been those of iron and manganese; but whether they
occur in sufficient quantities to be worked with profit,
remains to be ascertained.*

The true source of its wealth was discovered more
recently, and by a more plodding people. This exists
in its alluvial soils, of great depth and of almost inexhaustible fertility. They are commonly spoken of as
of three principal kinds; first, the stiff clay soil of the
shore and river lands,—rich in the inorganic elements
of plants, containing but little vegetable matter; secondly, the pegass,—a black peaty soil composed chiefly
of vegetable matter of the nature of peat, and like the
former it may be conjectured not formed where it is
found, but an alluvial deposit also, brought down from
the cool highlands by the rivers, and deposited on rest,
where they have overflowed their banks; thirdly, the
soils of the far interior, called loamy,—of which in

---

* In a letter lately received from my friend Dr. Blair, it is mentioned
that gold has been found in the interior, and that zinc is reported to have
been found.

truth little is known.* Minute research no doubt would show many varieties, many more than have ever been suspected,—and this both as regards the surface, and those beneath the surface, the subsoils. The subject,—one of great importance to agriculture, will, it is to be hoped, be duly investigated; it has already I know been entered upon by a very competent enquirer, Dr. Shier, the late colonial chemist.

Though abounding in water to excess,—a region truly of rivers and marshes,—British Guiana is

* The absence of lime, at least in any notable quantity, in the clays and soils of British Guiana is remarkable. Of the many specimens I have examined, not one has effervesced with an acid; nor could I detect lime after the action of an acid in the solution by the ordinary tests of this earth. In all of them, I found a comparatively large proportion of magnesia. The origin of these clays and soils, namely, from the decomposition of granite rocks, which it must be inferred is the case, and their having been long acted on by water, may account for this peculiarity.

The only samples of the soil of the interior I have seen, were brought from the Penal setlement on the Massarooni river; one was a brown loam; the other a red clay; neither of them yielded any lime,—the first only a trace of magnesia, and the second little more.

The pegass which I have examined, has left on combustion about 12 per cent. of ash; the vegetable matter consumed being about 88 per cent. The ash consisted chiefly of siliceous sand, with a trace of alumine, a very little sulphate of magnesia, and, as I believe, phosphate of the same earth with a little peroxide of iron.

The clays of the country, it is deserving of remark, are of a nature well fitted for making bricks and earthenware, so that if thorough draining, so much needed, should come into use, an excellent material is at hand for manufacturing the pipes requisite. Some of these clays are so very fine as to be probably fit for making earthenware of the red kind, similar to the antique,—the Greek and Etruscan. Under the microscope, a portion taken up by boring from a depth of 72 feet, was seen to consist of particles, the largest of which,—and these were few,—exceeded the blood corpuscles but little in their diameter; the majority were very much smaller, barely within the limit of distinct vision, using a high power (an eighth of an inch object glass.)

remarkably deficient in springs. I speak of course of the low country,—not of the mountainous interior. The few springs that there are, are due to intelligent enterprise, are Artesian, and the result of deep boring; and very useful they are, though the water of most of them is more or less impregnated with iron.

In climate British Guiana bears more resemblance to Trinidad than to the more distant West India Islands, especially Barbados. Like Trinidad, it is in great measure exempt from violent winds. I am not aware that there is any record of its having experienced the effects of any destructive hurricane. Like Trinidad, it is little subject to drought or to parching hot winds. Equability, particularly as regards temperature and degree of atmospheric humidity, may be considered its chief characteristic,—and that both by night and day. During the latter, a temperature above 85° is rare, as is also during the former, one below 75°. And the difference between the dry and wet bulbed thermometer is oftener below 6° than above it. These are qualities which might be expected from the situation and character of the country. As regards rain, the climate is most peculiar, but not so much in relation to the total quantity that falls, of which exaggerated accounts have been given, as to its seasons. Two rainy seasons are commonly described, the greater and the less. The former is said to begin about the middle of April and to continue through May, June, (the most rainy month) July and part of August; the latter is expected to begin in the latter end of November, and to end about the latter end of January. The table underneath,* giving the results of ten years' observations with the Pluviometer, kept at George Town,

RAIN,—GEORGE TOWN, DEMERARA,—PORT CANGE, BERBICE.

| | 1843. | 1844. | 1845. | 1846. | 1847. Dem. | 1847. Berb. | 1848. Dem. | 1848. Berb. | 1849. Dem. | 1850. Dem. | 1851. Dem. | 1852. Dem. |
|---|---|---|---|---|---|---|---|---|---|---|---|---|
| January | 7·99 | 0·20 | 5·70 | 2·06 | 9·30 | 6·38 | 6·30 | 11·25 | 5·40 | 15·16 | 3·93 | 6·75 |
| February | 8·16 | 2·29 | 5·91 | ·85 | 3·23 | 3·35 | 6·90 | 12·10 | 7·44 | 3·86 | 6·60 | 8·50 |
| March | 3·61 | 7·58 | 7·6 | 2·06 | 6·15 | 8·27 | 7·68 | 7·85 | 12·59 | 14·60 | 8·90 | 8·60 |
| April | 1·78 | 7·14 | 8·79 | 5·93 | 11·48 | 7·80 | 7·72 | 3·70 | 17·54 | 5·94 | 16·15 | 5·88 |
| May | 16·16 | 17·26 | 14·23 | 14·08 | 12·88 | 11·00 | 20·28 | 23·25 | 7·94 | 15·60 | 11·23 | 16·50 |
| June | 13·32 | 19·93 | 11·76 | 14·92 | 14·29 | 13·27 | 11·17 | 8·73 | 20·44 | 7·94 | 19·82 | 11·67 |
| July | 3·79 | 6·38 | 4·07 | 13·27 | 10·21 | 4·86 | 5·55 | 3·50 | 20·38 | 9·88 | 8·69 | 8·83 |
| August | 5·00 | 6·15 | 2·48 | 8·80 | 3·82 | 5·10 | 2·54 | ·75 | 10·80 | 10·41 | 7·52 | 10·11 |
| September | 3·52 | 1·09 | 1·23 | ·61 | 1·12 | 2·00 | 6·22 | 4·25 | 1·00 | ·63 | 2·90 | 1·17 |
| October | 0·72 | 0·06 | ·35 | 5·88 | 3·29 | 2·25 | 0·66 | No record | 3·55 | 1·86 | 2·70 | 0·52 |
| November | 3·58 | 6·31 | No rain. | 5·57 | 7·30 | 6·23 | 3·49 | | 10·36 | 5·15 | 12·00 | 5·25 |
| December | 6·07 | 10·04 | 6·78 | 11·23 | 10·29 | 4·00 | 18·82 | | 14·40 | 6·31 | 3·0 | 12·18 |
| | 73·70 | 84·43 | 69·14 | 85·26 | 93·36 | 74·51 | 97·33 | | 132·04 | 97·34 | 103·34 | 95·96 |

* This table is drawn up partly from observations made at the Military

Y y

Demerara, tolerably accords with these reckonings, shewing at the same time that they must be received with some degree of latitude, and that even in the dry seasons, an absence of rain for an entire month is very unusual.

Taking an average of these ten years, it would appear that 93 inches are about the mean quantity; the extremes being under 70 and above 100; both injurious to the crops, the one from deficiency, the other from excess of moisture, but the latter in the greatest degree. It need hardly be remarked that were rain gauges kept in different parts of the colony, marked differences probably would be observed in the quantity of rain. The limited observations recorded in the table made at Fort Cange, Berbice, accord with this supposition.

The prevailing winds are the land wind and the sea breeze; the one from the south and west, the other from the north and east: the former experienced most during the rainy seasons, the latter during the dry, and the days of fine weather and absence of rain. Dr. Blair, in whose monograph on yellow fever much minute information is given on the meteorology of this country, describes the sea breeze as exhilarating; the

---

Hospital, at the Observatory, and at the office of the Royal Engineers, all in George Town, with the exception of those made at Fort Cange, Berbice. Comparing the monthly results of the several gauges, in many instances, there is a remarkable want of accordance; but whether owing to the effects of partial showers, or to different forms of the instruments, or to careless observation or registering, I have no means of ascertaining.

land winds on the contrary as depressing, and as creating a feeling of chilliness and discomfort, and in many persons a sensation of nausea, not satisfactorily accounted for, as he remarks, by the mere meteorological changes. He says that these winds are universally believed to be unwholesome,—a remark which applies to land winds generally, at least with very few exceptions. Fortunately for the comfort and health of the inhabitants, the salubrious sea breezes greatly exceed in frequency the land winds—viz. as measured by their respective powers in the ratio of 313·59 to 34·65 ; the totals of the cardinal points for one year (1844) were 53·7 north, 260·42 east, 34·30 south, ·35 west, as determined by Dr. Whewell's anemometer.

Of the climate of this country in relation to health, very opposite accounts have been given ; by some authors it has been described as little worse than the western coast of Africa ; by others, its degree of healthiness has been as much overrated. I may have occasion to recur to this subject in the sequel ; now it may suffice to remark that a wide distinction requires to be made, between its influence on the newly arrived, and the long resident, on foreigners and natives, on the white race and the colored, and more especially on those arriving from, and inured to warm and tropical climates, and those coming from cold regions and habituated to them. As might be expected from the character of the country, the prevailing diseases are intermittent and remittent fevers, severe and often fatal to those unacclimatised ; frequent,—especially fevers of the intermittent kind,—but not severe amongst the Creoles, and of these, in a vastly larger proportion amongst the white, than amongst the colored races: a result also in accordance with general experience. The enormous quantity of sulphate of quinine, the febrifuge *par excellence*, imported and used in British Guiana, is a circumstance by itself sufficiently indicative of the nature of the diseases most common and rife.*

Whether favorable or not to the life of man, British Guiana is eminently so to life generally. Air, water, earth, are here teeming with life; it is, as it were a great laboratory of life, and especially of those lower orders of plants and animals of an aquatic kind or origin, and of the latter, more especially insects. These, especially those very troublesome ones, musquitoes and sand-flies, of which there are several species, are often in such excess as to darken the air, and almost constantly in such numbers as to require special means of protection from their attacks. Were it not for these plagues the climate might be too delightful, too much conducive to ease and enjoyment. They seem to perform here the part of the keen biting wind and cold blasts of winter in our northern regions, acting as stimulants to excite invention and care to exclude them; and in so doing administering

---

* According to a respectable druggist in George Town, the quantity annually exceeds 5,000 ounces. It is difficult to learn the precise amount, as drugs at the custom house are only entered in bulk, and as many planters import more or less, direct on their own account.

to home, in-door comfort, and even to health,—the same means which conduce to the one, contributing also to the other, such as scrupulous cleanliness, thorough ventilation and draining, and protection from the night air.

The population of British Guiana is hardly less miscellaneous than that of Trinidad, indeed putting in place of the descendants of the Spanish and French in the latter, those of the Dutch in the former, the difference becomes hardly appreciable, the great body of the people here as well as there, exclusive of the aborigines, the least important part, consisting of Africans, either emancipated slaves or their offspring, or of later immigrants from Africa or the West India Islands, and of other immigrants, Hindoos, "Coolies," from India, or Portugese chiefly from Madeira, making a total according to the last census, (that of 1851,) of 127,675. Now comparing this amount with the returns of the former census, that of 1841, (98,133 was the then total) it would appear that there has been an increase to the extent of 29,562; and this increase, it would further appear is partly,—in small part, owing to the births exceeding the deaths, but principally to augmentation by immigration. The former has been estimated at 1·26 per cent. annually,—about the same proportion as in Great Britain.*

The details of the population give the following results as to the races constituting it. Of the total 127,695, the natives of British Guiana, (including the

* See Reports &c., &c., P. P. for 1851, p. 145.

7,000 Aborigines,)* were 86,451 ; natives of Barbados, 4,925 ; natives of other West India Islands, 4,353 ; African immigrants, 7,168 ; Old Africans, 7,083 ; Madeiranes, 7,928 ; English, Scotch, Irish, Dutch, and Americans, 2,088 ; Coolies from Madras, 3,665 ; Coolies from Calcutta, 4,017 ; not stated, 17. If examined more in detail, it will be seen, and it is an important fact, that the male proportion of the population is in an abnormal and unhealthy excess over the female ; this excess not being less than 3,446, or 27 per cent. It is not found to exist in the Creole, the native part ; on the contrary, in the normal way with them, the females out-number the males to the extent of 3,993, or 4·6 per cent. ; but in the immigrant and white portion ; in them the difference is well marked, especially amongst the English, Scotch, Irish, Dutch, and Americans, and amongst the Asiatics—the Coolies. In the former the excess of males is the great one of 1,172, or 65 per cent. ; in the latter the very large one of 4,434, or 58 per cent. These disproportions hardly need any comment as regards their unsocial and immoral tendency, having in view an organised community. Truly it may be said, to repeat the emphatic conviction of the governor of Trinidad, that since the granting of emancipation too little attention has been paid to any legislation having for its end the formation of a society on true, sound, and lasting principles.†

---

* Arawak Indians, according to another estimate, amount to 10,000.

† The following Bounties are paid under an act of the Local Legis-

The agriculture of British Guiana, its almost sole dependence, has since it was first colonized by the Dutch, commencing about 1669, undergone great

lature, for the regulation and encouragement of Immigrants into British Guiana.

| Names of Ports or Places from which Immigrants may be introduced into the Colony. | Rates of Bounties allowed for the introduction of Immigrants. |
|---|---|
| Madeira ... ... ... ... ... ... ... ... ... | 25 dollars. |
| Azores or Western Islands ... ... ... ... ... ... | 25 |
| Canary and Cape de Verd Islands ... ... ... ... | 25 |
| Curacoa .. ... ... ... ... ... ... ... ... ... | 20 |
| Margarita and Spanish Main ... ... ... ... ... | 20 |
| St. Helena ... ... ... ... ... ... ... ... ... | 25 |
| Sierra Leone... ... ... ... ... ... ... ... ... | 25 |
| Brazil ... ... ... ... ... ... ... ... ... ... | 25 |
| Havana ... ... ... ... ... ... ... ... ... | 30 |
| United States of America and British North America | 30 |
| China or Chinese from any port east of Point de Galle in Ceylon, imported on board any vessel which shall clear for this Colony prior to the 31st day of March, 1853 ... ... ... ... ... ... | 100 |

Parties introducing immigrants are not only entitled to the above rates of bounty paid in cash, but would also have the preference of their services as indentured labourers, for terms not exceeding five years, on the gradual repayment of one half of the bounty, provided always, that it can be shown to the satisfaction of the local authorities, that suitable preparations have been made for their location, as regards food, lodging, and medical attendance. Bounty was paid under this Ordinance in 1852, upon 4,018 Immigrants, viz., from Sierra Leone, 267 ; from Calcutta, 2,779; from Madeira, 972. Upwards of 40,000 Immigrants have been introduced, upon whom bounty has been paid since the Emancipation Act." Note appended to the "Introductory Remarks," to the catalogue of articles transmitted from British Guiana to the Exhibitions of the Works of all Nations in New York and in Dublin in 1853.

The expense of bringing and returning a Hindoo Coolie (to be returned after five years) has been estimated at 150 dollars, or £32,—£16 each passage,—equal, it is said, "to the price of a second-rate slave in Cuba or Brazil;" and it is further said that "in the opinion of our planters, generally the powers of the Bozal negro are far superior to those of the Hindoo." *Eight years in British Guiana*, p. 271.

changes, but not so much as regards the means of
tillage employed, which has varied wonderfully little,
as the extent of land under culture, and the kind of
crops cultivated.

In its early history, its infancy, as in the other West
Indian Colonies, the attention of the planters was not
given exclusively to the raising of one kind of produce,
it was divided between the producing of sugar, cotton,
cocoa, coffee, indigo, and provisions. As in Trinidad
their advances were slow, till stimulated by British
capital, after three successive surrenders without blood-
shed, viz. first in 1781, next in 1796, and last in 1804,
since which time British Guiana has been a permanent
British possession.

In 1781 we are informed that Demerara, or Deme-
rary as it was then and is still officially called, and
Essequibo yielded 10,000 hogsheads of sugar, rum in
proportion; 5,000,000 lbs. of coffee; 800,000 lbs. of
cotton; and a certain quantity of cocoa, and indigo.*
Berbice then was little productive, not having recovered from the devastations effected in the slave
insurrections which broke out in 1763.

In 1827 there were exported 66,136 hogsheads of
sugar, rum and molasses in proportion; 8,279,653 lbs.
of coffee; and 15,904 bales, (400 lbs. to the bale) of
cotton. This amount was about the maximum produce
before emancipation. In 1850 the quantity of sugar
it exported was reduced to 37,351 hogsheads; of coffee
to 25,086 lbs. and of cotton to *nil*; its culture having

* Chron. Hist. West indies, vol. ii, p. 491.

been given up. This as yet is the minimum of production since the abolition of slavery, excepting that of 1846 described as a year of great drought when only 26,901 hogsheads of sugar were shipped.*

The causes of these changes, so great in so short a time, the most remarkable of them coming within the knowledge of the present generation, are nowise obscure, whether considered in their prosperous, or adverse turns. They are much the same as those already alluded to, producing like effects in Trinidad. First, the highly remunerative price of colonial produce, towards the end of the last century and the beginning of the present, tempting the investment of British capital, and exciting Dutch industry. Secondly, the great fertility of the soil, and, as to extent and extension, almost without limit. Thirdly, the facility with command of capital of obtaining almost any amount of slave labour. And, I might add, lastly, the advantage as to profit of maintaining these labourers cheaply, subsisting them chiefly on plantains grown on the estates.† During the prosperous period great fortunes were rapidly made ; a cane field was considered equivalent to a mine of gold ; and the value of estates was enormously enhanced.

The reverses commenced with the falling off of the supply of labour after emancipation, accompanied by a diminishing fertility of the estates under culture‡ and

---

* Reports &c. for 1851, p. 57, where is an official return of the exports of the colony from 1826 to 1851.

† Estimated at 25 dollars a year.

‡ The following is the reply of a manager to the inquiries of a proprietor, anxious to know whether the land on his estate was exhausted

by reduced prices of agricultural produce. The former was apparent in many of the properties first brought into cultivation*; the latter, especially in the instances of cotton and coffee, owing to their increased production with cheaper labour under more advantageous circumstances in other countries,—coffee for instance in Ceylon,† and cotton in the southern and slave states of America, gradually leading to the abandonment of the coffee and cotton plantations, and to the concentration as much as possible of the limited and costly labour on the cultivation of the cane, the produce of which paid best. The consummating reverse was

or not. "Here is an account of the yielding of the different fields for the last 15 years, extracted from our journal and compared by myself. The average, previously to 1838, was 1½ hogsheads per acre; from 1838 to 1840, 1¼; and from 1840 to 1842, inclusive, barely 1 hogshead." *Eight years in British Guiana;* by Barton Premium, a planter of the province, London, 1850. A melancholy narrative and worthy the attention of those who wish to become well acquainted with the colony, the causes of its distresses, and the character and feelings of the planters; provided allowance be made in the perusal for the bias under which it was written.

* We are told by a landowner that "an acre of newly planted land will give two tons of sugar for the first year, gradually falling off to not more than one fourth of that quantity as the stocks become old." *Demerara, after* 15 *years freedom,* p. 31. The practice hitherto generally in this colony has been to trust to the native vigor of the soil, and to fall back on the virgin soil, on that first cultivated shewing marks of impoverishment.

† In 1848, the coffee crop in Ceylon was estimated at 33,600,000 lbs. an enormous amount, considering that the cultivation of the plant, except in a small way by the natives, hardly worthy of mention, was not commenced till after 1821. Stimulated by a differential duty, capital was heedlessly, I may say recklessly, wasted in its extension, followed by disastrous losses to the planters, on the doing away of this duty, rivalling those of the West India proprietors, from the like measure affecting their interests.

the result of the alteration in the sugar duties in 1846, throwing open the English markets to foreign colonies, those having not only cheap free labour, as the East Indies, but also a command of slave labour, as in the Brazils and the Spanish West India colonies.

The fall in the value of estates after this measure was almost incredible. Lord Stanley's statements on this point in his letters on the West Indies to Mr. Gladstone are not I believe exaggerated, however much they may be in some others relating to the affairs and present condition of these colonies. I could state many instances corroborative, from the best authority. I shall limit myself to one or two. A friend writing to me on the 4th November, 1849, alluding to the then state of the colony remarks, "Another plantation has been sold by execution sale. It had been bought by the late proprietor about four years ago for 18,000 dollars, and was then considered very cheap. He expended about 30,000 dollars, it is said, on its improvement. At the time of last sale the estate was in perfect order, in cultivation, buildings, and machinery, and it was bought with twelve months' credit for only 6,800 dollars at public competition." He adds,—"This may give you some idea of the awful depreciation of property." He continues, "to shew you the downward tendencies of property in this colony, I am tempted to give you an account of one of the finest sugar estates of the whole colony situated about three miles from George Town on the west bank of the river, and with which I am connected as mortgagee. Its apprized value in 1840 was £40,000.—Two parties with myself bought it in 1843, at £25,000. About six months after I had purchased, finding the cares and anxieties of it, incompatible with my other pursuits, I sold my third to the other parties at cost price, and gave them ten years to pay it, securing myself by a mortgage. But originally there was a debt on the estate of £6,000, secured by the *first* mortgage, which remains. The two proprietors are ruined, my mortgage is not worth sixpence. But what is worse, as the first mortgagee considers that he has a personal claim against me for any deficit in the first mortgage claim, I am *forced* into negociation for the purchase of this mortgage in order to avoid the risk of law-suit; I believe I shall have it in a few days for £3,000." Resuming the subject in a following letter, one of the 3rd of December, 1849, my friend writes;—"The mortgage which I referred to in my last as being necessary for me in order to prevent a much larger liability, I have purchased for 14,500 dollars. I have had the estate placed under sequestration, with the full consent of the hopeless proprietors; and I have made a provisional sale of it for 20,000 dollars, payable in seven years. The capital sum for which I dispose of this magnificent estate on such long credit, would not dig the canals and trenches on it. My father-in-law, before the emancipation, derived a revenue of from ten to twelve thousand pounds sterling a year from it; and the free inheritance by his sons have encumbered them and been their ruin." He adds, "my apparent profit on the above transaction has been 5,500 dollars. But the security is not very good, and I have had to provide for the outgoing proprietors by letting them have a pasture estate belonging to me of about the size of the plantation they have left, at a pepper-corn rent. How are the mighty fallen! (I must not keep back his exclammation,) plantation S—— eagerly sold for 20,000 dollars." Well may we give credit after this to the statement of Mr. White, a planter of British Guiana, given in Lord Stanley's letter to Mr. Gladstone, that the value of sugar estates in that colony has been reduced since emancipation from 20 millions sterling to £66,000 which is less than the gross value of the yearly produce of the estates as estimated even so late as 1846, viz. £700,000 sterling; so that the landed property is not worth one year's purchase. Were any confirmation required of the disastrous changes which have come over British Guiana it may be found in the "Report of the Commissioners appointed by the governor to inquire into the state and prospects of the colony," in 1850, the time it is hoped of its greatest distress, and in the reports and despatches of the governor himself, Sir Henry Barkly, who was acquained with the country in its better days.

On the state of agriculture in British Guiana I shall be very brief. Two different feelings are excited in taking a view of it,—one of admiration,—the other of disappointment. The one at the manner in which the estates have been formed, the immense labour expended in raising dykes and embankments, in cutting drains, and excavating canals, essential to the reclaiming the land from the sea and marsh, and protecting it when reclaimed; works on which millions have been expended. The other of disappointment and regret at the little skill displayed in the cultivation of the estates thus laboriously and expensively made. It is no exaggeration I believe to say that in rudeness it even surpasses that of Trinidad. There the plough and some other instruments beyond the merely manual ones are beginning to be used : there manure is beginning to be applied : however backward, there ;the tendency is to an improved system of cultivation. The same can hardly be said of British Guiana; that it cannot, we are assured by a very competent observer, Dr. Shier, after a residence in the colony of many years and intimately acquainted with it. I shall quote his remarks; they form a part of an able letter addressed by him in 1852 to the governor in reply to some queries. "With respect to the cultivation of the sugar cane it has already been remarked that by reason of the lowness of the land, and the plan of draining in use, namely that known as the open drain and round-bed method, the system of cultivation remains exactly as in the times of slavery, every part of the operations of culture being performed by manual labour. The plough and other implements have been tried, but cannot succeed in effecting a cheap and effective tillage, till a system of close or covered drainage is resorted to and the open drains abolished.* Almost the only implements of tillage in use are the shovel, the hoe, and the cutlass. † The tillage in general is of the rudest kind, and were it not for the unparalleled fertility of the soil, ‡ nothing like the results actually obtained could be conceived possible. There is no such thing known as clearing or fallowing a field, so as to get it into good tilth and free from weeds before planting.

* " It is surprising how many horses, mules, and oxen have been sacrificed in the endeavour to establish this mode of tillage permanently. One of my neighbours lost sixteen oxen in ploughing about twenty acres, and after all, some hands were obliged to go over it with the shovel." *Eight years in British Guiana.*

† The basket may be added,—the rude substitute for the wheel or handbarrow and cart. I remember being impressed strongly by the waste of time and strength in seeing a party of labourers, highly paid, employed in repairing a breach in a sea embankment—carrying clay on their heads in small baskets, and trudging through the adhesive clay,—bare footed of course,—where labourers in this country would be working and walking easily, using wheelbarrows for carrying and planks for walking on.

‡ In a letter recently received from a friend resident in Demarara, in reply to some queries of mine, relative to the soil, and whether supposed to be exhausted,—he states his belief that none of the estates are worn out; adding, "Plantation Thomas, the estate next Town (George Town) is I believe the oldest sugar estate in British Guiana ; it must be, at least, 50 years in cultivation, and every acre of it still under cane crop without manure." He further adds, "when estates have plenty of land, as soon as the quantity per acre of sugar declines much (although the quality always improves) the old land is abandoned and fresh land, though at a distance, taken in. On our east coast the process was so extensive that the cultivation (and the buildings also) have receded often a mile from the public road. But latterly a retrograde movement is going on, and cultivation is in many places brought down to the very edge of the railway by the repairing of old fields. A few years of the splitting and cracking of the soil by the sun disintegrates and renders soluble enough of fresh mineral elements, and the wild vegetation is so rank that both organic and inorganic materials are soon supplied for fresh crops."

There is no alternation of crops; no manuring; but a field once planted, although at the time full of weed and the seeds of weeds, continues to be cultivated for a great number of years, new plants being put into the blanks that appear at the time of its being examined immediately after being cut. The tillage properly so called is very slight, and consists of a little digging in the neighbourhood of the cane rows, called shovel ploughing, and this is not always given every year. In addition to the digging, the grass and weeds are cut off by the hoe, and laid in heaps on the surface, two, three, or four times according to the rapidity of their growth and the nature of the weather. In addition to this the decayed leaves are stript from the canes once, twice, or three times during the season to admit the sun and air. When the canes are ripe and when it is found necessary to cut them, they are cut by cutlasses, and the immature upper part being rejected, the lower part is cut into lengths and carried out on the labourers heads to the nearest point of the navigation canal that bounds every field on three of its sides. The canes are then piled up in heaps or at once put into the punts by which they are conveyed to the sugar works to be manufactured. Plant canes, that is to say the first crop after planting, ripen on an average at fourteen months on coast estates, and sixteen months on river estates. Ratoon canes, i.e. canes of the second or any subsequent crop from the same stoles, ripen on coast estates at twelve months and on river estates at fourteen or fifteen months. But protracted periods of rain or drought, and these occasionally occur to some extent, influence the period of maturity."*

In their manufacturing processes the planters are more advanced than in their agricultural, and this by means of an almost unlimited supply of capital. No cattle, wind, or tide mills are now in use for crushing the canes and expressing the juice: the superior power of the steam engine is exclusively employed. Instead of about 50 per cent. of juice, the old proportion, in many instances now as much as 65 per cent. is obtained. Improved methods too are being employed in the evaporation of the juice and the making of sugar: the vacuum pan is pretty generally used, and by means of it a superior article obtained. These are redeeming circumstances and no doubt they will be carried further. And there are others which may be briefly alluded to, such as bringing into use improved stills, for the making of rum, heavier rollers at a slower speed for crushing the canes, and machinery for drawing off water, which may lead to thorough and covered draining, and this to the use of the plough and horse hoe. And these improvements no doubt will spread and be followed by others, as confidence is acquired in the resources of science, in which at present, owing to want of information, there is too little faith.

As to the general management of the estates and their concerns, the system here is much the same as that followed in Trinidad. As there, and in most of

* Reports, &c., P. P. 1851, p. 159.

the other islands, a large number of the proprietors are
absentees. No properties are let for rent.* There is
little or no community of interest, too often an opposing
one between the persons in charge, the resident managers, and the attornies non-resident, and those employing them.

The supply of labour too is much the same as that in
Trinidad, and the description of labourers,—the majority of them unskilled,—similar.† These as already

* That is, money-rent; on some estates the metayer system or share and
share system of cultivation has been tried, and with partial success:—but the
reports on it as a whole have been unfavorable.

† The following extract relating to wages is from the note of a friend in
the colony, of the 26th March, 1852, to whom I had applied for information, "The price of labour varies somewhat in the colony; perhaps it is
cheapest on the east coast, being the most densely peopled and of easy
access to the city. The following rates I have obtained from a planter
who is acquainted with both the east and west coast of Demerara and the
banks of the river.

| | |
|---|---|
| Weeding and moulding (east coast) ... ... ... ... | ·32 |
| "      "      " (west coast) ... ... ... ... | ·40 |
| Weeding and trashing (west coast) ... ... ... ... | ·40 |
| Cutting canes and carrying to punt 500 cubic feet (east) | ·64 |
| "   "   "   "   "   "   " (west) | 1·00 |
| Relieving and tying trash ... ... ... ... ... ... | ·33 |
| Throwing out small drains ... ... ... ... ... ... | ·50 to ·70 |
| Shovel ploughing (east coast) ... ... ... ... ... | ·33 |
| "      "      " (west coast) ... ... ... ... ... | ·40 |
| Jobbing ... ... ... ... ... ... ... ... ... ... | ·24 to ·32 |
| Boiler men, per day ... ... ... ... ... ... ... | ·40 to ·48 |
| Digging trenches ... ... ... ... ... ... ... | ·64 to ·100 |

In addition to these wages, the labourers get houses, garden ground,
water carrier (while at work) and generally medicine and medical attendance,—the wages are stated in cents and dollars."

Two or three years ago they were somewhat higher. Governor Light,
writing in 1848, states that "the labourer of ordinary work in the field
completes his task in four or five hours, hitherto paid at the rate of four,
five, or six bitts per day," i. e., 1s, 4d., 1s. 8d., 2s. 1d.

mentioned are principally Africans, either natives or imported, Portuguese, or Coolies. Each class has more or less of a distinctive character. The Creole Africans, as they have greater facilities in living and are exposed to greater temptations even than in Trinidad to idleness and dissipation, so they are even less to be depended on for field labour; a large number have withdrawn themselves from regular estate-labour, have become squatters or small freeholders, and are fast degenerating, so they are described, into savages.* Of 82,000 estimated as

* The following is from the reports of the commissioners relating to this class. After stating the resident labourers on plantations as amounting to 39,375, of whom about 19,436 are immigrants,—reflecting on the large proportion of Creoles who have withdrawn, they remark, " The system of freeholds (as it is called here) appears one of the crying evils of the day, and is indeed little better than a licensed system of squatting. Where whole districts present but a scene of abandoned estates it is very easy to purchase land for a trifling consideration; and thus numbers combining, deserted plantations are bought up and villages quickly formed on their sites. There are great numbers too, who strictly speaking *squat* up the rivers and creeks, that is, settle themselves on crown land without any title whatever. The forest teeming with game and the rivers with fish, afford them plentiful subsistence; and the ground with very little tillage yields them an abundant supply of provisions. They carry on a small trade in fire-wood, charcoal, &c., but by day the greatest part of their time is spent in absolute idleness. The accounts your commissioners have received of the demoralization going on in the negro villages is calculated to excite the deepest alarm, and rioting and debauchery seem but too prevalent among them. In many of the most populous villages in the most thriving parts of the country, very significant signs of actual retrogression are plainly perceptible. Formerly the Creole had a taste for luxuries in food and dress, and would willingly work to earn the means of gratifying his desires; but now he seems content to go about with the least amount of clothing consistent with decency and to be satisfied with the coarsest fare."

Governor Barkly in a despatch of the 18th July, 1849,—after pointing out some of the failings of this class remarks with great point and just-

composing the rural population in 1850, it is stated by
the commissioners that only about 42,000 contribute
in any way to raise the staples of the country.*

ness,—"another still more deplorable taint generated by slavery and
conducing to the same result (an almost stationary population in circumstances favorable to a natural increase) is the wide spread immorality
which prevails in the intercourse of the sexes, notwithstanding an outward
deference to religious ordinances." He adds, "To the casual observer
the emancipated peasantry seem the most pious of devotees; their oldest
friends often accuse them of being the deepest of hypocrites. To me they
always appear like the early converts to christianity, compelled by con-
science to avow a faith the effects of which the depraved nature of their
previous habits incapacitates them from exhibiting in practice." Report
1849, P. P., p. 244.

An anonymous writer who takes a very unfavorable view of the African
character, reflecting on the same retrogression, remarks "Let not the
blame however be entirely laid upon the black man. From the very
beginning, the policy of the planters would seem to have been a great
mistake. The freed negro was treated as a wayward child, but never
as a free man; he was coaxed, but never compelled to fulfil the duties
of his station. That stimulus to exertion, which exists in other countries
was carefully removed in Guiana; houses and grounds and hospitals
were placed at his command without charge; food and clothing were
equally within his reach, by the slightest exertion; schools were established,
to which he was entreated to send his children; and churches were
erected for the support of which he was never required to contribute.
When inclined to work, his services were zealously competed for on the
neighbouring estates; and when disposed to be idle, no laws restrained
him from wandering at will over every man's property." *Demerara after
fifteen years Freedom*, by a landowner. London, 1853.

* Since emancipation, owing to abundance and cheapness of land and
the natural love of possessing it and independence, the number of freeholders and of persons residing on freeholds has increased very rapidly.
These, it is estimated, now exceed one third of the total population.
According to Mr. Harfield, commissary of population, the first conveyance by "transport" of such lands was in 1838; in 1844 the number
of such holders, including their families in the number, amounted to
about 19,000, in 1847 to 29,000, at the end of 1848 to about 44,443.
The number of freeholds he states to be about 446, on which are
erected 10,541 houses,—containing 44,403 inhabitants, averaging sometimes more than 4 persons to each house. Report 1848, P. P., p. 255.

The imported Africans, it is said, are more industrious, but less skilled, and that as they acquire some degree of skill, they fall off in industry. The Portuguese are considered the most hard-working of all; but being very much intent on money-making, they prefer trade to field labour as the most profitable: very many of them have turned hucksters and small shopkeepers, and have become very useful in this capacity, going wherever they are needed and doing away the necessity before indispensable on the part of families of having large stores to meet every day wants. The Coolies have the repute of being civil and easily managed, and intelligent and neat though not strong labourers, and consequently better fitted for light than for heavy work. The climate appears to be best adapted to the Africans and Coolies, and less to the Portuguese. The latter are most subject to the fevers of the country and to yellow fever. In a despatch of the governor accompanying the Blue Book for 1851, he remarks. "The mortality of the different classes of immigrants, under indentures or working on plantations, was 364, out of 15,200, or rather under $2\frac{1}{2}$ per cent. but it was unequally distributed, the Portuguese from Madeira losing 5 per cent. the Africans less than 2 per cent. and the Coolies, very little over 1 per cent.," shewing, as he observes, "an unprecedentedly small rate of mortality, and proving how admirably they are adapted to the climate; it may be added, if taken due care of, for here, as in Trinidad, when they have been neglected, there has been great suffering and a fearful mortality amongst them.

The remarks made on the labourers in Trinidad are applicable, I believe, without exception or extenuation to those of British Guiana, of whatever class they may be composed, and also to the state of its society; that, here, as there, being little moral, ill assorted and ill organized ;—partly from the nature of its discordant and low elements, and partly from the very inadequate means hitherto taken to educate the people. According to the returns on the subject in 1851, of 25,467 children between the ages of five and fifteen, only about ten per cent. were able to read and write, and only 7,486 were attending any kind of school.* The evils of this social condition seem to be keenly felt in its consequences, such as those described by the commissioners in their report, and given in a foot note to a preceding page. A new

---

* Of the existing schools and the progress of the scholars a very unfavorable account is given. Governor Light in a despatch of the 3rd of May, 1848, states, " I see little improvement in the writing of the best of them : the petitions got up by the influence of the London missionaries are as deficient in penmanship as formerly ; the chief number of signatures is by cross with their scholars as with those of other sects. I have had before me a petition to the combined court of this year, where the elders, signing after the minister, show miserable specimens of writing ; the petitions I receive from labourers are generally marked with crosses, or if signed, it is in the lowest grade of writing : if the petitions have the benefit of original writing, they present the most extraordinary jumble of bad orthography, grammar, penmanship, and composition that ever emanated from a human being under the influence of education." He states that " the last resort of a man who has failed in other attempts at earning his bread, is to become a schoolmaster." Correspondence &c., P. P. 184—, p. 569.

system of popular education is now under the consideration of the authorities. May it be established and work well,—which it can hardly fail to do, if well constructed on the broad basis of usefulness, and well directed; and then, and not till then the colony is likely to prosper.* It is to be hoped that in this new scheme of education there will be means afforded not only for the training of the labouring youth in industrial and moral habits, but also for the instruction of the youth of the higher classes, so as to fit them to perform their part in life with credit to themselves, and advantage to the community. There have been at times good impulses acted on for promoting science in the colony, but for no long continuance, as is too clearly shewn in the brief history of the Colonial Laboratory, and Meteorological Observatory, both of them institutions, especially the former, of the greatest promise, considering merely the material interests of the planters. It is true that the expenses of the one were considerable, and that the support of both was withdrawn at a time of great disaster, during the commercial panic, and agricultural distress of 1847—8, but surely had there been confidence in the resources of science, that confidence which knowledge imparts, and a due regard for it, the closing of the Laboratory when its aid was most wanted would not have been

* I regret to learn that owing to the clamour raised against it, this proposed system of popular education, similar to the national system in use in Ireland, but more comprehensive, has been laid aside, and the old and imperfect plan of education—system it cannot be called, is continued with the abuses.

considered an economical measure. The like neglect of science, is indicated, in the disregard shewn to the formation of a public library in the principal town of the colony, and to the publishing of any periodical, except it may be newspapers, fitted for instruction, and the state, I may add, of the agricultural society of the colony; in reply to the last enquiry I made respecting it, I was informed that though "there was great activity amongst the planters, the agricultural society had degenerated into a mere reading-room society."

As regards the future of this colony, there seems to be opening a gleam of hope and a brightening prospect. The produce of sugar last year (1852) so much increased, that the quantity exported, almost reached the maximum shipped in the times of slavery, being about 60,000 hogsheads. Another favorable circumstance is, that the produce per acre now, even exceeds what it was in the times of slavery; then, we are told it was under a hogshead, —now, we are assured it is over a hogshead an acre.* This increase is probably owing, not to any improvement in the tillage of the land of which there is no proof; but to the improved manufacturing machinery and processes of which there is convincing proof. Moreover, now it is admitted that on encumbered estates, most of them belonging to new proprietors who have purchased them at their lowest depreciation, the present market price of sugar (about 33s. per cwt.)

* Dr. Shier, letter already quoted, p. 358.

more than clears the cost of production, leaving some profit to the producer.

For the success however to be permanent, many circumstances, here, as in our sugar colonies generally, must conspire, the conjunction of which cannot but be uncertain and create anxiety; such, to mention a few, as the having a larger and more industrious population, and a not over costly rate of labour,— a more economical, and better instructed and public spirited proprietary body,* with fewer absentees,— an increased demand in the markets of the world, from increased consumption, of the staple produce, sugar, with no unduly increased extension of the cultivation of the cane; keeping out of the account as sources of anxiety the accidents of seasons and the various contingencies to which every interest, every branch of industry is exposed. The hope is more encouraging and the prospect brighter looking forward to the time, which sooner or later surely must come, of universal slave emancipation and of free and fair competition in tropical agriculture. Then it will be hard and strange indeed, if British Guiana, (and the remark applies equally to Trinidad,) cannot contend successfully with any of the slave colonies, when no longer such, which are now most successful. These have not greater natural advantages, not even Cuba, either as regards climate or soil, fitting them

* The anonymous writer already quoted, strongly points out the want of public spirit as altogether interfering with the advance of the colony. " That utter want of unanimity and concert, that intense love of self and self alone, that wretched plea that things will last their time." p. 97.

for rich and varied production. The produce even now in Cuba, of sugar per acre, we are assured, is less than either in Trinidad or in British Guiana. Of the productive powers of the latter, even in its present state, we have proof afforded in the catalogue of the articles contributed to it by the Great Exhibition; the number amounts to 156. According to the arrangement adopted, the whole are comprised in three sections,—first, including raw materials and produce ; the second, machinery; and the third, manufactures. The two latter, in their poverty, are remarkably and very characteristically contrasted with the first, in its richness,—under the section of machinery no article having been contributed, and under that of manufactures only 39, and these entirely examples of Indian work, chiefly specimens of their implements, instruments, and utensils. The first section too, as divided into the productions of the mineral, vegetable, and animal kingdoms, exhibits marked contrasts, in relation to wealth and poverty,—belonging to the first* there being only 4 articles, (clays and sands,) to the last only 2, whilst to the second no less than 113 are assigned, viz. first, under the head of substances used chiefly as food, or in its preparation 47,— many of them, exclusive of the staples, very valuable, deserving of being made commercial articles, especially

---

* Of the clay and sands of the colony, of which there are many varieties, probably some will be found of value in the manufacture of glass and porcelain. No. 1 in the catalogue is described as " a white sand," no doubt siliceous,—" that has been exported to the United States of America for the purpose of glass making."—No. 4, " a decomposed rock, supposed to be valuable in the manufacture of pottery."

various starches, meals, rice, and dried fruits ; secondly, under the head of materials used chiefly in the chemical arts or in medicine 29,—a number very inadequately representing the richness of the colony in these pro- ductions ; thirdly, materials for building, clothing, &c. III, of which, all but 20—specimens of cotton and certain grasses and vegetable fibres,—are samples of the woods of the colony. It is in these woods that British Guiana is peculiarly rich ; the forests of the interior are described indeed as inexhaustible,—as abounding in timber trees of excellent quality, yielding woods fit for various uses,—for the cabinet maker, as well as the ship builder, and as easily reached by navigable rivers. Sir Robert Schomburgk, the enterprising and intelligent traveller, speaks of the fitness of the timbers of this country, as unparalleled for naval architecture, for strength and durability, some of them it is believed even surpassing in these qualities the oak and the teak, and in such abundance, he thinks, as to be competent to sup- ply all the ship-building establishments in Great Britain.

I shall not attempt to enter on the natural history of British Guiana ; a vast subject opening wide fields of research, and on which volumes might be written. To the enterprising and exploring traveller, especially if a naturalist, few countries are better fitted either to ex- cite curiosity or to gratify it. If he wishes to see nature in the luxuriant grandeur of wild tropical mountain forest scenery, he has only to ascend one of the noble rivers to the highlands of British Guiana. If his desire is to study man in different stages of society and progress, he will have ample opportunities afforded

him, in the interior amongst the native tribes, and
on the coast in the several races strangely brought
together from various quarters of the globe, and with
difficulty kept together, the majority of them in a state
of transition either from the rude condition of the
savage or slave, or the low one of the Hindoo. If
agriculture be his favorite pursuit and political economy,
and they can hardly well be kept apart, he will have
much to see and examine and reflect on, wherever he
goes, whether it be over a sugar estate in prosperous
cultivation, of which we believe there are still a few
owing to peculiarly favorable circumstances, or one
going to rack and ruin from a state of high prosperity,
or a coffee or sugar plantation, once blooming like a
garden, now become a wilderness.

To explore this magnificent colony, as it has been so
often and justly called, even moderately, would be a
somewhat arduous task, needing a good deal of time
and ample means. I shall mention one or two excursions which may be made with profit, in a short time,
and without much difficulty; and one or two spots near
the principal town within walking distance, that may
be worth visiting.

New Amsterdam, the chief, indeed the only town in
Berbice, situated on the shore of the river of the same
name, is about 80 miles from George Town. The journey by land from one town to the other is commonly
made in a rude conveyance, called the mail-cart from
its conveying the mails, which, when I was in the
country in 1847, went and returned three times a week.
Without stopping on the road except to change horses,

it made out the distance in from about 12 to 16 hours, according to the state of the roads. Now, the time may be somewhat shorter, as since I was there, there is a railway constructed as far as Mahaica, the first stage, about 6 miles distant.

For the day's journey some preparation must be made, some provisions laid in, as there is no stopping to dine, or any place where refreshment can be had by the way. And what is more important, precautions must be taken in regard to clothing. First a good large veil should be procured, capable of being well tucked in round the neck; next gloves, two pair, one to be worn over the other, that the seams may not meet, and further some defence from invasion of chest and legs by close buttoning or tying and tucking, allowing no chink or aperture to remain unclosed, wherever such may be from the fashion of the dress. These precautions are necessary against the torment of mosquitoes which infest these regions, and to such a degree, that it is even necessary to use smoke pots, when changing horses, to keep off these insects, and prevent the animals from becoming furious from their attack.* Of the incidents of the journey, judging from my own experience, some few accidents may be calculated on, such as an upset, or a threatened one from a wheel sinking into the mud to the axle, on a piece of road, through an encumbered or deserted

---

* A smoke pot, which is often necessary even in the dining and drawing room in Demerara, is an earthern vessel, in which there is a smouldering fire kept up by means of half-dried grass, which burning slowly without flame, gives off a very disagreeable smoke,—a lesser evil only than the pest of insects.

estate, neglected in consequence of the want of means as well as of inclination of the proprietor, whose duty then was to keep it in repair; a sudden stop from bush obstruction, the off or near horse getting entangled in and unable to force his way through the thick shrubs encroaching on the grassy track,—such the road being in many places; a delay here and there from some bridge broken down, or in a dangerous state from decay, or from some ferry boat deserted.

Of the country itself through which one passes, it is not easy to give an idea in a few, or an adequate one in many words, so peculiar is it in many respects, so varied in its minute details, and yet monotonous as a whole. The first part of the way, so far as the military post Mahaica, and beyond it to the Macomy River, there is some cultivation, — sugar estates occurring at no great intervals. After leaving that river, traces of cultivation almost entirely disappear; one enters a wilderness, open spaces, savannahs, deserted cotton estates now become cattle farms, marsh, jungle, or bush, (as the wild coppice and woodland are here called,) meet the eye in succession or altogether, strangely intermixed. What is most striking is the excess and luxuriancy of vegetation, especially of those lower forms, such as the aquatic plants, and I may add the parasitical vines and creepers. Wherever there is a surface of water it is hid by aquatic plants,—varied and bright tinted confervæ, and more conspicuous water lilies, some bearing white, some yellow flowers. Wherever the ground is sandy, somewhat dry, and a little raised and bearing trees, there the parasitical plants and vines abound, spreading over and hiding not only the branches, but even the natural foliage of their support, and giving to them forms totally different from their real, often resembling the effects of art, as if made by assiduous clipping,—some in nearly square masses, some conical, some nearly pyramidal, others like walls or lofty trimmed hedges. Where neither covered with water nor dry, but wet approaching to marshy, there the bush and low jungle spread over the ground,—a tangle of these parasites with reeds and broad leaved ferns of a gigantic size. Not unfrequently an air of peculiar desolation is imparted to the landscape by dead trees in masses. The excess of animal life seen in passing is hardly less remarkable than of that of vegetable. Birds abound as well as insects, and help to give some animation to the dreary way; paroquets, with gay plumage and rapid flight, stoical waders of different species, stolid vultures, almost tame, "the witch," a species of pica of forbidding aspect, these are the most common.

Of the termination of the journey in New Amsterdam after crossing the Berbice river, not less than two miles wide, I can say nothing laudatory. Judging from my own experience, it is likely to be found in character with the journey itself, especially if arriving at night, as in my case, and without a friendly house to go to, under the necessity of seeking shelter and finding it with difficulty in some poor inn or lodging house, and there to be received almost as a matter of favour.

If not pressed for time the stranger will do well to ascend the Essequibo, the largest of the three great

rivers of the colony, and visit the penal settlement,
established on the Massaroni, near its junction with the
Cayoonee, one of the principal, if not the greatest con-
fluent of the Essequibo, that mighty stream with its
two or three hundred islands, and which when joined
by the above rivers expands into a lake, described as
eight miles in width and navigable for ships of the
largest size. The voyage—a distance of about 70
miles (about 50 from the mouth of the Essequibo,) is
easily made in a steamer ; and from what I have heard
is a peculiarly agreeable one, not only on account of
the beauty and magnificence of the scenery, increasing
in impressiveness with the ascent, but also from the
freedom from insects, even from mosquitos, an exemption attributed to the quality of the water, but
owing more likely to other circumstances, especially
the elevated, dry, and rocky nature of the country.*

In a short time, and even in the neighbourhood of
George Town, much that is interesting may be seen
deserving the attention of the traveller. I remember
with pleasure two visits which I made whilst there.
One was to La Penitence, an estate where I witnessed

* The color—a clear dark brown—of the waters of the Essequibo and
its great tributaries, by some has been attributed to iron, by others to
manganese, oxides of both these metals having been found on its shores.
Chemically considered, neither conjecture seems to be probable. Is it not
more likely to be owing to the presence of vegetable matter of a peaty kind
similar to " Pegass," suspended in part and in part dissolved ?

The banks of this river are esteemed peculiarly healthy and fit for
European immigrants, as is also the interior highland. On account of
this presumed quality and the coolness of its atmosphere, depending on its
elevation, it has been strongly recommended as a site for an Imperial penal
settlement—a substitute for Norfolk Island. See report, 1849, P. P.,
p. 266.

the first attempt in the colony to introduce thorough deep and covered draining as a substitute for the open drain system, under the direction of Dr. Shier, the results of which were of a very satisfactory kind. The field, thus drained, presented of course an uninterrupted level surface allowing of the use of the plough and was remarkably contrasted with an adjoining field with its open drains permitting only the use of manual labour. In the thorough drained field a steam engine was at work, raising water from the canal in which the deep pipe-drains terminated. The expense entailed by this mode of draining seems to be the only impediment to its coming into general use. That it will be remunerative, its advantages being so many, appears to be the opinion of those most competent to decide on its merits, amongst which, no inconsiderable one is the improved quality of the sugar, the produce of its canes, from the abstraction of salt, always in excess in the soil, when tilled in the ordinary way. Those wishing for full information on the subject, can find it in Dr. Shier's published report on the operation.

The other visit alluded to was to an estate, also very near to the town, one of about 1,500 acres, of which about 500 were in canes, belonging to a successful and wealthy planter, a native of Barbados, who commenced his career as a manager. His house, his grounds, his works, everything, in brief, that I saw, denoted a flourishing condition, and the presence and control of a master's mind. The road to the house, broad and straight, was through an avenue about a mile in length, of stately cabbage palms, of uniform height and size, all planted by the owner,—suggesting an approach to some stately temple, as in ancient Egypt. Each palm was hardly less than thirty feet in height, beautifully tapering from a broad base, and each with a magnificent mass of leaves gracefully arranged in a dome-like form. The house, a well-constructed one, and large,—it might be called elegant, was surrounded by neatly-kept and pretty shrubberies, an unusual sight any where in the West Indies. The works there for making sugar, including the distillery, were in high order and provided with superior machinery; steam was the main power employed. The whole was a gratifying sight, and a hopeful one, I thought, for the colony, as proving, that even in the worst times, a sugar estate may be kept in cultivation without loss, even with some profit.*

* From a recent communication, (the 10th of March, 1853,) I have learnt with regret that this gentleman is dead, and died insolvent. My friend from whom I had this information, adds some particulars,— very characteristic of the reverses of the colony. " He was not successful latterly, though pre-eminently so in his early days. He was once manager of the estate he afterwards owned, plantation H——. That estate then gave him between salary and perquisites more than £1000 per annum, and the proprietor £15,000. He was discharged from it for having the audacity to offer for it £100,000. He purchased two other fine sugar plantations and retired to England with £40,000 besides. After the emancipation was complete by the termination of the apprenticeship, the proprietor of plantation H—— got into difficulties; and Mr. —— who was at home, again offered for the estate and got it for £36,000. But some of his friends upbraided him for taking advantage of the proprietor, he Mr. —— voluntarily came forward and added £4,000 to the price."

# CHAPTER XII.

## ANTIGUA.

ANTIGUA may be considered as bearing very much
the same relation to the leeward islands, that Barbados
does to the windward, being the seat of the central
government and the head quarters of the military
force,—that however detached from Barbados. Like
Barbados too, compared with the islands subordinate
to it, Antigua is the most populous, productive, and
wealthy;—distinctions it is believed, owing chiefly
to similarity of circumstances; some relating to the
soil and climate, some to the people inhabiting it,
and some to the manner in which it was originally
settled and planted.

By consulting the map, it will be perceived how
it is situated in relation to the adjoining islands,
and how it is nearly at the same distance from Guadaloupe and St. Christopher's, and to the windward of
each; both of which are to be seen from its higher
grounds when the atmosphere is clear, as is also

Montserrat, which is nearer than either of them,—
Guadaloupe being about forty miles distant, and Montserrat only about twenty-five.

In form Antigua is circular, about 20 miles long,
and 54 in circumference, comprising within its area
about 108 square miles, equivalent to 69,277 acres.
Its coast, in many places skirted and defended by coral
reefs, is remarkably indented, abounding in excellent
harbours, superior in this respect to any other of the
West India islands. In relation to its general aspect
and scenery, it is of mixed character; an irregular surface, hilly rather than mountainous, gentle in its features, and pleasing rather than bold and picturesque.
It is in a great measure destitute of wood, as much so
as Barbados, and even more destitute of springs and
streams.

As regards its geological structure, and in accordance
with that the character of its surface, it may be divided into three portions, viz. the eastern, western, and
middle or intermediate; the former including the coast,
extending from Willoughby Bay to near Dickenson's
Bay; the second from a little to the eastward of English
Harbour to Five-Islands Harbour; and the last, the
intermediate portion, extending from one shore to the
other.

This last-mentioned portion is little raised above the
level of the sea; is comparatively level;—a plain, bordered on both sides by hills. The eastern division
flanking this plain is moderately hilly, the hills commonly gentle and rounded, separated by flat and basin-

like hollows, in which small collections of water, ponds are of frequent occurrence, and marshy or swampy ground not infrequent. The western division also bounding the plain, is formed of higher hills, (the highest of these not exceeding 1000 feet) and these more connected, and ridge-like, with depressions between them constituting vallies of gradual and in some instances rapid descent, without any basin-like hollows, retentive of water.

In these three divisions, marked contrasts are exhibited in their geological relations. On one side, the western, the rocks are of an igneous character, denoting violent action and protrusion from beneath, akin to the volcanic, but without actual eruption or explosion,—rocks belonging to the fletz-trap family, varieties of basalt, green stone and trachite, with conglomerates intermixed, composed of fragments of these. On the other side, the eastern, the character of the rocks is totally different, such as belong to the tertiary aqueous formations, being chiefly calcareous freestone and limestone, containing organic remains distinctive of the class. In the intermediate space, the plain stretching across the island, both kinds of action, the igneous and the aqueous, may be said to be exhibited ; the former in the indurated clays and siliceous cherts forming the skirting western declivities, the latter in the numerous petrifactions, varieties of wood and coral, principally if not entirely siliceous, met with as fragments, scattered over the lowland and imbedded in its soil.

The soils of the island are not less varied than its

rocks: stiff clays may be considered as predominating in the western division, lighter ones and calcareous marls, in its eastern and in part in the intervening lowland. These are generally productive, and some of them, especially the marls, of extraordinary fertility.* Exceptions, however, are not uncommon; they are most remarkable where incumbent on the indurated clay and chert formations. Another exception, and it is a singular one, occurs in particular spots, which without any obvious cause, are unfavorable to the growth of the cane; these spots are called "gall patches:" they are met with, I believe, only in connection with marl. Such information as I have been able to collect respecting them, is given in the subjoined note.†

* In none of the marls of Antigua or in any of its calcareous formations have I detected the remains of infusoria, such as are abundant in some of the chalk-like formations of Barbados and Trinidad. The calcareous marls of Antigua such as I have examined, have contained variable proportions of carbonate of lime and clay, with siliceous matter in fine angular particles. The poorer soils have proved to be chiefly siliceous, composed principally of water-worn and angular particles, as seen under the microscope.

† The "gall patch" most commonly occurs in light marls; is not distinguishable by any appearance of the soil, by any peculiar odour, or luminous exhalation, or by being fatal, or, as well as can be ascertained, by being anywise noxious to animal life. It is most injurious to the cane, less so to the maize and guinea corn. The late Dr. Nugent of Antigua, a man of a very inquiring mind, designated it the opprobrium of agriculture, neither the cause of it having yet been discovered, or any empirical means tried found of use in correcting the evil. It is said that it may be cured for a year by penning cattle on the spot. An enterprising planter, I am informed, had a gall patch excavated to the depth of twelve feet and filled up with mould from fertile ground. The spot did well for about twelve months and then the disease reappeared. It is said to show itself most in its withering influence, in wet and showery weather. The soil

The climate of Antigua is in most respects similar to that of the other islands. It differs most, and that in a very important particular, in the uncertainty of its rains, their average less amount, and in being occasionally subject to long droughts.* In accordance, I may

from a "gall-patch" may be removed, it is said, to another spot and there be fertile. Some specimens, at my request, were sent to me when I was in Barbados with a bottle of air, which, at my suggestion, was collected over one of the "gall-spots." They were accompanied by a note from the manager of the estate from which they were taken—Thibou's,—of which the following is an extract. "They (the specimens) are from a piece of land, which, excepting three yellow spots, (yellow, I apprehend, from the blighted canes, one about 40 feet square, one of 30, and one of about 20,) is perhaps as productive of canes as any land in this division. The three yellow spots nourish cane sprouts to the height of 18 or 20 inches, at which period they become yellow and die, but produce yams, potatoes," (according to other information the sweet potatoe suffers almost as much as the cane,) "and other vegetables abundantly. No. 1 is taken from the surface. No. 2 is the subsoil taken from the depth of 11 inches. There is no rock nor grit near the surface, nor could I perceive any difference between No. 2 and the subsoil at the depth of 3 feet. The bottle was filled with water and turned down in the soil, at the depth of 12 inches from the surface in which it remained for 24 hours, was taken up and sealed, being kept in the same position." These specimens I examined, but with negative results : I could detect nothing in them or in the air that was peculiar, so as to account for the injurious effect. The subject is too important to be given up in despair. So important is it, owing to the considerable loss these " gall patches " occasion on some estates, that I have heard a planter say he would willingly give £1,000 for the discovery of a remedy, and another present add that he would willingly contribute £500. I need hardly remark that careful enquiry on the spot affords the best chance of discovering the noxious cause, and that known, it is probable there would be little difficulty in finding an effectual corrective. Barrels of the soil from the patches, it is said, have been sent to England for analysis but without successful result, though submitted to the searching skill and science of Mr. Faraday.

* The Rev. Mr. Smith in his " *Natural History of Nevis &c.*," published in 1745, states that when he visited Antigua, it was during a period of great drought; when water was brought from Guadaloupe and Montserrat;" and "sold for eighteen pence the pailful."

Bryan Edwards, adverting to this peculiarity of climate, remarks that

mention that I have even found the sea washing its
shores salter than any other part of the Archipelago
of the Antilles, the water of which I had any opportunity of trying.* The average yearly fall of rain is
stated in the yearly almanack to be 45 inches. I believe
the deduction is founded on limited experience. The
following Table formed from observations made at "the
Ridge," the military station, about 300 feet above
English harbour which it adjoins, would seem to indicate a much less average amount, at least for that part
of the island.

|  | 1846. | 1847. | 1848. | 1849. | 1850. | 1851. | 1852. |
|---|---|---|---|---|---|---|---|
| January................ |  | 2·77 |  | 2·92 | 1·57 | 3·75 | 0·81 |
| February ............ |  | 1·50 |  | 1·30 | 1·95 | 6·38 | 2·85 |
| March ............... |  | 1·72 |  | 1·00 | 2·60 | 2·25 | 0·40 |
| April ................. |  | ·57 | 2·92 | ·85 | ·43 | 2·75 |  |
| May ................. |  | ·86 | 1·17 | ·53 | ·25 | 5·43 | 4·36 |
| June ................. |  | 1·92 | 1·37 | 1·99 | 2·64 | 8·84 |  |
| July ................. | 8·31 | 2·33 | 2·91 | 3·68 | 7·53 | 1·81 | 2·85 |
| August .............. | 8·69 | 6·38 | 4 00 | 4·26 | 7·52 | 7·31 | 6·24 |
| September ........... | 3·29 | 1·91 | 5·47 | 1·37 | 3·78 | 1·23 | 5·09 |
| Ootober .............. | 2·46 | 3·98 | 7·49 | 2·96 | 1·31 | 3·99 | 0·88 |
| November ........... | 9·17 | 6·25 | 4·69 |  | 0·31 | 1·82 | 4·34 |
| December ............ | 13·47⅛ | 2·42 | 1·79 | 3·30 | 3·24 | 5·67 | 4·28 |
|  |  | 36·51 |  | 24·16 | 33·13 | 51·23 | 32·10 |

" in consequence it is difficult to furnish an average return of the crops,
which vary to so great a degree that the quantity of sugar exported from
the island in some years is five times greater than in others : thus in 1799
were shipped 3,382 hogsheads ; in 1782 the crop was 15,102 hogsheads ; and
in the years 1770, 1773, and 1778, there were no crops of any kind ; all
the canes being destroyed by a continuance of dry weather ; and the whole
body of negroes must have perished for want of food, if American vessels
with corn and flour had been at that time, as they now are (i. e. during the
war) denied admittance." *History of the West Indies*, vol. I, p. 447.
    * The difference in the degree of saltness was very slight as is shewn by

Drought is not its only infliction, it has suffered from hurricanes, only in a less degree than Barbados,* and from earthquakes more than any other of the British Antilles: the earthquake of 1843 was unusually severe and destructive, it overthrew most of the churches, many of the public buildings, and a large number of the dwellings of the inhabitants.†

As regards salubrity, the climate of Antigua has, from all the information I have been able to collect, nothing to boast of. Fevers of the intermittent kind are far from uncommon, especially amongst newly arrived Europeans when residing in the country; and

the following result. The specimens of water tried were taken up near shore, between the 25th and 28th of February, 1846, and kept till weighed in bottles well secured with glass stoppers.

Water from the harbour, Granada was of specific gravity 10,273.
Off St. Vincent, about half a mile from shore    ...   ... 10,265.
——— Dominica   ...   ...   ...   ...   ...   ...   ...   ... 10,265.
——— Martinique  ...   ...   ...   ...   ...   ...   ...   ... 10,267.
——— Guadaloupe ...   ...   ...   ...   ...   ...   ...   ... 10,267.
——— Antigua outside the English harbour ...   ...   ... 10,274.

* During that of the 12th of August, the barometer, it is stated, fell 1.5 inch,— the sea rose above its usual level, depositing a great quantity of marine organic remains and of decaying vegetable matter, to exhalations from which an epidemic fever that commenced in the following month in the town of St. John has been attributed.

† In Antigua, the loss of life was inconsiderable, but the loss of property great. It is worthy of remark as an effect of the earthquake, the only one that I have heard of that was well marked,—apart from damage to artificial structures,—was a partial sinking of the causeway leading to that island, an islet adjoining the town of St. John.

The catastrophe was even more severely experienced in Guadaloupe. There the town of Point à Pitre was engulfed, and 5,000 of its inhabitants perished.

3 D

dysentery, particularly during periods of unusual drought, when unwholesome water is used, may be considered as a disease, always under these circumstances, more or less endemic.

Its population though proportionally the largest of the leeward islands, is small in comparison with that of Barbados, being 340 to the square mile, about half that of Barbados similarly estimated; and is still more remarkable in not exhibiting any increase for many years even in the colored races. It is stated in a valuable work on these colonies, first published in 1761, that Antigua was then judged to contain about 37,000 inhabitants, of whom 7,000 were whites, 30,000 blacks.* In 1787, the total was returned as 40,908, of whom 2,590 were whites, 1,230 free people of color, 37,088 slaves. In 1805, the total was given as much the same, viz. 40,300, of whom 3,000 were whites, 1,300 people of colour, and 36,000 slaves.† In 1844, according to the census taken in that year, the total was 36,178; according to the last census, that of 1851 it was 36,799. In neither of these are the races distinguished. Comparing these two last totals, the difference we perceive is only 621, and this we know is partly owing to increase by immigration.‡ So stationary a population, such a marked want of increase, has not passed unob-

---

* An account of the European settlements in America, vol. II, p. 92, 5th edition.

† The estimates for the last two periods are from Sir. W. Young's *West India Common Place Book*," p. 3.

‡ In 1849 the number imported was 132.

served; but has not I think been satisfactorily accounted for, especially by those who are disposed to refer it to the immoral and licentious lives of the colored races, and to the neglect by them of their offspring in infancy. To me, the cause seems more likely to be in the climate, such as it has been already noticed. And this inference appears to be confirmed, on comparing the numbers of the sexes at different ages and how the proportion of the women increases with advancing years, starting from an almost equality, the men who are most exposed to atmospheric and other noxious influences dying off in a higher ratio than the women. Thus, whilst the adult females exceed in number the adult males by 2,112; the females under 14 years are less in number than the males by 208,—the adult females being 13,441, the adult males 11,329,—the females under 14 years being 6,079, the males 6,287. Nor are such differences, in relation to the sexes, confined to the results afforded by the last census. Sir Charles Fitz Roy commenting on those of the preceding census, that of 1844, points out even a greater disparity as to numbers, viz. comparing the men and women above 60, that the number of the latter was double that of the former.*

One of the earliest settled and in part from the overflowings of Barbados, led by Colonel Codrington, the father of the great benefactor of the West Indies; the population of this island has been and still is very similar to that of Barbados, and with the exception of

* P. P. 30th June, 1845, p. 17.

the small number of Portugese immigrants lately introduced, not more compounded or mixed as to races, being limited to Creole whites of British, and to negroes of the African stock.

During the time of slavery, nowhere in our West Indian possessions were the negroes treated with more humanity than in Antigua. Of this there are various proofs, greatly to the credit of their masters, and in its influences not less to their advantage. They were the first to allow their slaves to receive any religious instruction; the first to permit them to marry; the first not only to tolerate but to encourage the labour of missionaries; the first to grant the slave the privilege of trial by jury; and lastly, and it was the crowning measure, when the act of emancipation was passed, they, and they were a solitary example, graciously and generously granted them their liberty, sparing them the galling and irritating trial of apprenticeship,—a second and hardly lesser bondage, as the disappointed considered it.

The missionaries who had first the confidence and encouragement of the planters (the methodists followed) belonged to the society of *Unitas Fratrum*, better known by the name of Moravians,—a body of christians acting on truly christian principles, proving their faith by their works, seeking and labouring most where most needed and likely to be most useful.* In Antigua their labours were eminently successful. Com-

---

* Some interesting particulars respecting this society, their principles and labours are to be found in Bryan Edwards's *History of the West*

mencing in 1756, we are assured that in 1818 they had nearly 12,000 followers, and now we are informed that at least two thirds of the labouring class belong to them.* The effect of their teaching was made manifest even during slavery. The answer of the House of Assembly, when after the slave insurrection in Barbados in 1816, it was exhorted to take measures of safety, affords a very satisfactory proof,—being to the effect, that the members did not consider any such measures necessary, having confidence in their people from the kind treatment they had received, and expressing conviction that the "increasing influence of moral and religious principles would effectually prohibit any rash or desperate attempt."†

Since emancipation, if the peasantry have not made all the progress that the ardently hopeful looked for, I think it cannot be denied that they have improved, and are an improving people. One Governor, the present, Mr. Mackintosh speaks of them as a "well-ordered community,"—as admitted to be distinguished for

*Indies*, vol. I, p. 449. He states that the number of converted negro slaves under the care of the brethren at the end of the year 1787, was,—

In Antigua, exactly ... ... ... ... ... ... ... 5,465.
— St. Kitts, a new mission... ... ... ... ... ... 90.
— Barbados and Jamaica, about ... ... ... ... 100,
— St. Thomas, St. Croix, and St. Jan, about ... ... 10,000.
— Surinam, about... ... ... ... ... ... ... ... 400."

The exertions of these missionaries at present, both in Antigua and in the other West India Islands, appear to be well sustained. A training school recently formed by them in Antigua is favorably mentioned in a despatch of the Governor. P. P. April 5th, 1849, p. 217.

* Parliamentary Paper.
† Letters on the West Indies, by James Walker. London, 1818.

" cheerfulness and courtesy:"—that there is "no increase of crime, if such a term is applicable in definition of the petty offences arising out of the hasty facile temperament of the negro." Another Governor, Mr. Higginson, the immediate predecessor of Mr. Mackintosh, adverting to the small number of prisoners in the gaol of the island, and that amongst them there was not one female, remarks—"It seems almost superfluous to repeat what has so often afforded subject of gratulation, the peculiarly rare instances of the commission of grave or sanguinary crimes amongst the emancipated population of these islands. In orderly demeanour, in observance of the laws, in submission to constituted authorities, in respectful deportment towards their superiors, and in the discharge of many of the obligations of social life, the people of Antigua are eminently conspicuous ; and it gives me sincere gratification to record a fact (the few prisoners and their light offences !) which redounds not only to their credit, but to that also of their pastors and others, whose precepts have contributed so materially to its accomplishment.*

Of their improved state, as regards their material wants, the evidences are of a very satisfactory kind. Three fourths of the labourers, we are informed, have cottages of their own, generally near the estates on which they work, forming villages or hamlets, of which there are as many as 87, all built since emancipation, and each possession a small freehold, the land

* Parliamentary Papers, 1849.

attached seldom exceeding half an acre, oftener under that. They are described as having a pride in the erection, and adornment of these cottages, in the possession of property of their own, in striving to raise themselves in the ranks of social intercourse and in promoting the advancement and welfare of their children. Sir Robert Horsford, reporting on their condition in 1845, remarks, speaking of the labourer,—" The calabash or gourd in which he was accustomed to carry his provisions to the field have in all cases been replaced by the neat covered tin saucepan; his cottage, in addition to articles of furniture far beyond his condition of life, is now supplied with plates, dishes, knives and forks, drinking glasses, and all the many domestic appliances which were wholly unappreciated in his days of bondage." He adds, "a coat of superfine cloth, a black hat of the best description, waistcoat and trowsers in good keeping, gloves and an umbrella or light walking cane are all indispensable on Sundays or holidays; and the female portion of this class of society are not behind their sisters of the superior ranks in throwing into shade the extravagance of the male, and walk forth on these occasions of festivity, in all the costly garniture of the latest fashions."*

Further we are informed, and it is more to their credit, and marks a different feeling from these little excesses in dress, that many friendly societies are established amongst them, with which 12,588 persons are connected; and also that they are beginning to avail

* Half-yearly report of stipendiary magistrate, 1845. P. P.

themselves of a Savings' Bank lately instituted. Their attention to the education of their children is not so satisfactory, nor so clearly increasing.* On my first visit to the island which was in 1846, I was told that the number of children attending the schools was about 3,000, half the number those of the Church of England, the other half those of the Moravians and Methodists.† In the latest report of the Governor that I

* Having a pleasure in recording what has come to my knowledge favorable to the African character and considering it a duty, I cannot withhold from myself the gratification of mentioning what was told me at the time by the late Mr. Cunningham, then the acting governor,—how in Anguillas, a dependency of St. Kitts,—the natives, the greater portion of them people of color, had exerted themselves to establish a school there, where a few years ago, the inspector of schools had reported that the measure was impracticable, and how they effected it without any aid,—converting a ruinous building which they had repaired into a school-room, and engaging a master,—they allowing him a salary and the parents of the children contributing something in aid, and this at a time when there were only a few dollars in the island treasury and in an island the yearly public revenue of which did not exceed £200. Would that we could see similar exertions in our country villages! Mr. Cunningham said he found many of the children pretty well advanced in reading, writing, and arithmetic.

† The following from my note book relates to the schools in St. John's, which I transcribe the more willingly, as according to the latest report, the church-school was actually closed for some weeks,—and its continuance it may be inferred in danger, "owing in a great measure to unhappy discussions between the authorities lay and clerical," and this "after the stipends of the masters, at the time long in arrears, had been defrayed by the proceeds of a public subscription." From a despatch of Governor Mackintosh of the 12th of April, 1852, P. P.

"March 4th, 1846,—after breakfast went into town, St. John's, and accompanied the bishop and archdeacon to see the schools. First visited the infant school, where there were about 100 young scholars from 4 or 5 years of age to 12 or 13. About 38 of them could read very tolerably. There was singing and counting, and in conclusion, marching out to play and luncheon, many of them having come provided with something—such as a bottle of water,—a bag with some solid food. As they marched round and each took up his own, there was much of order, propriety, and cheer-

have seen, the number of scholars is stated as nearly the same, viz. 3,004, or about 1 in 12 of the population.

After noticing their good qualities, I must not pass over some of the bad ones attributed to them, and it would be extraordinary indeed if they did not possess any. By some they are called intractable and capricious in their conduct to their employers. Certain of them may be so, but the character of the whole, marked by respectful bearing, as already described, is hardly compatible with the charge. Licentiousness, it is said, is a crying sin amongst them; and I believe truly. In a sermon which I heard preached in the cathedral by the archdeacon, he raised his voice loudly against it, stating that half the children brought to be

fulness. The children, with one or two exceptions, were negroes, as were also their teachers.

Next, went to a higher school, where there were about 100 boys, also mostly of color,—who had been advanced from the lower, the infant school. Here reading, writing, and arithmetic are taught. A payment is made by the parents, from about half-a-crown to one and sixpence a quarter, according to what is taught. They read pretty well, and gave intelligent answers generally to the questions of the bishop, as did also the readers in the infant school, indicating some reflection.

We concluded with a visit to the classical or grammar school (since closed.) Here there were about twenty-five boys and two masters. Two from the higher classes, were tried in Sallust and Sophocles and translated well. They had been at the school hardly three years. The eldest, about 14, was a half caste; the youngest about 12 or 13 was of white parents. The opinion here seems to be, from what I could learn from the bishop and archdeacon, both of long experience, that the children are quick and easily learn, readily make an advance to a certain stage. As to their capacity for the sciences and the higher branches of knowledge, they say it is difficult to judge from want of experience, none here having had more than elementary instruction. Their ease of manner, absence of shyness, seem to be equally admitted."

baptized were illegitimate. More than two thirds of his large congregation were people of color. Perhaps many of them felt their weakness, and resolved to lead a purer life. Be this as it may, we are assured that the brave and conscientious preacher, (it was the Venerable—truly venerable Archdeacon Holberton) doing his duty, did not give offence, insomuch as on his leaving the island shortly after on a visit to England on account of his health, he was requested to accept a piece of plate, two pieces, one from each of two friendly societies, with "a suitable inscription,—expressive of their affectionate regard for him as their pastor and director, and of the sincere estimation in which they held his untiring administration to their spiritual and worldly necessities." Of these societies one was established in 1828, composed of the former slave and free colored inhabitants of the city and neighbourhood of St. John:"—the other, "in 1832, exclusively of the black agricultural labourers who attend the parish church."* If I may offer a suggestion, I would remark, that those who in their ministerial capacity may have to deal with the failing just alluded to of the African character emerging from barbarism and slavery, would do well, were they, whilst imitating the zeal of the archdeacon, to follow the prudence of the Moravians and consult the rules laid down by them for their guidance in the matter, (so far as they are compatible with our civil laws) founded on the apostolic instructions addressed to the first frail christian converts.

* Parliamentary Papers, 1852.

Of the other classes of the community in Antigua, were I to speak from my own impressions, I should describe them very favorably, and yet I believe justly. The manner in which the planters first treated their slaves, and afterwards the emancipated, betokened more than ordinary good feeling, good sense and foresight. In no part of the West Indies where I have been have I experienced more kind hospitality, or met in society more agreeable and well-informed people, or witnessed more of the decencies, courtesies, and elegancies of civilized life. The houses of the upper classes here I found even better furnished than of the same class in Barbados; their floors carpeted, the windows provided with curtains, with more of prints on the walls; the roads good and kept in repair as in Barbados by a general rate, and carriages in use. My favorable report, I would not limit to the white portion of the society, I would extend it to the colored, to the comparatively few who are liberally educated. Of such there were three members of the House of Assembly. At one of the meetings at which I was present, the subject brought forward was the public grammar school and the religious instruction to be given there. According to the original motion, it was to be restricted to the doctrines of the church of England. These gentlemen moved as an amendment the substituting of sound christian teaching so as to open it to the children of dissenters, many of the rate payers being such. None who spoke appeared to me to speak so well, whether logically, or as regards information, and I may add manner. But forming a minority, their amendment was negatived. The school as might have been expected, established on so narrow a basis, did not flourish; and according to the last report, it has ceased to be. I wish I could in terms of praise advert to the higher attainments,—those of science and literature, so essential to the wellbeing of society. Neither scientific or literary tastes seem here to be cultivated; not that they are despised, but rather neglected, especially the former. There is a library, belonging to a society, but it is ill furnished with books; instruments I am informed were provided for meteorological observations, but they have not been used, except for a very short time;* agricultural societies have been formed, but they have had a short existence and have accomplished nothing; and as in British Guiana and in all the other colonies, excepting Barbados, the only periodicals that have been edited have been newspapers.

In its agriculture, whether we consider its past history or present state, Antigua approximates more to Barbados than to any other of our West Indian colonies, except it may be St. Kitts. In its early period it was not exclusive; it yielded cotton, indigo,

* They were purchased, I am informed, by the Antigua library society. The short time they were used and the circumstances leading to their disuse, such as the observations being interrupted by Sunday,—the rain gauge overflowing on one occasion when 7 inches of rain and more fell in the 24 hours,— the self-registering thermometer becoming useless,—circumstances referred to as irremediable,—are not a little indicative of the good intent and weakness of purpose too commonly displayed in West Indian undertakings.

and other crops besides sugar. Latterly, and even to a greater length than in Barbados, all but the sugar cane have been given up, and now to it alone, as their staple, the attention of the planter is directed even to the neglect of growing provisions, and for reasons already assigned,—mainly, the greater return that it affords, or is supposed to afford.

In relation to the interests of the planters, the first circumstance I shall mention, and it is a fortunate one, is, that there is no land in the island unappropriated; no crown lands unoccupied,—none to which labourers as squatters can withdraw and live in a manner independently. Of the 69,277 acres, estimated to be included in its area, at least 52,503 belong to sugar estates, or are private properties. Of these the number is about 158, of the average extent of about 332 acres, the extremes being for the largest 1,931 acres, for the smallest 17.* Another fortunate circumstance,—fortunate for the planters, is, that of the large number of the labouring class who have purchased freeholds and provided themselves with cottages, the space of ground they have provided has generally been so small, and its qua-

---

* According to the return of estates given in the Antigua Almanack the number of these estates, and their total acreage in the several parishes,—the six into which the island is divided,—are as follows;—

| | | | | | | | |
|---|---|---|---|---|---|---|---|
| St. John's | ... | 45 | containing... | ... | ... | ... | 14,967 acres. |
| St. George's | ... | 17 | " | ... | .. | ... | ... | 4,849 " |
| St. Peter's | ... | 22 | " | ... | ... | ... | ··· | 7,570 " |
| St. Philip's | ... | 31 | " | ... | ... | ... | ... | 7,895 " |
| St. Paul's | ... | 17 | " | ... | ... | ... | ... | 8,261 " |
| St. Mary's | ... | 26 | " | ... | ... | ... | ... | 8,961 " |

lity so inferior, (the poorer lands being commonly sold to them) as not to suffice under culture for the support of themselves and family, being little, if any more than they were allowed to have during slavery, and consequently their labour is not lost to the estates. Bearing on the same point, as I think it does, I will mention another circumstance, only indirectly connected with the landed interest, viz., that the revenue of the island is mainly derived from duties on imports, none being exacted on exports,—thereby making the labourers contribute a large portion towards it, and so much the larger, the higher their wages and means of procuring comforts and foreign luxuries.* Other advantages might be pointed out which the planter enjoys in Antigua: and first, and not least, is the excellence of a large proportion of its soils, especially of the calcareous marl ones,—very productive in themselves,—well adapted as they commonly are for implemental husbandry, of great depth, and, with care, inexhaustible. The amount of produce that some of them yield is almost incredible, and yet it is well authenticated, such as three, four, and even

---

* For this wise measure—for so I cannot but call it, Antigua is most indebted to its late respected treasurer, Dr. Musgrave,—a man I am proud to have called my friend, able and estimable as he was in all the relations of life. In advocating this mode of raising the revenue, he mentioned that when there was little competition, the duty did not enhance the price to the purchaser, but diminished the profit to the seller, checking exorbitant gains : and his experience of different rates of duties, tried during the many years he filled the important office of Treasurer—the island Chancellor of the Exchequer, as he sometimes in pleasantry called himself,—accorded with this view of the subject.

five hogsheads of sugar per acre. Even the disadvantages of its climate, the droughts to which it is subject, taking an enlarged view, can hardly be considered such; on the contrary, rather advantages, as tending to industry, frugality, and foresight. When man can live without care, we are sure he will live idly; ease and indolence are everywhere associated, and most of all in a tropical climate, where rest is considered luxury.

Of the agriculture of Antigua, that is, the culture of the sugar cane, to which its agriculture, as already observed, is mainly confined, a very brief notice may suffice.—With straightened circumstances and increasing difficulties, it has been gradually, slowly, improving, becoming more skilled, I cannot say more scientific, more economical and more productive, that is, as to the amount of crop. The hoe has ceased to be, what it was so long, the only implement employed in tillage, wielded by the hand of man. Really brute force is now beginning to be brought into action, and the drudgery of hand labour to be shifted to the horse, the ox and the mule. In the recent reports, those of the stipendiary magistrate, the following are some of the implements which he states are getting into use and are likely to come gradually into general use, such as, besides ordinary ploughs and harrows of English construction, some of American, of moderate cost, for instance, a light plough for moulding ratoons, the hill-side plough, the ox-shovel for clearing ponds or ditches, the horse-rake for gathering in the trash, the bush-scythe, and a light three-pronged fork. A strong proof of the necessity felt of economising labour, substituting brute for human, was given in the formation of a horse-hoe and harrow company,—which however well designed, I know not for what reasons failed, like too many new undertakings of great promise. It was proposed and I believe established in 1845 with the object of aiding in the culture of "such estates as cannot afford, or may be desirous of escaping the risk of keeping up a constant supply of implements, labour, and stock equivalent to their demands," and to such "as may from a pressure of circumstances require occasional extraneous aid." With the introduction of implemental husbandry, the area of culture was in very many instances contracted, from the experience as old almost as the annals of agriculture of greater profit accruing from a few acres well cultivated than from many negligently.

With the exception of Barbados not one of our West Indian colonies has had a better supply of native labour (there being about as many labourers as there are arable acres) or at a more moderate cost, or more continuously, and with more content, and consequently with less need if any, of immigrant aid. Since emancipation the wages here of the day labourer have fluctuated from 6d. sterling to 1s. 6d., and from that to 9d., about which it is at present; the labourer located on the estate being commonly allowed a cottage, a small portion of land, and medical attendance, under contract of a month's notice, if to quit, on either side.

The amount of produce which, cæteris paribus, should be the indication of the skill and industry applied, has, since emancipation, though perhaps less than before, been subject to some great variations, but more it is believed owing to the seasons than to any other circumstances; and so far of good omen. A return of crop is given in the Antigua Almanack year by year from 1828 to 1850 inclusive, from which it appears that the average of the whole was 13,272 hogsheads; that the greatest for any one year was 20,921; the least 5,434, and this followed the very next year by one of 18,534 hogsheads. The average may seem small, and hardly in accordance with what has been said of the excellence of the soil, making all due allowance even for dry and unproductive seasons; but if the quantity of land actually in cane cultivation be taken into the account, the result will not appear unfavorable or contradictory. In 1833 the land under canes was returned as 8,138 acres.* Now supposing this to be about the average—it is more likely to be in excess—comparing it with the average of crop, the produce per acre would be rather more than a hogshead and half (1·6.) This surely is nowise unfavorable, especially looking back to what the estimated produce was during the period of slavery. Sir Wm. Young writing in 1807, cal-

* Tables of revenue &c., P. P., (supplement to part iv, colonies, 1833,) p. 19.

culated the average amount at no more than 900 hogsheads, and these of only 13cwt., and per acre at only one third of a hogshead! and this at a time when the number of slaves was not short of 36,000,—according to him one to each acre! Surely these are encouraging results, and full of promise for the future; and how different from the prospect which opened before the Author just referred to, (one of the most accurate on West Indian affairs,) in which he viewed the soil of the leeward islands, and especially of this island, as becoming less and less productive, either from exhaustion, a very commonly received notion, or from increased tendency, as he imagined, to drought from the clearing of woods, or from both combined.

In the manufacturing part of their business, the planters of this island appear to have made less progress than in some of the other colonies, especially in British Guiana. The old methods seem to be generally kept in use; the windmill for crushing the canes, rather than the steam engine, with a very few exceptions; and the open boilers or teaches rather than the vacuum pan, or other more refined methods for evaporating the juice, without any exception, that I am aware of. This backwardness has been attributed, and probably justly, to straitened circumstances and the want of capital necessary to provide costly apparatus.

To the defects and evils bearing on the welfare of the planter and of the island I shall be very brief in adverting: it is not right to pass them over entirely.

Here as in Barbados, too little regard is shown to stock. On this subject I cannot do better than quote the words of a gentleman well acquainted with the island and not likely to take any exaggerated view of the failings of its planters. Sir Robert Horsford, in his capacity as stipendiary magistrate, in his half-yearly report, from June to December 1844, after expressing doubt of the proper application of guano as a manure from ignorance in those using it of scientific principles,—remarks, "another grand defect in agricultural economy, as I conceive, arises from the little care and attention bestowed upon the management of cattle, as well in the manner of feeding as the improvement of their breed. Except during the period of crop, at which time there is an abundant supply of fodder, through means of the cane-top, nothing is afforded beyond what the animals can collect, in their wanderings over the dry and sun scorched pastures, or an occasional bundle of dried trash at night. The cattle therefore become reduced in flesh, poor and weakly in constitution, and the manure which they yield is proportionably deteriorated and profitless. Few attempts have been made to improve their breed by the introduction of other stock ; and from the same causes these attempts have generally ended in failure ; their diseases, if not wholly disregarded, are but little understood, and they are left to live or die, as the ordinary course of nature may prescribe. These observations equally apply to sheep, asses, pigs, and all other domestic animals."* In the Antigua almanack, under the head of the vegetable kingdom, forming a part of a well written topography of the island, it is stated that Guinea grass is extensively cultivated; but, as it is added, that together with an indigenous species, the *cent. per cent.* (panicum colonum) it, with the top of the sugar cane, constitutes the principal green food of the stables, it may be inferred to be of very limited use, and grown chiefly to be given to carriage and saddle horses, on the good condition of which the gentlemen of Antigua pride themselves. The neglect of the extensive pastures, is I believe notorious, I never heard of their being manured, or in any way cultivated; or of any artificial grasses being sown in them; or of the scythe being ever applied, or hay, or dried forage obtained from them: a neglect it is true, common to all the West India colonies. Another defect is the little or no attention paid by the planter either to horticulture or the growing of field provisions,—a neglect

---

* Poultry, in which the island abounds, bred by the peasantry, may be mentioned, I believe, as an exception. They are not only plentiful but good, especially the fowls and capons, the art of making which has probably been taken from the neighbouring French islands. There is one peculiarity in the management of their poultry which was new to me, and may be worth mentioning, viz., their taking them out to feed on the pasture lands where insects abound and small seeds. They are conveyed in large conical wicker-work cages,—taken out in the morning and brought home in the evening,—when a feed of corn is given to them, to attach them as it were to home. A number of these, forty cages, scattered over a hill side, has a singular appearance. The fowls so treated are commonly hens with their broods: the hens are let out of the cages, each tethered by a long string, to prevent her straying,—the chickens are allowed unrestricted freedom.

gradually increasing since the time of slavery. Gardens with all their innocent delights are almost unknown there; flowers and fruits are almost equally rare; excepting the pine apple, for which the soil and climate are peculiarly favorable, scarcely any fruit is cultivated. The labourers grow some provisions in their small plots of ground, but far from sufficient to supply the wants of the inhabitants generally, not enough even to supply their own. Consequently the island is mainly dependent for supplies of food from abroad which is furnished in a great measure from the United States and paid for in dollars, constituting a species of one-sided traffic more to the advantage of the seller than buyer,—a traffic not likely to terminate so long as the planter is embarrassed in complicated transactions, and almost in bondage to the home merchant; or as long as the American can dispose of his dollars to more advantage by laying in a mixed return cargo at the Havannah or any of the adjoining rival ports.

Some other evils and not of less moment, as they are commonly and I believe truly considered, I shall do little more than name. Standing in the front of these is absenteeism: three-fourths of the properties it is said are abandoned to what has been called a vicarious cultivation, that is, intrusted to the care of attornies and managers. Next may be mentioned the large number of encumbered estates, and the ruinous rate of interest. And next in connection, the difficulties arising from legal forms and costs to the transfer of landed property, such as were experienced in Ireland before the encumbered estates act was passed, and are still experienced in England and Scotland.

How hopeful would the future of Antigua be were these impediments and other difficulties got rid of, and improvements commenced in agriculture, carried further, with less restriction as to crop, and extended to the manufacturing processes. Then indeed might the island afford an example not only of a well ordered but also of a prosperous community.

In concluding, I shall mention a few excursions which may be made with ease by those visiting the island and wishing to have some general idea of its scenery and cultivation.

Arriving by the packet, one of the West India mail company's steamers, the stranger is landed at English Harbour, a most secure little port, attached to which and protected by the military station on the heights above—"the Ridge," is a naval dockyard, the only one in the command. A well appointed carriage, an omnibus which is always in waiting, will convey him across the island to the town, or city, as it is now called, of St. John, and its capital,—indeed its only town. He will be fortunate if he has a friend there to receive him under his roof, unless the inn is greatly improved since I was there in 1846 and 1849. Even in this drive of six or seven miles, which is about the distance of the town from the harbour, he will see a good deal of the country, of its nakedness, nice cultivation, neglected pastures, and gently hilly and undulating surface, with pretty villas belonging to the estates scattered here and there, wind mills and negro villages.

Gun Hill, two or three miles from the town, a great mass of trap rock rising abruptly out of the low ground, spoken of before as the central intervening plain, is worth visiting, for the view it affords from its rugged summit of the country spread out on each side, depressed in many places, seemingly even below the level of the sea, and dotted with ponds reminding those acquainted with Barbados of many parts of that island, and rendering it easy to be understood how like an Italian maremma, it may be productive of malaria.

To see some of the most fertile parts of the island, the best cultivated and most productive, and I may add, some of the prettiest, a visit should be made to Willoughby Bay, to "high windward," and to Mount Josua, both situated in marl districts; the one about 14 or 15 miles, the other about 4 from St. John's. Never, anywhere in the West Indies, have I seen finer canes than on the latter, or seen a spot reported so fertile. It is said that four and five hogsheads an acre are not an uncommon yield on this property; and that the proprietor residing on it, in a few years accomplished the paying off a debt of £60,000; but be it remembered, this was before the equalization of the sugar duties was commenced, yet, if I have been rightly informed, since the act of emancipation.

The Shekerly hills or mountains, the highest portion of the elevated western district, may be explored in a morning's ride. Where loftiest, the elevation is said to

be about 1,200 feet. If the ascent is made by Windmill
hill, returning to St. John's by the valley, " the Bermu-
dian valley," much will be seen that will repay for
any little difficulties, for there are no dangers, to be
encountered in threading the steep ascents and de-
scents by a narrow and rugged bridle path. Nowhere
in the island is the scenery so bold and picturesque,
or the views near and distant finer, the one comprising at least two thirds of the island, the other
Guadaloupe, Montserrat, Nevis, and St. Kitts. Nor
is the pleasure confined to the prospects; there is an
enjoyment in the cool air and pure atmosphere, in
the sound of running streams, of which two or three
are to be seen, small indeed, but living waters, and in
the sight of some fine trees and wooded declivities.
At Green Castle, an estate which is passed after
descending the hill, there is a geological appearance
worthy of notice, viz. a bed of volcanic ashes con-
solidated into a tufa, cropping out from beneath a
mass of dark crystalline trap rock, differing but little
from basalt in its character, and near it a bed of white
clay, not unlike St. Stephen's clay in Cornwall, similarly overlaid.

A stranger should not leave the island without
paying a visit to one or more of the new villages
which have sprung up since emancipation, so creditable
to the emancipated negroes. Judging from what I
saw, he will not come away with an unfavorable
opinion of their inmates, or in doubt of the sound
policy of encouraging such establishments.

# CHAPTER XIII.

## MONTSERRAT.

ABOUT equidistant from Antigua and Nevis, the islands nearest to it, Montserrat, as its name implies, (a name given by its discoverer Columbus,) is truly a mountainous island. No island in these seas is bolder in its general aspect, more picturesque, and I think I may add without exaggeration, more beautiful in the detail of its scenery,—indeed, one might be tempted to say, considering its fortunes, that it has the fatal gift of beauty.

Never accurately surveyed, it is conjectured to be 34 miles in circuit, about 12 miles in length and 7 or 8 in width, and to comprise about 48 square miles, or 35,000 acres, of which at least two thirds are steep declivities, uncultivated and covered with wood; a circumstance to which, with its varied mountain forms, a great deal of its beauty is owing.

In structure it appears to be entirely volcanic, and composed of igneous crystalline rocks, of clays, probably derived from the decomposition of these rocks, and of tufas, conglomerates and volcanic ashes; the latter often exhibiting a stratified arrangement.

Two spots are pointed out as craters of eruption; one the Soufriere, about 800 feet above the level of the sea; the other a hollow, now a lake of clear water, at the summit of one of the highest mountains, described as about one hundred yards, round and overflowing after heavy rains.

Its soils are various, mostly light and porous, and are commonly fertile,—for instance when under canes, (plant-canes,) and carefully cultivated, yielding, it is said, often 4 hogsheads of sugar per acre, and on an average not less than 2 hogsheads.

Its climate is esteemed healthy, that is comparatively, for though called the Montpellier of the West Indies, it is not altogether exempt, no more than Barbados, from the complaints peculiar to these regions. From its mountainous character, covered in great part with forest, and from the circumstance of its rising out of a deep sea, and having no indented bays or sheltered inlets, being entirely destitute of harbours, its atmosphere is comparatively cool and equable in its temperature. During the three days I passed in the island in March, the thermometer ranged between 76° and 73°, and this very little above the level of the sea. There is reason to believe that it is less subject to droughts than Antigua, though like Antigua and the other adjoining islands, liable to suffer from hurricanes and earthquakes. The earthquake of 1843, which committed so much havoc in the island first mentioned, was scarcely less destructive in Montserrat.

Neither are springs frequent nor streams, excepting in the northern portion of the island, where clay is said to abound. Of hot or mineral springs, I am not aware of any excepting those of the Soufriere, and one in a marshy spot, near the town of Plymouth, the temperature of which when I visited it was only a few degrees above that of the atmosphere. A petrifying spring was described to me, but which I did not see, situated in the side of one of the higher hills, the water of which, from such information as I could obtain, is strongly impregnated with carbonate of lime, dissolved by means of carbonic acid gas, like the water of like quality at the baths of San Fillippo in Tuscany. Specimens were shown me of its petrifactions, as they were called, though not strictly such, for they were parts of vegetables, chiefly leaves included in the deposited calcareous matter. Some of them were so perfectly preserved in the marble incrustation, on which their forms were delicately impressed, as to have retained even their color.

According to the last census, that of 1851, the population of the island then amounted to 7053, of whom only 150 were whites. Referring to its past history, this shows a remarkable diminution. It was first settled in 1632 by a party from St. Kitt's, chiefly Irish Roman catholics who had quitted their native country when under the severe rule of Cromwell. It is stated by Oldmixon that in the short space of 16 years, such accessions had been made to them, that they numbered 1,000 white families and had a militia of 360 effective men. In a work already quoted on the European settlements in America, published in 1761, we are informed that the total of the inhabitants rather exceeded 15,000, of whom about 5,000 were Europeans and ten or twelve thousand slaves. From this time its population appears to have rapidly decreased, and to have continued decreasing till it reached its minimum, about the time of emancipation; after which the tide turned in favor of increase, but an increase limited to the colored race: a circumstance, especially when compared with what is noted of the stationary numbers of the same class in Antigua, which may be adduced in proof of the salubrity of Montserrat; for the difference in favor of the latter certainly is not owing to any superiority in the morality of its peasantry. The following numbers taken from authentic sources worthy of credit, show the amount of decrease and increase during the period adverted to.

| PERIOD. | WHITE. | COLORED. | SLAVES. | TOTAL. |
|---|---|---|---|---|
| In 1787 — | 1300 — | 260 — | 10,000 — | 11,560* |
| " 1805 — | 1000 — | 250 — | 9,500 — | 10,750† |
| " 1828 — | 315 — | 818 — | 5,986 — | 7,119‡ |
| " 1834 — | 312 — | 827 — | 5,026 — | 6,165 |
| " 1844 — | — — | — — | — — | 7,365 |
| " 1851 — | 150 — | 6903 — | — — | 7,053 |

* Reports of Privy Council, 1788.
† The West Indies Common Place Book, p. 3.
‡ P. P., 30th June, 1845, giving the results of the census 1828, 1834, and 1844.

But, to give a just idea of the increase during the latter period, the efflux from immigration should be added to the numbers for 1844 and 1851. This certainly has been considerable, owing to the temptation of higher wages in the neighbouring colonies as well as the bounty paid in Trinidad. Four vessels, it is stated have been constantly employed in this trade, and no less than 16 dollars were given for a considerable time, at length discontinued from the abuses it gave rise to, for every labourer landed. The number thus withdrawn from the population between 1834 and 1844, allowing for those who returned, has been estimated at 2,507,* and between the latter year and 1851, at 3741!

Neither of the state of the population nor of the agriculture of Montserrat can the same favorable account be given as of those of Antigua. The contrast as regards the latter—the agriculture, is very striking and truly melancholy, and the more so, as its wretchedly depressed state appears to be more owing to mismanagement, carelessness, and neglect than to any incidental circumstances, though occasionally aggravated by such circumstances, as by a French invasion, if we may go back so far, in 1712, when extensive depredations were committed,—by a later French invasion in 1782 and the capture and occupation of the island in 1783; and more recently by the earth-

* See reports, P. P., July, 1846, p. 109. Imposition was one of the abuses; such as shipping backwards and forwards the same individuals for the sake of the bounty.

quake of 1843 and the epidemic small pox in 1849
—50.

Of the small waning white portion of the population,
reduced so low, it may be inferred chiefly owing to
absenteeism,—making all due allowances for diminution, the effect of climate, (the healthiest climate of
the West Indies, be it remembered, being unfavorable
to the northern constitution,) I would not wish to
speak disparagingly. That there are amongst them
deserving men I have no doubt, and that there are
in their little society some very agreeable ones I know
from my own experience, during the short time I was
in the island in March, 1846, when hospitality and
kindness were shewn me which I shall long gratefully
remember.

Of the colored inhabitants constituting the great
majority of the population, the reports made officially
by those in authority, viz., the president administering
the government under the governor general, and the
stipendiary magistrate, are not on the whole unfavorable, indeed they are more favorable than might be
expected under the circumstances in which they are
placed,—having few of the advantages in regard to
education possessed by those of the same class in
Antigua, less inducement to labour and greater
temptations to idleness, or that low degree of labour
little more than sufficient to supply their most necessary wants.*   This will best be shewn in bringing to

* Remarks, deducible from the reports of the authorities, unfavorable to
these people, I am disposed to think are hardly just, except when applied to

notice the state of agriculture, such as it has recently been, and is (perhaps a little improved, it is to be hoped) at the present time.

Of the 39 estates belonging to the island, and which, as there is no crown land that I am aware of, and little colonial,—may be considered as comprising almost its whole surface, whether cultivated or waste, four only are conducted by resident proprietors, eight by lessees, and the remaining 23 by attorneys acting for the absentee owners; and, what is more remarkable, as many as twenty-three are, or were in 1847, under the charge of one individual, in his different capacities of owner, lessee, executor, attorney, or receiver. Of the whole number of estates twenty are reported as being in ordinary cultivation, ten as in imperfect or semi-cultivation, leaving nine which, it is believed, are abandoned. Cultivation when carried on on the estates is restricted to the cane. The mode of conducting it and the manner of manufacturing the sugar are the same as in the time of slavery, with the exception of the partial use of guano as a manure, and the erection of two steam

them in their depressed state, under the circumstances to be hereafter related when they were so severely tried,—and what is so much to their credit, when, though there was no police or military force in the island, they committed no outrages. When I was in the island a large number of them contributed each, that is, for each family, a dollar yearly for medical attendance; a certain number of them paid a sum, though it was small, for the schooling of their children; and some had joined a friendly relief society,—which broke down through the mismanagement or dishonesty of its directors. In advance of some of the other islands,—they are not altogether without manufactures, however rude; they make bark-ropes for sale and articles of earthenware.

engines for crushing the canes. In tillage the only implement used is the hoe; two ploughs it is said were imported, but were allowed to decay without having a trial. The canes when cut are conveyed in panniers, of rude construction, on the backs of horses, mules and asses to the yard, and from thence on the heads of labourers to the mill; no carts I believe are used in the town and immediate neighbourhood of Plymouth and not even a wheelbarrow on the estates.

Possessing a labouring population amounting, as it was estimated in 1849 to 3,742, we are assured that less than a moiety, not more than 1,200 are actually employed on the estates of the proprietors, and yet that no less than 1,327 families of labourers are located on the estates as tenants at will,—that is without any contract, liable to be ejected at any time without redress, and nevertheless fancying they have a claim to remain in possession even when doing little or nothing for the proprietors. This tenancy-at-will may be considered as one of the main roots of the existing distresses and is therefore deserving of attention both as regards its origin and its operation.

On emancipation, either from straitened circumstances, or false views of economy, the planters, instead of engaging the freed labourers to work for them, paying them fair money-wages,—instead of encouraging them to form free villages, and to establish themselves in small freeholds, preferred allowing them to retain the cottages they previously had, and an ample portion of land for their own use, on the condition, a mere nominal

one, without legal validity, of their giving one or two
or more days labour to the estates at a low rate of
wages, on an average not exceeding 4*d*., and according
to one authority, little exceeding 2*d*. sterling a day.
The consequence of this was, the diversion of labour
from the estates of the owners to the grounds of the
labourers; and to such an extent, that were it not well
authenticated, there might be some difficulty in giving
it credit. The following (which is not described by the
stipendiary magistrate as an extreme or extraordinary
case) may afford some insight into the system. I shall
quote the whole paragraph in the end of which is the
instance. "I am of opinion, (states the magistrate
writing in 1845,) that to establish a proper system and
an understanding beneficial to all parties, the present
gratuitous arrangement and small rate of wages, should
be abandoned; the landholders should lease and the
labourer should rent their domiciles and land to yield
them its fruits of increase, and pay for the agistment
of their cattle, and thus learn to be more honestly
independent; and let those who require their labour
pay a remunerative consideration of the same in money:
each party would then be placed on a sound and lasting
footing and the law of *meum* and *tuum* would be better
understood and respected, and their proper relative
situation being thus established, mutual benefit would
result; whereas, in the present state of things, the
labourers, in a general point of view, have in proportion
greater benefit from the lands than the owner thereof.
I will give an instance, which may be considered very

far from being a solitary one: a peasant with his family plants as much as they can well take care of, in cassava, a very prolific article, which meets with ready sale, and not only in this island, but in Antigua at profitable returns, and other roots which yield a good increase; rears horned cattle, sheep, goats, and hogs on the property, which sold alive or as butcher's meat give good prices, (a steer will sell at from 20 to 30 dollars;) and they cut *ad libitum* wood, grass, &c. for sale. From these resources the labourer is placed in a very comfortable position; he is raised not only above the necessity of daily labour, but obtains a degree of wealth; the only return he gives is to work a day or two, or a week or two; then for many days, if not weeks, he absents himself from estates' labour. If estates' work be undertaken per task, they take three times as long as requisite, labour two or three hours a day, devote as many more to their grounds, and apply as many more to rest their skins."[*]

In the same report he remarks, " It is evident that the only method of enforcing labour at present is to eject the parties with their stock; but even in few instances where this course has been pursued, the more ignorant consider it an act of oppression and injustice, because it is a deviation from accustomed indulgencies. Several of their number have recently complained to me, apparently under the impression that I had the power to insure the retention of what they deemed their right; I endeavoured to make them

[*] West Indies Parliamentary Papers, 8th August, 1845, p. 180.

understand that they were mere tenants at will, and have no right to the many indulgencies granted them ; and that unless they so conduct themselves as to merit the approbation of their employers, they ought not to be surprised if such favors were withdrawn ; that as they pay no rent and give such uncertain labour, both as to time and amount, their employers have good reason to be dissatisfied, and that sugar estates must have continuous labour, &c."

Under this system of labour, scant and inefficient,— nominally cheap,—really dear, the produce of the estates has gradually fallen off, and matters from bad have become worse, terminating in the ruin of the greater numbers of the proprietors,—a ruin impending before, the affairs of the island having been long deteriorating, and consummated in 1846, the year of the great alteration of the sugar duties.

In 1789 it would appear that Montserrat exported 3,150 hogsheads of sugar; in 1799, 2,595; in 1805, 3,000; in 1834, 1,608; and in 1848 only 426. The shipping employed diminished in like proportion, till in 1847—48, not even one vessel came for freight; all direct communication and trade with England ceased ; the little sugar that was made had to be sent in small craft, droghers, to Antigua and St. Kitt's, to be exported from thence ; no less than fifteen estates were sold for arrears of taxes, and thirteen houses in the short period of three years ;* and in one

* An estate valued at £ 27,700,—and that considered now more than half its value,—was sold lately for £ 295 ; another valued at £ 18,000, was

year that of 1847, as many as seventeen sugar estates were either abandoned, or the cultivation of them greatly reduced; on few of them, only on eight, it is said, were the labourers settled with with regularity; on all the others they were paid at long and uncertain intervals, from six to twelve and even eighteen months. The large amount of £600, is reported as due from one proprietor alone to his people. As a consequence the truck and barter system, prohibited by the laws of England, sprang up and spread daily with all its attendant evils and abuses.

The labourers in their turn suffered much; pinched in their circumstances, they rapidly deteriorated. In 1845, they were described as having more comforts in their houses, as consuming better food, and dressing more cleanly and neatly than formerly;* in 1851, they are spoken of as filthy in their dwellings and dress, more so than in the times of slavery,—careless more than before about the education of their children, and as falling back into concubinage.† Wretched as all

sold for the amount of taxes—£181. A house valued at £400, was sold in execution of taxes for £3 1s. 9d. A house in town rated £200, was sold three times within two years; first for £23,—next for £63,—and last for £6 8s. 5d. The island revenue is in part raised by a tax on land and houses,—and whether productive or not, cultivated or abandoned, the tax to be paid is the same. The tenants-at-will, contribute nothing in the way of land tax, and the duties on imports being light, not much through indirect taxation.

* So they appeared to me, and such they were described in the early part of 1846 when I was in the island. On a Sunday, I attended the chapel of the Wesleyans; the congregation was large, almost entirely of colored people, well dressed, and apparently devout; the officiating minister was a young man of color, fluent, earnest, and impassioned.

† In the instance of West Indian peasantry, the temptation to concu-

this is, and I have only partially described the wretchedness, it did not reach its culminating point till 1849—50, when small pox unknown for 60 years, nor guarded against by vaccination, invaded the island and spread, infecting almost the whole of the inhabitants: and, to add to the affliction, a drought prevailed about the same time—an unprecedented one of from nine to ten months, occasioning the destruction of at least one half the ground provisions, as well as a great loss in canes. Though the epidemic was mild, proving fatal only to about 200 of the attacked, or about 3 per cent., and though assistance was liberally afforded from Antigua in a grant of money and provisions, yet misery and destruction are described as having prevailed to a frightful extent, and horrid pictures are given of suffering; some labourers, it is stated, even perished from starvation, though large sums were due to them for wages at the time.* Other details might be given to show the sad state to which the island was reduced, owing—apart from the natural visitations—to " want of capital, want of credit, want of confidence between planters and labourers and want of energy ;"—the

binage is great, especially amongst the emancipated; and from several circumstances, but two especially,—one the almost equality of sexes as to means of livelihood—the wages of men and women, field labourers, being much the same,—the other that a lawful wife arrogates to herself more than a mere help-mate unmarried, considering herself entitled to better clothes, more respect, and greater privileges. Amongst a people emerging from the worst of barbarism, from slavery, the tendency must be, unless they are advanced by education and their condition improved, especially their domestic comforts, to fall back again to the brutal sensual state—living on from day to day without thought of the morrow.

* Reports, 1851, P. P., p. 117.

causes assigned, and, I believe, truly by one of the authorities, reporting in 1848 on the state of the island.\* It might be mentioned how the public officers were in arrears of pay for two years; how the churches injured by the earthquake, for the repairs of which and other public buildings, a loan of £3,000 had been granted from home, were left almost in ruins; how another loan of £20,000 granted to assist in relieving the losses of the planters occasioned by the same catastrophe, had been misapplied and misappropriated, with other particulars; but I willingly avoid entering into them; what has already been related being more than sufficient to exhibit the effects of a bad system, beginning no doubt, in extravagance, akin to what has been so largely witnessed in Ireland, and ending here as there in want, in the sale of incumbered estates; and here as there, it is to be hoped, in the transfer of the properties to a different class of persons, solvable, circumspect and with some energy—with the chance of improvement and the coming of better times,—of which there appears to be already a dawn. To the auspicious circumstances indicating this change I willingly turn.

A hopeful circumstance and one seeming to indicate that the distresses so terrible were owing to the causes assigned, is that the few resident proprietors escaped them in a great measure; thus, it is reported that "the properties on which the owners are resident have not suffered so much as those intrusted to the

---

\* Copies &c., relative to West India distress, 1848, P. P., p. 385.

management of attorneys, but are almost invariably in a comparatively flourishing condition;" they, it may be inferred, being of the small number who paid their labourers regularly, and thereby secured better and more continuous labour. Another of the hopeful circumstances is in the change of owners, from embarrassed to unembarrassed parties. Already, we are informed, that some abandoned estates have been restored to cultivation and that the labourers on them being punctually paid, there are more applicants for work than can be employed, even at the rate of $5\frac{1}{2}d$. a day for hard work, and $4d$. for lighter. Other encouraging circumstances are the fertility of the land, especially such as has from being abandoned, lain fallow; and the cheapness and abundance of labour; the profitable returns of judicious cultivation; the fitness of the soil and climate for various crops, and the many sources of profit open to the intelligent and enterprising planter.*

Ample proofs of these are afforded and are to be

* In the official reports of 1851, an example of successful cultivation is given, viz., a net profit of £400 from a small property rented at £150 sterling per annum. The lessee was Mr. E. Semper, and we are assured that the statement was voluntarily made by him.

The profits of a corn and cotton estate have been estimated here as high as 120 per cent. The coffee of Montserrat is esteemed as of excellent quality. It is supposed that the climate of the higher grounds is well adapted for the olive; and that the silkworm, for making silk, might be bred here with advantage. Both, it is said, have been successfully introduced into Martinique. Three species of mulberry are at present in the island and grow luxuriantly. Probably the nutmeg and cinnamon might be cultivated here with profit, and other spices. One large tree of the latter I saw in a garden at Richmond Hill, belonging to Colonel Shiell.

found in the official reports half-yearly and quarterly made on the state of the island, confirmed as they are by an increase of produce and the additions made to the staple produce. Thus the sugar crop of 1851, was nearly triple that of 1850, being as 926 hogsheads to 381. The same year amongst the new articles of export, the following are mentioned;—firewood, 628 cords, value £328; timber, 321 cords £56 18s.; arrowroot, barrels 107, boxes 11, £218; Indian corn, bushels 201, £54; cotton, in process of picking, 18 bales, £144; moreover, in the same year it is stated in evidence of a favorable turn, that the direct trade with the mother country was reestablished, and that two vessels had arrived from England for cargoes, and in proof of the same, that three well-built vessels made of native wood had been launched, and that nine new " stores" had been opened in the town.

One or two other circumstances, pointing favorably, must not be passed over, especially as they relate to the root, or at least to one of the roots of the past distresses;—I allude to the relation of planters to labourers. In a report of 1852, it is stated that a contract act has been passed of a mild kind, putting a stop to the paying of wages in truck;—but how much more it secures, is not specified. It is also mentioned, that the proprietors long averse to the letting of land to labourers for money rent, from a feeling of jealousy of making them too independent, are beginning to be sensible of their mistake, and that two have set the example, one letting land at the rate of 10 dollars a year for two acres; the other at the higher rate of 10 dollars for one acre, and two dollars more, if houses be included,—in each instance the land let being mountain land.

Whether the favorable change thus apparently commenced, continue and make progress, or the contrary, must depend on the conduct of the inhabitants, and especially the landed proprietors. If they reside and exert themselves in the management of their lands, turning them to the best account in the cultivation of the cane, cotton and coffee, for each of which and for other productions, different situations in this mountainous island seem to be peculiarly favorable, they can hardly fail of success, with such a command of labour as they have; but to secure this labour and make it efficient, they should not expect to succeed and surely they will not, unless they attend to the true interest of their labourers, connecting it with their own, and whilst endeavoring to reclaim them from their failings so as to make them more industrious, more honest, truthful, and to be depended on, giving them a motive for being so. It is said, that at present, about one tenth of the sugar crop is abstracted by the pilfering of the negroes, and that little less is devoured by the rats;*—the same carelessness, no doubt conducing to the double loss.

As to the government of the island in relation to its

---

* So numerous were they in 1846 that on one plantation, it is reported, that a cane piece of 10 acres was so ravaged by them as to be useless, and was abandoned.

future welfare, whether any alteration is required in its form, is a question not undeserving of consideration.

Self government, the grand desideratum, the imagined panacea of all abuses in our colonies, which this little island possesses,—its president representing the crown, its council the aristocracy, its house of assembly the people,—is no doubt the best that can be, where the qualities, the elements, are found which are essential to its success. What are the elements here at present? Of the landed proprietors we have seen how few are resident; the number of freeholds we are told is only 150; of persons paying direct taxes only 170; of these able to read, above the age of sixteen, only 111, or, were the age extended to twenty, only 85. These surely are scanty elements,—whether for the purpose of electing or being elected, especially taking into account that the officials, that is, the members of the two houses and others officially employed, are not less than 77. And, if possible, the incongruity is increased, even *ad absurdum* by there being, as we are assured, no law regulating either the qualification of voters, or of their representatives, the members.* Moreover, looking to the working of this complicated piece of machinery, we find it, if in motion and acting

* The members of the House of Assembly are twelve, two for each of the five parishes into which the island has been divided and two for the town of Plymouth. The limited duration of the Assembly is the same as that of the House of Commons, septennial; but like it, it may at any time be dissolved. Of the members last elected six were white, the other six colored; not long ago, a colored man who had been a slave was elected, but he declined taking his seat, it is said, from a feeling of modesty.

at all, commonly only jarring and making a noise, doing no good work, rather destructive than productive. In reflecting on the subject, and in reading the speeches of the president administering the government, and the addresses of the council, and the house of assembly in reply, and his honor's replies to each of them, in set form, language, and length—sometimes exceeding in length,—very like what we have been accustomed to hear delivered in the imperial parliament, it is difficult not to refrain from a smile, or to avoid the idea that the proceedings are a burlesque. And, the same remark applies to the law courts, with their vice-chancellor, attorney, and solicitor general, queen's council, &c. The late president by whom so many elaborate speeches were made at the opening of the annual session, during his long rule of eleven years, seems to have been fully convinced of the nullity of the existing forms, though from his seriousnesss, hardly of the absurdity of them, and that however fitted the inhabitants might at one time have been for self-government, that time is past.

Of what is to be seen in Montserrat, I shall be very brief. Those visiting the West Indies in the regular steam packets, may see a considerable portion of the island on the passage from St. Kitt's to Antigua, or in returning, it being necessary to steer along it for the purpose of landing the mails. They will have an opportunity of observing, if by day, its bold headlands, north and south, one at each extremity, its nearly central three-headed mountain range, reaching, where highest, about

2,200 feet above the level of the sea,—the finely wooded
and broken declivities with their transverse gorges and
gullies, and the lower slopes and cultivated grounds.
No one, however, without a special object is likely to
stop, there being so few inducements tempting, and not
a few discomforts repelling. Even the landing is not
easy, and is seldom effected on the naked beach without
a wetting. If induced to stop, and under favorable
circumstances as to weather and introductions, the traveller may here pass a few days very agreeably, were
it merely in the enjoyment of its very delightful scenery. He will do well to devote one day, or a part of
a day, to the principal sight in the island, its Soufriere;
another to the crossing of the mountain to the windward coast; and perhaps a third to the exploration of the
northern portion of the island. He will be fortunate
indeed, if on his landing, he should be met, as I was,
by a kind host, invited to take up his quarters in a
pleasant villa, with the promise and not a false one, of
"a cool bed," and to find when least expecting it, not
only the civilities but many of the refinements and
luxuries of polished life, a remnant of a more flourishing time,—it is to be feared in its last stage.

The Soufriere is distant from the town about four
miles and is easily approached to its very edge by a
bridle path. Passing through a sugar estate finely
situated,* then through a charming wilderness of

* An estate of about 300 acres with a good house and works complete,
and 60 head of cattle ; it had produced 200 hogsheads of sugar. The
erection of the windmill alone, I was told, cost nearly what the estate recently sold for.

shrubs and trees, you suddenly come upon it, situated
in a recess or hollow—a crater like hollow, with a deep
chasm opening below and another yawning above.
The scene is everywhere impressive : I will not call it
horror in the lap of beauty ; and yet in some particulars
it calls up the idea, somehow, in its varied and contrasted aspects,—the surrounding and hanging woods,—
the ascending fumes of sulphur and steam, — the
scorched, charred trees, the boiling, gushing springs,—
the black streams and pools,— the broken naked ground
within and around, formed of varied and richly colored
clays, sparkling with sulphur in crystals, and salts ;
and in addition—the pleasant cool mountain air and
the partial hot obtruding vapour. It would be tedious
and useless to dwell on these peculiarities, as however
singular they may appear and striking in their effect
on the eye,—the phenomena are very much the same
as those of any other Solfoterra best known. I shall
merely remark, that the temperature of the hot springs
was the boiling temperature, and that they are consequently to be avoided ;—a Roman Catholic bishop I
was told got scalded by his incaution ; that the waters
are strongly impregnated with sulphuric acid and alum ;
that the black streams and pools owe their hue to sulphuret of iron suspended in a finely divided state,
mixed with earthy matter, and derived from the oxide
by the action of sulphuretted hydrogen ; and the varied
colored clays, to the oxide of the same metal and
of manganese. At one time sulphur-works were established here ; it was during the interruption of commerce

with Sicily. A good deal of sulphur was collected and sent to the United States; but owing to the impurity of the article, having been obtained by smelting, it had not a ready and profitable sale, and so the undertaking proved a losing speculation. Probably had the sulphur been obtained by another process, that of sublimation, the result might have been different.

The excursion across the mountain ridge to windward is by a path, a good part of the way steep and rugged, and occasionally, where winding along the brink of a precipice, somewhat hazardous, prompting the caution of dismounting and leading one's horse. The great charm of it is, the tropical mountain forest, in all its wildness and variety, according to the height and nature of the ground ; the air so cool and refreshing; and where emerging from the forest, the extended prospect of earth, sea and air, the two latter, as it were, blending, hardly distinguishable one from the other,—seeming to be an atmosphere to the near and strongly marked land ;—then a succession of descending terraces, ending abruptly without any apparent shore. Never was I more impressed, if I may venture to express my feeling, with the solemn beauty of such scenery, especially in the higher regions, where in the damp rocky gorges, wide spreading livid lichens and dark mucors incrusted the wet dripping rocks, and where a luxuriant vegetation, all of a peculiar kind, shut out the sky, producing a dense shade totally impervious to the direct rays of the sun, the tree-fern springing up to the height of thirty feet, and the bamboo occurring in masses fifty or sixty feet high. In winding downwards, after quitting the mountain forest, the first traces of cultivation seen, are of a rude kind, the provision grounds of the labourers, reclaimed by the fire and the hoe from the forest itself, often in small patches, hardly bigger than garden beds, with fruit trees and plantains, the latter most abundant in the sheltered hollows. Descending further, a sugar estate is passed, standing apparently on the margin of the sea, but in reality at some distance, and much above its shore. Here, there was a pretty good house and works, and a large negro village of thatched huts, occupied by tenants-at-will,—huts as first seen a mile or two off, not unlike bee hives. From this estate to the next, that of Hewitt's, the road partly paved, leads through a cultivated country; and so also from this estate back into town. Hewitt's consists of 300 acres; it has yielded as much as 300 hogsheads I was assured, and a clear profit in one year (probably exaggerated) of £20,000; when I visited it, it yielded about 80 hogsheads. The house a large one, with its marble floored galleries, bore marks of former opulence.

Of the excursion I have recommended to the northern portion of the island, I cannot speak from my own experience, not having had time to undertake it. The country there was described to me as hilly rather than mountainous, such indeed as it is seen from the sea in passing, and not without streams as already observed, of which one is sufficiently powerful to turn a wheel, the only water mill in Montserrat.

# CHAPTER XIV.

## ST. CHRISTOPHER'S.

Form and extent of the Island.—Its geological structure.—Climate.—Historical notice.—Population.—White and colored Creoles contrasted.—State of society.—Labouring class, their circumstances and improving state.—Land in cultivation.—System of agriculture.—English immigrants.—Remarks on manure.—Steam engines and sugar making.—Diminished production and profits.—Grounds of hope for the future.—Excursions which may be made by the passing traveller.

In passing from Montserrat to St. Christopher's, as from Antigua to Montserrat, we have an example of difference and contrast, not indeed as in the latter, in the island itself physically considered, nor in the inhabitants, (except perhaps in the absence of the Irish element, in regard to races,) but in condition, which if not prosperous is at least not distressed, tending further to show how much "conduct is fate."

St. Christopher's, deriving its name from the great Admiral, and by himself, its discoverer, so christened, is somewhat singular in its form, not unlike a guitar. It is equally remarkable for rugged boldness, and soft beauty, for wildness and cultivation; the former in the chain of hills gradually rising into mountain, which traverse it from south to north, reaching where highest, as at Mount Misery to an elevation of

3711 feet above the level of the sea ; the latter in the
flanking and lower hills and slopes gradually descending almost without interruption even to the water's
edge. Its mountainous and hilly portions are commonly
wooded ; enhancing its beauty, its declivities are either
cultivated or in pasture. Its low coast is little indented
with bays, and, excepting very partially, neither defended nor girded by coral reefs or abrupt cliffs.
The total circuit of the island is estimated at 73
miles, comprising within its area about 68 square miles,
or 41,851 acres. Its main body exclusive of the
narrow portion, that is, the long spit corresponding to
the handle of the instrument with which it has been
compared, is about seven leagues in circuit, and is
traversed by a road the whole way ; a pleasant road to
travel, bordering on the sea which is seldom out of
sight, passable for carriages, and when kept in repair
a very tolerable carriage road.

St. Kitt's, as it is mostly called for the sake of brevity,
is undoubtedly of volcanic origin, and principally
formed, (as well as can be judged from its surface and
the sections displayed in the many gullies, here called
"guts," by which its sides are transversely intersected,) of erupted materials, such as ashes, scoriæ,
lapilli and lavas. The only exception I am acquainted
with is that which occurs at Brimstone Hill, an offset
of Mount Misery, about seven or eight hundred feet
in perpendicular height.* It is remarkable as consist-

* The summit, Fort George, the citadel is said to be 750 feet above the
level of the sea.

ing of a mass of igneous rock a good deal resembling
basalt, but not columnar, and of coral and shell limestone, the latter flanking the former, and evidently
uplifted, exhibiting in its perpendicular and worn
strata, strange, fantastic and picturesque forms. Perhaps were diligent search made, other instances of
the same kind might be discovered, especially in the
higher ridges, which as yet have been little explored.
I am induced to think so, having met with a portion
of limestone of a crystalline structure, little differing
from marble, loose amongst fragments in a gully washed
from the heights adjoining, and this in a spot where it
is not likely that it could have been dropped. One
crater of eruption, if it may be so called, and the
only one tolerably certain in the island, is situated
in Mount Misery, in its side, at some distance from its
summit, which rises gradually above it. It is difficult
of approach, and not without danger. It has been described to me by a gentleman who succeeded in reaching and descending into it, as a deep hollow, longer
than it was wide, without an outlet, walled in by precipices and steep declivities, from fissures in which
steam and the fumes of sulphur were emitted. Sulphur
he said in crystals abounded, encrusting the rocks and
loose stones, and a stagnant pool, as he supposed of rain
water, occupied the bottom of the Soufriere.

The soils of the island are generally light, easily
worked, and in consequence peculiarly fitted for implemental husbandry; as are also the gentle declivities;
and they are so porous as seldom to stand in need of

draining. None of them that I have examined have effervesced with an acid, and yet all have been found to yield lime when acted on by an acid, as well as the other inorganic elements essential to fertility. They are remarkable for the small proportions of vegetable matter which they contain; it probably on an average does not exceed one per cent. In their mechanical texture they greatly differ, some being extremely fine, composed of an almost impalpable powder, whilst others are very coarse, composed chiefly of gravel, lapilli, or scoriæ in small fragments. These latter are comparatively barren.

In its climate St. Kitt's is esteemed healthy; yet its inhabitants are not exempt from those diseases which are peculiar to the West Indies, or from those influences which so much impair the constitution of Europeans, and shorten their lives. The white troops have suffered severely in garrison at Brimstone Hill from yellow fever, and even the colored Creoles at times have not escaped fever, especially those inhabiting Basseterre, the principal town.

As regards temperature and other sensible atmospheric qualities, it differs but little from the climate of Antigua. From being more mountainous, it is probably a little cooler. It has suffered hardly less from hurricanes and earthquakes. Though less subject to droughts than Antigua, it is not exempt from them; indeed notwithstanding its mountainous and wooded character, it appears to me doubtful that it is more favored with rains than the low, and almost woodless Barbados. The remark applies to the lower grounds,—its cultivated parts, not to its hilly and mountainous region. There, no doubt, rain is more frequent and probably falls more heavily. This is indicated not only by the mists and clouds which so frequently envelop the summits, but also by the streams, three or four of them perennial, and not inconsiderable in their volume of water, which descend from these heights.

The following table shews the results of observations with the rain gauge for a period of six years, at the military hospital, Brimstone Hill; where from its proximity to Mount Misery it is understood that showers are more frequent, than in the lower and more distant grounds.

RAIN.　ST. KITT'S

|  | 1846. | 1847. | 1848. | 1849. | 1850. | 1851. | 1852. |
|---|---|---|---|---|---|---|---|
| January ... |  | 6·06 | 3·10 | 6·68 | 1·38 | 3·93 | 1·64 |
| February ... |  | 2·50 | 2·11 | 3.41 | ·75 | 9·64 | 2·52 |
| March ...... |  | 1·90 | 1·62 | 1·81 | 2·00 | 1·51 | 0·82 |
| April ........ |  | 1·70 | 4·18 | 2.94 | 2·42 | 3·18 | 0·28 |
| May ......... |  | 1·10 | 1·68 | 2·71 | 1·69 | 6·70 | 3·30 |
| June ........ |  | 2·90 | 1·74 | 1·81 | 6·06 | 5·58 | 2·77 |
| July ........ | 7·44 | 4.17 | 2·70 | 2·43 | 7·73 | 3·56 | 4·50 |
| August ...... | 7·75 | 5·73 | 6·38 | 5·15 | 6·51 | 4·63 | 6·25 |
| September... | 4·80 | 1·92 | 2·53 | 3·53 | 4·92 | 4·31 | 11·50 |
| October ... | 3·90 | 2·78 | 4·87 | 7·59 | 2·58 | 5·48 | 4·48 |
| November... | 6·90 | 8.25 | 3·43 | 3·16 | 4·76 | 4·34 | 2·25 |
| December ... | 7·90 | 3·65 | 6·58 | 5·26 | 4·53 | 4·40 | 7·48 |
|  |  | 42·60 | 40·92 | 46·48 | 45·33 | 57·26 | 47·79 |

St. Kitt's has proudly been called the mother of the Antilles, the first attempt by Europeans to settle in the West Indies having been made here, and that about the same time by the French and English.

The accounts which we have, of the rapid progress of the two colonies, of the increasing numbers especially of the white inhabitants, and of their wealth even amidst almost incessant wars, fiercely carried on between the two people, are not a little marvellous,—the English occupying the middle portion of the island, the French, the two extremities, Basseterre in the South, Capesterre in the north.* Amusing details on the subject, strongly illustrating the incidents and manners of the times and the habits of the French and English colonists are to be found in Pere Labat, who more than once visited St. Kitt's, on one occasion meeting, in company, the distinguished officer General Codrington, who shortly after, — namely in 1702 defeated and expelled his countrymen.† After the island had become altogether a British

* I heard it asserted, and that there was historical authority for it, that when the island was divided between the French and the English, the latter could muster 16,000 white armed men. This was related at a dinner table, and was not doubted by any of the intelligent company, residents, some of the best informed in the island!

Of the high price of colonial produce, one of the causes of attraction to, and of the rapid prosperity of the first settled West India colonies, an instance is given by Pere Labat,—how M. D'Enambuc the conductor of the first party, on his return to France from St. Kitt's, sold his tobacco at ten livres the pound. He attributes to the culture of tobacco, and the free commerce in it, the rapidly flourishing state of the island. At one time he states, before the introduction of the sugar cane, in the French part of the island alone, there were 10,000 men fit to bear arms, whilst after the introduction of the cane, it was difficult to find 2,000. (vol. vi, p. 330.)

† It was this year, 1702, of the conquest of the French portion of St. Kitt's, that General Codrington, commanding the troops on that occasion, bequeathed those estates in Barbados, constituting the endowment of the college there bearing his name, instituted for missionary purposes.

It is worthy of remark that a greater undertaking of the same kind

possession in 1713, ceded by the treaty of Utrecht, it may be inferred, from such information as is available, that its white population, though for many years considerable, gradually declined, much in the same manner, and doubtless from the same cause, as in Montserrat, though not so rapidly. In 1761 the whites were estimated at about 7,000, the negroes about 20,000; in 1787, they had diminished to 1912; in 1805 to 1800, when the negroes had been augmented to 26,000; in 1826 the whites were 1600, the free colored people 1996, the slaves only 19,885; in 1831 the total population is stated to have been 21,608; in 1844, according to the census then made when races were not distinguished, 23,177, shewing an increase of 1,569, and this notwithstanding a considerable emigration of labourers had taken place, as from Montserrat, to the southern colonies. This increase is supposed to have been altogether of the colored people, and is happily contrasted with their constantly declining numbers during the period of slavery.

That the whites continue to diminish in numbers can hardly be doubted; it seems to be the natural tendency from the effect of climate. A highly intelligent gentleman, with whom I conversed on the subject, men-

might have been accomplished resulting in a manner from this conquest, had the minister, Sir Robert Walpole been true to his word. I allude to the college, which the great and good Bishop Berkeley, then Dean of Derry,—it was in 1726,—left England to establish in Bermuda, mainly by means of promised funds, to have been raised from the sale of crown lands which had belonged to the French in St. Kitt's. An attempt of a like kind, it is said, is about to be renewed in the same island: may it be more successful.

tioned some facts strongly in proof. One was, how, about a century ago, a party of white labourers, consisting of about a hundred were brought to the island; that he had tried to discover their descendants, but in vain; he had been unable to detect any traces of the stock; such were nearly his words. Another was,—it was elicited in asking him what he thought of the value of life here in the instance of whites,—that old age is uncommon, that he knew only one person, a merchant who had lived quietly and carefully, turned of 70; that the majority after 40 laboured under some chronic disease and died before they attained the age of 50. He said that men of his profession,—he was a lawyer, a barrister fully employed,—specially suffered; men in the exercise of their calling, nowise necessarily exposed to the inclemencies of a tropical climate, either to its noonday heats or to its night dews and chills.

I fear too, that as to condition, the white portion of the population, especially those of the lower class, are little better off than the poor whites of Barbados. From the same gentleman I had information illustrating this. When conversing on education, he remarked, that there is no grammar school in the island in which the children of the middle class can be tolerably taught, which he said, is felt as a great evil. He added,—the conversation turning on the condition of the people, that there is a good deal of distress amongst a considerable number of the white inhabitants reduced from better circumstances. A white manager, he instanced, has a large family; they are brought up easily; are in the habit of associating with gentlemen and ladies, and of considering themselves as such. He may have two or three hundred a year, a house to live in, and some other advantages. He saves little or nothing, dies and leaves his family destitute. The sons of man's estate, may be able to provide for themselves, as overseers and managers. The unmarried daughters have no means of earning a decent maintenance. Their pride will not allow them to go into service. By their needles they can earn very little, there are so many sempstresses. They are not qualified by education to become governesses. They sink to the miserable state of paupers.

The state of society here, (I allude to the higher class, a very limited one,)* formed of a few landed proprietors, attorneys, professional and official men, does not differ much from that in Barbados or Antigua; it is, perhaps, applying to it a home term, more provincial,—more of the older period, some-

* A gentleman, whom I met in Montserrat, speaking of the diminished number of the white population in St. Kitt's, and especially of the upper class, said, that the grandfather of a friend of his remembered, when Sandy Point was the principal town, 60 couples of ladies and gentlemen joining in a country dance, all belonging to the town itself and the immediate neighbourhood.

Even so late as 1745, there appear to have been many whites, proprietors of small estates. The Rev. Mr. Smith who in that year published a Natural History of Nevis, &c., speaking of an excursion he made in the neighbourhood of Basseterre says, " At first we rode through many sugar plantations till we came to thick woods, where now and then we passed by a small cotton settlement, whose humble and temperate possessor, (hermit like) lived by virtue of his own, and three or four slaves labour."

The same author mentions that when he went into the wilder parts of the island, it was necessary to go armed, " in case of need against runaway negroes." His party consisted of four white men and six negroes well armed with pistols and cutlasses.

what less refined than that of Antigua, and less luxurious. The men generally appeared to be superior to the women in manners and information, a remark commonly applicable in the West Indies, but of course, not without exceptions. There being pretty good roads and carriages in use, social intercourse seemed to be carried on in a very easy, agreeable and hospitable manner, much to the advantage of the casual visitor if well introduced. For my own part, I remember gratefully the many kind attentions, and, with pleasure the many worthy and pleasing people whose acquaintance I formed there. One of the advantages of the place is a well supplied reading room, and a respectable library, formed and supported, and well supported, by subscription,—the library I believe the largest and best in these colonies.

Of the labouring class a favorable account can be given, little less so, than of the same class in Antigua, the labourers here, having had some of the same advantages as there. It is true that they were not exempted from the apprenticeship trial,—a trial that led to much disaffection and disagreement, and to the partial withdrawing of labour from estates at a time when, in Antigua, there was an ample, willing and cheap supply, after emancipation, and when in consequence, with good seasons and good crops, and high prices, the profits there of the planters were great.*

* The opinion expressed above is that entertained by persons who witnessed the working of apprenticeship and were competent to judge of it. The gentleman already mentioned, speaking of it said, it was a bad

As in Antigua a good deal of attention has been paid to their instruction, especially by Moravian and Wesleyan missionaries. And here, as there, they have been rather encouraged than checked in the forming of free villages, and in establishing themselves in dwellings of their own, either on small plots of land purchased or rented. When the latter, they have the privilege of placing dwellings of their own on them, as in Barbados, with the power of removing them, which is easily done, being altogether of wood—walls, roof, and flooring, should they chuse to change their place of abode.

The official reports relating to them are satisfactory, as to their condition and conduct. Their dwellings are described as "rapidly increasing since emancipation, in comfort, extent, value, and durability." "The establishment of a village," says a magistrate, "was formerly predicted as a focus for crime. The contrary has been satisfactorily proved, and the most enlightened proprietors are now convinced of the advantage to be derived from the settlements, which contain an assemblage of well-disposed and industrious peasantry ; while on the other hand, the old collection of negro

system and ought never to have been introduced, as tending to destroy the good feeling which had sprung up from the kind treatment the slaves had experienced from their masters, during the latter period of slavery ; —but when commenced, it ought not to have been shortened, the shortening of it without an equivalent, being held by the planters to be an act of injustice. He added, that even the appointment of magistrates to protect the emancipated, however right in itself, had an unfavorable influence ; they were applied to on all occasions, whether of real or merely imaginary wrongs,— whether regarding points of law or matter of feeling ;—suspicion and mistrust in the minds of the labourers taking the place of confidence and reliance.

huts, or wretched hovels, (belonging to estates, and occupied by tenants-at-will,) has too commonly since emancipation, proved a refuge and a lurking place for the indolent and profligate of the labouring class."*
Another magistrate in his report, adverting to similar results,—the decrease of tenants-at-will, or those having "gratuitous occupation," and of the increase of the lease and rent holders, remarks,—" I cannot but feel peculiar gratification in hailing the changes which they present as an earnest of a better, and more wholesome state of things. I have often, and at much length dwelt upon the evils which have sprung up from the old system of permissive occupation without abatement of wages. The oft-recurring questions of litigated rent and disputed tenancy, were fruitful sources of dissatisfaction and complaint, and much of that spirit for emigration, which at one time gave a wandering character to the young peasantry of the island, was to be attributed to the irritating consequences of this most objectionable system. The gradual progress therefore of free tenancy, which now promises the total annihilation of the last remnant of a slave system cannot but be a source of congratulation to all lovers of social order and contentment ; and too much praise cannot be bestowed upon the labouring population of this island, who alone by their own industry and perseverance are bringing about this most desirable end."†    Other traits are mentioned in accordance, and

---

* P. P. West Indies, 8th August. 1845, p. 191.
† Loc. cit., p. 193.

of a confirmatory kind, indicative of good conduct, and of improvement; as the rareness of drunkenness, the diminishing number of convictions, the formation and well-doing of friendly societies, and the bearing contentedly a reduction of wages, when, owing to the change made in the sugar duties, the profits of the planters were suddenly, and so greatly reduced.* One circumstance, which I shall have more fully to advert to further on, viz. the introduction of immigrants by the planters, from a deficiency of labour, that is, of the continuous labour required in cane cultivation, may seem to be opposed to this favorable account, but it is so more in appearance than in reality. According to one authority, the labouring population, if it could be put in motion, is adequate to the cultivation of the soil. This opinion seems to justify the planters. Another remarks, and justifying, I cannot but think, both the planters and the labourers, more especially the latter,—"If labour has been de-

---

* In a recent letter from a friend in the island he states, "As regards wages, a woman's daily hire is nominally 6d., but she works only for six hours willingly. A man receives 8d. or 1s., giving more of his time." He adds the following remarks on the subject of labour, which I transcribe, not because I admit their justness, for I think, experience proves that they are not just,—men's wants not being limited,—wants growing with indulgence, but because they are views commonly entertained in the West Indies. "A native tropical population don't seem to earn as much as they can. They find out the sum they must have to supply their wants, and work no more than is necessary to obtain that. Hence it follows that doubling or halving wages is just halving or doubling the labour effectiveness of the population; a reduction of wages is, in fact, an equivalent for so many labourers, proportionally, and would augment labour according to its degree, until they permanently gave its maximum of exertion, which would be at least a third more then they contribute now."

ficient, and I believe it has been so, I do not gather that the people generally speaking are less orderly or well disposed, but on the contrary, I believe they will safely bear comparison with other communities." He goes on to explain,—"The very small profits with which the people are content to rest satisfied from the growth of potatoes, yams, tous-les-mois, and other provisions on patches of land hired by them at the rate of a dollar a month per acre, render the fact of less continuous labour at command, whatever may be the regret with which it is received, no matter of surprise." He adds, " land is cheap, and the returns from an infinitely small amount of labour, (a great desideratum in the tropics) are rapid, as well as apparently ample for the enjoyment of an existence whose mere wants are very few."*

I cannot dismiss this subject, the colored labouring class, without offering a few words relative to the capacities of the African mentally, intellectually considered. The same gentleman whose remarks I have given respecting the whites, said, while we were conversing about the negroes, that he considers them quick to learn, not inferior in aptitude to Europeans, and fond of learning. He witnessed this in his own house; servants, for instance, learning to read from his children. I transcribe the words from my note book. It is well to have cumulative evidence on a point of this kind, and from persons such as I met in Antigua, whose opinion, similar to the above has already been given, and

* Lieutenant Governor Drummond Hay, reports P. P., 1851, p. 106.

this of my St. Kitt's acquaintance, who from their thorough acquaintance with the people are competent to judge correctly of them.

St. Kitt's has been boasted of as the garden of the West Indies. Its fertility certainly has been remarkable; it is still respectable, if at all diminished; and with care may be pronounced to be inexhaustible. Like Antigua and Barbados it has the advantage of its land being all appropriated, and there being none, as in Trinidad and St. Vincent, and some of the other Colonies belonging to the Crown, offering a temptation to squatters. The whole surface comprising 41,851 acres divided into nine parishes*, contains 143 separate estates,—the number given in the Almanack of the island, in which the dimensions of each in acres is to be found, and the name of its proprietor. In size, these vary from 2,314 acres to 19, the average being about 300 acres, including pasture or waste land with that fit for canes; but limiting it to the latter, vastly less, averaging little more than 130 acres, the whole of the cane land returned as capable of culture, if not actually cultivated, being under 19,000 acres. The estate of maximum size, that of 2,314 acres, as large as some of the parishes, is called "Salt Pond;" it occupies, I infer, the greater portion of the narrow promontory

* These parishes are the following, those marked with an asterisk belonged to the English, the others to the French, when the island was divided between them. St. George, Basseterre; St. Peter, Basseterre; St. Mary, Cayon; Christ Church,* Nichola-Town; St. John, Capisterre; St. Paul, Capisterre; St. Ann,* Sandy Point; St. Thomas,* Middle-Island; Trinity, Palmetto-Point. These nine parishes send 24 members to the House of Assembly.

extending southward from Basseterre, corresponding—keeping the figure of the island in remembrance, and its similitude, the guitar,—to the handle of that instrument. Belonging to this property, the pasturage, that is, the waste land, is as much as 2065, the cane land being no more than 249 acres. These figures, whether relating to the number of estates or their average, are of course to be considered only as approximate, the properties being liable to constant change; the tendency, now being on one hand, to enlargement by addition, joining one estate to another; on the other, to reduction of size by selling or leasing off portions for labourers to form their settlements on. When Bryan Edwards wrote, in 1793, the extent of land in cane cultivation was, according to his stating, 17,000 acres.

As to the system of agriculture followed here, it is in some respects peculiar and much to be commended, especially in the large use made of implemental husbandry, and the frequent use made of green dressings.* More than one circumstance conduced to the former, the more important; first the troubles about continuous labour, and a due supply of it after emancipation, and the abstraction of many of the labourers, tempted to emigrate by the higher rate of wages in Trinidad and British Guiana; next the nature of the soil and ground

* These so called "green dressings" are effected by ploughing in the the pigeon pea (cajanus indicus) a plant possessing deeply penetrating roots well adapted to extract the fertile elements from the soil,—and leaves abounding in these elements especially the vegetable alkali, and silica; the minute spicula constituting the down of the under surface of the leaf consist principally of silica.

so inviting; and next, and not least, the intelligence
of some of the planters, intent on improvement, and
of one in particular who took the lead. Now, nowhere
in the West Indies is this kind of husbandry more followed, if anywhere so much, or with greater satisfaction and success. Now, we are informed that the
common plough, the bedding plough, the hoe-harrow,
or weeding machine are in common use; that a plough
with two or three horses, a man and boy, will displace
eight persons with the hoe, doing the work called
"holeing," and that a weeding machine with one horse
and a man and boy will displace sixteen persons in the
weeding of canes. An intelligent manager, with whom
I had conversation on the subject assured me, that on
the new plan, the estate he had the charge of was cultivated after the English method of agriculture with a
saving of half the labour, and half the expense, and in
a superior manner compared with the old, and with
increase of profit. Within the last few months I am
informed by a friend residing in the island, a proprietor, and deeply interested in its agriculture, that
the same system is continued, the same as when I was
there, that "all our fields," these are his words, "are
cultivated by plough and harrow and not more by hoe
than can be helped." He adds, "as in England young
turnips are hoed, so here young canes are hand weeded,
but wherever the brute can be walked, the man is not."
In carrying out this system here where the soil is commonly so light and easily worked, the horse is preferred
to the ox. A gentleman residing on his property

which was carefully cultivated, speaking of the advantage of employing the more active animal, said he found that one man, an Englishman, ploughing with three American horses, could do more work than could be accomplished by ploughing with eight oxen and two men in addition as drivers. With the attempt at this system of cultivation another measure was connected, that of the introduction of English labourers, which appears to have been partly successful, partly a failure; the former so far as the object of the planter was concerned; the latter as regards the people themselves. Of the whole number imported, little exceeding two hundred, many were ill-selected, ill-conducted, ignorant, infirm, totally unworthy of the name of skilled labourers. Some proved useful, continued to be so, and have given satisfaction; more turned out useless; many sickened and died; of the total, now, I believe, very few are remaining, and a century hence in all probability not a descendant of them will be discoverable, and all traces of them will be lost, as of the party of the same kind that preceded them already alluded to. *

* The following simple petition addressed to the colonial minister, then Lord Stanley, expresses their woes.

*To the Right Honorable the Secretary of State for the Colonies, &c., &c.*

St. Christopher, 20th August, 1845.

We your humble petitioners as immigrants, that is, lately arrived at the island of St. Christopher, humbly showeth ;—

That ever since we placed our feet on these shores, we are continually ailing,—a complication of diseases or distempers, some with headache, others debility, complaint in the bowels, a rash all over their bodies, which makes them very uncomfortable, accompanied with uneasiness of mind, &c.

We humbly beg to observe to your lordship that the climate is so

3 M

During slavery a good deal of attention was paid miserably close and warm altogether that we can scarcely withstand it, perspiration to such an extreme, and then dreadful cold ensues. We, your memorialists, can assure your lordship, that we can fully maintain the truth of our assertion. We further beg to observe to your lordship, that if your lordship afforded the honor and pleasure to this island of your presence our statements would fully convince your lordship as to our veracity. We assure your lordship that there is not the slightest reluctance on our part to continue in the island for an honest livelihood, by pleasing our employers by our industrious labour if the climate agreed with us, but unfortunately it do not, and we are much afraid if we continue longer in this injurious hot climate (the West Indies) death will be the consequence to the principal part of us, which will be an irreparable loss to what parents and families we have left at home at our native country.

We would therefore most humbly solicit your lordship's kind, strict and compassionate consideration in this our present case, and relieve us accordingly as your lordship may deem fit.

For we must candidly assure your lordship that we would rather exchange our present situation, and return to our native homes. And your memorialists, as in duty bound, will ever pray.

(Signed,) William Reynolds, Immigrant, on the part of the whole that is in the island, and consent of all parties.

(Return, P. P., 26th August, 1846, p. 111.)

A sum of one thousand pounds was granted in aid of bringing out this party. The terms on which they came were, according to articles of agreement subscribed to in England, that "they bound themselves to a service of three years for wages at the rate of 10s. 6d. a week, are disposed of on their arrival by lottery, and are provided by their employers with house, garden, and medical attendance." In the report of the stipendiary magistrates respecting them, amongst other objections, fears are expressed that these immigrants from England may "corrupt the otherwise sober and moral character of the negroes." Alluding to some of the difficulties they have to contend with, it is remarked, " The immigrants who have been imported into this island complain generally of the food which they are able to obtain. The supply of meat is scanty and precarious and costs 9d. per pound. Vegetables, except sweet potatoes, are procured with difficulty, and at a price beyond their means. They are therefore compelled to consume the cheaper and coarser sort of food, consisting of salted provisions, such as American herrings, and cod fish. There are no markets contiguous to the estates in the interior, no shops are at hand, and to procure the ordinary necessaries of life a messenger is commonly indispensable." Return, part 1. P. P. 26th, Aug. 1846, p. 280.

Some other immigrants, African and Portuguese have since been intro-

here to manures. Dr. Grainger in his poem, "The Sugar Cane," composed in the island says,

"Planter, would'st thou double thine estate,
"Never, oh never, be ashamed to tread
"Thy dung heaps, where the refuse of thy mills,
"With all the ashes, all thy coppers yield,
"With weeds, mould, dung and stale, a compost form,
"Of force to fertilize the poorest soil."*

Another, the author of letters on the cultivation of the cane, an old and experienced planter of St. Kitt's, writing in 1801, enforces the use of manure with the adage, as he calls it, "that the planter who makes the

duced, of whom the Lieut. Governor writing in 1850, thus reports. "The African immigrants amounting in number to 97 who arrived in the month of April, (1849.) have given general satisfaction to their employers, both in rendering comparatively steady, continuous labour, and in deporting themselves peaceably. They have all been indented under yearly contracts. The Madeira peasant also continues to be of great service in submitting to the call for regular and sustained labour which the Creole negro finds so irksome; but with him the tendency to withdraw from prædial labour for the purpose of establishing some retail shop is still on the increase. Many Portuguese moreover have of late left the island altogether. The desire of bettering their condition, so remarkable in their character, renders them restless; and a calculation of the numbers present at given periods would render very fluctuating results."—Reports (for 1849,) P. P. Part I. p. 46.

Those planters in the West Indies, or in any of our colonies, desirous of introducing implemental husbandry, would do well to follow the example of a great agricultural improver, a Mr. Gordon, who is thus spoken of by Arthur Young, in his Tour in Ireland,—he "imported a man from Norfolk, to whom he gave 40 guineas a year, with board, who brought ploughs, hoes, &c. with him;—gave him a guinea for every boy (Anglice man) he taught to plough; and every boy who could fairly plough, had a shilling a day wages. By this means he has collected a set of excellent ploughmen, who have been of infinite use, so that he has to this day ploughed with Norfolk and Suffolk ploughs worked with a pair of horses, and no driver, except the first and second ploughing of fresh land, which, and dragging, he does with great drags of 18 cwt. and drawn by bullocks."

* *The sugar cane*, p. 20, 4to, London, 1764.

most dung will make the most sugar," and still more
by an instance he adduces of extraordinary success,
following its very liberal employment, how for instance
" from some pieces of canes the scarce credible quantity of six hogsheads per acre had been obtained."*
Whether manuring has had due attention of late
years is doubtful. From what I saw and heard I am
disposed to think it has not; that tillage and labour
have been more absorbing subjects, as felt to be more
immediately pressing. And a like remark is applicable
I believe to the time and manner of planting, to rotation of crops, the growing of provisions, and some
other points of agriculture. Whilst travelling through
the islands I could not avoid seeing the neglect of
articles of manure possessed of great fertilizing power,
such as Mr. Caines alludes to, mentioned in the foot
note, collected by his successful planter when performing the part of a scavenger, and also, it was only in
one instance, the waste of cane-ashes, they having

* The following is an extract from these interesting letters,—the whole
of which are well deserving of perusal, and not less by the tropical agriculturist than by the statesman, and the advocate of emancipation.

" Attention to this subject alone (manuring) constituted the basis of a
fortune, which exceeded any that has been acquired in the leeward islands
for ages. It amounted to 80, or £100,000 in the funds, and four or five
considerable estates. The person who accumulated this wealth had the
lease of an estate, contiguous to a town where large quantities of dung
were always to be met with, and were always neglected. He turned
scavenger, and covered all his land with the nasty and precious heaps.
In a very short time his industry and judgment were abundantly
rewarded. From 60 acres of land he has often made 240 hogsheads of
sugar; and from some of his cane pieces, the scarce credible quantity of
six hogsheads an acre."

Letters on the cultivation of the Otaheite cane &c., by Clement Caines,
Esq.; London, 1801.

been thrown on the road instead of having been returned to the cane field.

The planting of the cane appears to be more irregularly conducted than can be right. I was told that on many estates the mill was almost monthly at work throughout the year, piece after piece ripening in accordance with the time of planting; a circumstance it may be complimentary to the climate, but hardly to the judgment and foresight of the planter, as we are sure, that even in St. Kitt's with all its irregularities as to rain,* there is a wet season, taking the average of years, best fitted for putting the canes into the ground and a dry one specially auspicious to their ripening and the reaping of them.

The growing of provisions is now very much left to the labourer. The planter probably would consult his interest were he to attend more to their culture, whether with a view to an immediately profitable return in growing a marketable crop, and supplying food for his live stock, such as sweet potatoes; or to the more remote one of preparing the ground for canes after the manner followed by some of the best planters in Barbados.

In the manufacturing process a start has been made here in substituting the steam engine in many

---

* An author, already quoted, (Caines) speaking of the weather, says,—
"Older planters than myself have told me that the changes were regular, that it was periodically wet and periodically dry. But my own experience does not give entire confirmation to what they say. I have known it frequently dry at the periods supposed to be most rainy; and very wet at the periods deemed the driest."

instances for the windmill, for the purpose of crushing the canes, and expressing the juice. When I was there first, now seven years ago, I was informed by the civil engineer in the service of the colony, that 33 were then in use ; that there, where the winds are uncertain, (and where are they not ?) they are decidedly more economical than windmills, and even apart from the uncertainty of the winds, the saving by the greater work they can do, and the fewer hands required, more than covers the cost of the coals consumed. He adverted at the same time to a mistake sometimes made in pulling down the windmill, or neglecting it, on the erection of the steam engine, as if not aware, the owner, that the latter is subject to derangement, and that workmen competent to repair it, are not always at hand or easily procurable ;—a difficulty this, it may be remarked more or less felt in all our West India colonies, and discouraging to the introduction of any complicated and delicate machinery.* Even in St. Kitt's where the steam engine is so much used, I saw on one estate, and that near the principal town, a cattle mill employed,

* Ignorance of the mechanical arts in these colonies generally, and the scarcity of artificers, especially of skilful ones, is very seriously felt, and is a source not only of vexation but also of great expense. A proprietor in this island, whose estate, not a large one, is near the town, assured me that his blacksmith's bill for the year amounted to 500 dollars, and this for ordinary repairs and work. The blacksmith told him that heavy articles, with most trifling derangement of machinery were often brought to him from a distance which ought to have been repaired on the spot, and a very unnecessary expense in consequence incurred.

The same proprietor made mention of a millwright, who had come from Barbados with two or three apprentices, and was rapidly making a fortune. For repairing my friends windmill he had received £ 50.

and it was apparently of an antique and rude construction.*

In the actual making of sugar, that is in the processes of the boiling house, no improvements of any moment so far as I could learn have yet been made. I did not hear, either of the vacuum pan, or of any new method having been introduced. And of late years the quality of the sugar of St. Kitt's, it would appear has somewhat deteriorated and fallen in value.

Of the amount of produce, the criterion at least of the extent of cultivation, there appears to be a great falling off; and it is remarkable that the decline has been from a somewhat distant period, commencing and continuing whilst the produce of Antigua was increasing. When Bryan Edwards wrote, viz. in 1793, the average yield of the island according to his stating was 16000 hogsheads.† From 1822 to 1830 taking the

* It consisted of three perpendicular rollers of small size, worked by six mules, three abreast, driven by two boys;—the mules in a very sorry condition. On an adjoining pasture,—if pasture it could be called, so numerous were the bushes in it, was a herd of mules of the same lean quality, some of them of very wretched appearance. On the same estate I saw three men employed in ploughing a light soil (one ploughman and two drivers) with eight oxen. Examples these of the old system. Extract from journal.

† His statement relative to the fertility of the soil in accordance with some of those of Mr. Caines, is worthy of the attention of the struggling planter at present. Speaking of the peculiar fitness of the soil for the cane, he says,—"Canes planted in particular spots have been known to yield 8,000 lbs. of Muscovado sugar from a single acre. One gentleman, in a favorable season, made 6,400 lbs., or four hogsheads of 16 cwt. each per acre, on an average return of his whole crop. It is not however pretended that the greatest part, or even a very large proportion of the cane land throughout the island is equally productive. The general average produce for a series of years, is 16,000 hhds. of 16 cwt., which, as one half only

average of eight years, (exclusive of 1828 for which no return,) the yield was 7855 ; the maximum during that period 9197; the minimum 6006.* Coming to a later period and almost to the present time, the produce of 1843 was 4,789 hogsheads; of 1844, 7,507 ; of 1849, (the returns of intermediate years are not reported,) 5,357 ; of 1850, 4,708, (two years of drought) of 1851, 7,270, (a seasonable year) giving for the whole an average of 5,926 hogsheads.

This decline of produce has been attributed by some to the worn out exhausted state of the soil; an inference the justness of which I cannot but doubt. I am disposed to refer it to the diminished extent of land under cane cultivation, especially since the abolition of the slave trade, stopping an unlimited supply of efficient labour, and since emancipation, occasioning a further loss of continuous labour. The produce per acre is the best proof of the degree of fertility of the land. Bryan Edwards states, and with the expression of astonishment, (see the preceding subjoined note,)

of the whole cane land, or 8,500 acres is annually cut, (the remainder being in young canes)—gives nearly 2 hhds. of 16 cwt. per acre for the whole of the land in ripe canes;—but even this is a prodigious return, not equalled, I imagine, by any other sugar country in any part of the globe. In Jamaica, though some of the choicest lands may yield, in favorable years, two hhds. of 16 cwt. per acre, the cane land which is cut annually, taken altogether, does not yield above a fourth part as much." He adds, " I am informed however, that the planters of St. Christopher's are at a great expense for manure ; that they never cut ratoon canes, and though there is no want in the country of springs and rivulets for the support of the inhabitants, their plantations suffer much in dry weather, as the substratum does not long retain moisture." *History of the West Indies,* I. p. 431.

* M. Martin's West Indies, vol. II, p. 332.

that it then averaged 2 hhds. an acre. When I was in the island I was informed by a gentleman of great experience, and well informed that it is still about the same, not exceeding, (these were his words,) 2 hhds. an acre, and he spoke of it unfavorably, in comparison with, in his own opinion, the more fertile soil of Antigua. Another gentleman, at the same time, informed me that from some of his fields he had made from 3 to 4 hhds. an acre, and these belonging to an estate, perhaps as long in cultivation as any in the island, it being at a short distance only from Basseterre ; but, be it remembered, it was well cultivated and well manured.

As regards the returns of estates, whether profitable or otherwise, there is no inconsiderable difficulty in coming to a conclusion. Before the sugar duties act of 1846,—that " crushing" shock to all West India interests, as it has been designated by a late governor of the leeward islands,—the cultivation appears to have been carried on with moderate profit. Since that event, the opinion of the most unbiassed seems to be, that the costs of production, and the other expenses incurred in bringing the produce to market, exceed its market value at the late low prices, leaving the producer, *nil*, excepting,—and fortunately exceptions are made,—in the instance of resident proprietors taking into their own hands the management of their estates, and conducting the business of them with some skill and attention to economy.

Other and more unfavorable views are entertained.

The following is an example. "You ask me," (says a friend, writing to me on the 31st of March, 1853,— a gentleman resident, deeply interested, and well acquainted with the agriculture of the island) "how we have got over the difficulties of the act of 1846. We certainly have not got over them, though we seem to exist in defiance of them. How, none of us exactly know. As a whole, the island has, unquestionably, been cultivated at a loss ever since. The proprietorship has therefore once again become nominal. Those that did not owe, now do, and those that did, owe more. We are always hoping and ever disappointed. In several instances, small properties have been taken into larger ones, and more engines (steam engines) have been erected. The price of labour too is reduced, so that variously, the cost of the production has been lessened, but the value of the product has fallen from year to year in greater proportion, so that the premium of profit has never been realized."

As preceding difficulties,—those immediately following emancipation, led here to improvements which in a great measure mastered those difficulties, may we not reasonably expect that the present will have a like effect. This we are sure of,—that in tropical agriculture there is great scope for improvement, and in various ways; that other crops besides sugar may prove profitable, and more so than the exclusive one of the cane; that greater economy may be exercised, and savings made, as instanced in Barbados in the general management; that more attention to

live-stock, both to their breeding and keeping is
requisite for success; and to be brief,—if these
things cannot be accomplished by the old proprietors
now resident, of whom there are so few,* or by those

A late Lieutenant Governor, Mr. Mackintosh, now governor of the
leeward islands, in a letter of the 5th of March, 1849, adverting to the
subject of absenteeism, states that the number of resident proprietors,
superintending the cultivation of their estates, does not exceed eight.
Reverting to the subject in a letter of the 6th of March, 1850, he
remarks. "The absentee system, as I have previously had occasion to observe to your excellency, has been allowed to develop itself in the island
to an excess, which has produced its usual concomitant evils. I need not
stop to particularize such as are always, and everywhere observable as
such. But there are many estates in this island which are the property of
opulent individuals residing in England, (some of whom have never seen
their properties,) who apparently, utterly forgetful of the peculiar responsi-
bilities, which can never be separated from the ownership of land, are
content that their estates shall be cultivated through the agency of mercantile houses at home. So great is this inveteracy of habits of thought on
this subject, that there appear in the evidence given before the committee
of the House of Commons, which sat in 1848, on West India distress, the
names of gentlemen, who though, owning and residing on their estates in
this island, were in the habit of leaving them daily for their mercantile or
professional avocations, classed argumentatively with reference to the
absentee system, as resident proprietors. The effect of this disastrous
system has been, that a body of men have grown up in consequence, who
though calling themselves merchants, derive all their profits by simply
reducing absentee landlordism to a profession. It is almost impossible to
exaggerate the proportion of embarrassment which this apparently hopeless
struggle to reconcile tropical profits with residence in a temperate climate,
has contributed to West Indian difficulties. As one consequence of it,
simultaneously with the depression and disappearance of the old families,
has increased the influence of the class of attorneys of estates, who still carry
on a lucrative business under a system which concentrates powerfully in
support of itself, the personal interests of the few whom it has ruined. It
is impossible I should think, but that with our comparative amenity of
climate, and many social advantages, there must be many in the mother
country, with small capitals, who would gladly invest their means at
present prices, in land of such prodigious fertility as ours. Some indication of such wishes indeed already exist. But all intending purchasers,

encumbered with debt, of whom there are so many, they may be effected by resident ones, and new ones resident free from debt.

Comparing the condition of the island generally with that of the one last considered, Montserrat, recently so depressed, seemingly ruined, yet presently shewing marks of revival, how much more encouraging are the prospects of St. Kitt's. Here there is no distress prevailing amongst the population generally. It is reported on officially as a well ordered community; the peasantry as more than commonly easy in their circumstances, and well-doing; no public colonial debt, one that was contracted having been recently liquidated; a revenue exceeding the expenditure, raised chiefly by indirect taxation, affecting most the working class, and yet not severely; roads undergoing repair; a new prison recently constructed, and other public works, as reported, in a respectable state. Were the prices of sugars to rise only a few shillings per cwt.,—and what is more likely, if the consumption of the article continue to increase as may be calculated on, and that in a higher ratio than the supply, especially of slave grown sugar,—the effect would be to afford an

---

meet on the threshold a great obstacle in the cumbrous and expensive system, which governs the tenure and alienation of real property, and which so needlessly enhances the cost and precariousness of acquiring it. I am not sanguine in the expectation that this great evil can ever be successfully grappled with, by a local legislature. But if the Imperial Parliament could be induced to apply a measure on the principle of the Irish encumbered estates bill, to the West Indies generally, I am deeply impressed with the conviction, that it would confer, on this island at least, a very substantial benefit." Report, (for 1849) 1850, part I, p. 47.

immediate profit to the planter; and the same result would follow, should our government choose to discourage slave labour, and encourage free labour, and that of our colonies, by reducing in just degree the duties on the sugars of the latter, and exempting those of the former from any further reduction than the promised one which is in progress, and to stop next year, at 10s. the cwt.;—a measure that could not, whilst it benefited the planter, render the article dearer, but rather cheaper, to the people of this country. These and other possible and not improbable contingencies should encourage hope of better times for this, and our other West India colonies, but most of all, the great consummation to be looked for, the total abolition of slavery, which now seems to be more and more called for, and urged, by all reflecting persons, who have the real interests of man at heart.

Many pleasant excursions may be made by the traveller in St. Christopher's, and easily made, even in the interval of the few days, from the arrival of one packet to the departure of the following, the distances being short, the roads pretty good, and conveyances easily procurable.

Basseterre, where he is landed in an open boat, at the risk of getting a wetting, as at Montserrat, should be his head quarters. There he will find a tolerable inn, as inns are in the West Indies, and be best able to gain information and be put in the way of seeing what is most worthy of notice.

Even in the neighbourhood of the town, whether

taking a walk, or a ride, a good deal is to be seen that is interesting. The valley, or rather plain of Basseterre, surrounded with hills on all sides, except where opening out on the sea, is itself a charming scene. It reminded me of the plain of Zante, being not unlike it in form; as generally cultivated and as verdant; the cane fields rivalling the vineyards, the cocoa-nut palms the cypresses.

Monkey-hill, so called from the monkeys infesting it, overlooks the plain; it is only a short distance from the town and will repay the labour of the ascent, which, as it is somewhat arduous, if attempted, should be made early, at least before the heat of the day. The prospect from thence is extensive and instructive. The hill itself, the upper portion is a good example of the wild pasture of the island; it is encumbered with brushwood, composed chiefly of fruit-bearing guava bushes.

Either in going or in returning, a visit might be made to the Monkey-hill estate, and the Olives', (indeed the hill is part of the former.) The estate-houses belonging to them both, pleasantly and coolly situated, standing high, two or three hundred feet above the level of the sea, are favorable instances of the dwellings of the planters, especially that of the Olives'. It stands on a well raised stone terrace, paved with marble, and when in its best days, it is now out of repair, had spacious open galleries and verandahs. One large finely proportioned room extends the whole length of the front, with a handsome deep cornice, and ample doors, both of dark mahogany, and if I recollect rightly a half wainscot of the same wood. It was in the by-gone and convivial times, the great reception and dining room. The estate comprising 283 acres, viz. 151 of cane-land, 132 of pasture, was spoken of, in proof of the depreciated value of landed property: the gentle-man then residing there said that the whole might be purchased for £3,000, a less sum than the original cost of the house.

One day may be profitably employed in visiting the north-west portion of the island, a portion of it, which in time of slavery and during the later French wars, was far more extensively cultivated than at present. In St. John's alone, in which there are eighteen sugar estates so named, nine I was told were abandoned, that is, the cultivation of the cane had been given up, and the land had either been let to labourers, for growing provisions, at the rate of, from 9 to 14 dollars a year the acre, or had been left neglected, or used for pasture; a neglect indicated by many of the sugar works in ruins, and some of the estate-houses.

Going by way of Middle-island church and Sandy Point, the latter the old principal town, to Dieppe, much of the coast will be seen, several of its villages, and much of its most pleasing and remarkable scenery and objects of interest, as well as an opportunity afforded to observe the dwellings, the manner in which cultivation is carried on, and some of the habits of the people. Where the road passes under a steep sea cliff, the only cliff by the way, traces, in a line of old prickly pears, were pointed out to me of the boundary established when the island was formerly divided between the French and the English. The banks of the East river and Bloody river, two of the four principal streams of the island, which have to be crossed, may tempt a halt. They have much beauty, and are finely shaded in spots by cocoa-nut groves, flourishing unblighted,—unfortunately a rare exception. Further on, little more than half a mile from Sandy Point, Brimstone Hill, called the Gibraltar of the West Indies, rises conspicuously, immediately above the way. It should not be passed by: the ascent is easy, by a good carriage road, the best in the island. Its peculiar structure, its singular forms already alluded to, the fine and extensive views it commands may be mentioned as its principal and ordinary attractions. To the military and medical officer it will have other points of interest, of no ordinary kind,—to the one on account of its position and defences, to the other on account of the ravages committed in its garrison by repeated, sudden, and solitary outbreaks of yellow fever,—a locality remarkable for its dryness, and other apparent circumstances of situation and climate, which it might be supposed would be favorable to health.*

Dieppe a large village at the N. W. W. extremity

* The barracks of this fortress are a striking example of defective construction in a sanitary point of view: the worst of them have undrained and unventilated ground floors, the flooring of boards, pervious to exhalations from beneath and to all liquid impurities from above : this is, or was the case even in the elevated citadel.

of the island, stands on the shore of a little bay, a
basin-like port within a coral reef, in which there is
a barely passable opening, not without risk for small
craft. Whilst standing on the heights above, we saw
a small vessel attempt the entry, and succeed. It
was an interesting sight, on account of the dangers,
and the skill required and shewn to avoid them. At
times she seemed amongst the breakers and all but
lost. Even in the narrow channel she did not advance
regularly; she had to tack this way, and that,—
always advancing; had she once missed stays, she
would have been a wreck, the coral rocks on which
the waves broke with violence, being inevitable.
Excellent and most steady steerage is requisite here,
and none but the very experienced attempting the
passage. Accidents, it is said, are rare. This district,
I was assured, has always been peculiarly healthy,
never having been visited by any epidemic, and
having escaped yellow fever, when other parts of the
island were suffering under it. Its climate is reported
cool and pleasant, and though subject to frequent
showers, yet not moist. The character of the country
is that of a continued ascent from a low coast to
the steep wooded acclivities of Mount Misery. The
surface is more or less irregular and broken. The
declivities are traversed by ravines, exhibiting much
variety as to depth and extent; they have not the
regularity of the gullies of Barbados, and have more
the aspect of having been formed by water, acting
on loose volcanic ashes, of which the ground consists,

than of being rents or fissures, the result of sudden
and violent force. Some of them, the higher ones,
are beautifully and most luxuriantly wooded, the
tree-fern abounding: some, lower down, are cultivated, well stocked with plantains; others are naked
and barren, strewed with loose detritus, with deeply
cut sides, exhibiting successive layers of ashes,
marked by their different hues, most of them horizontal.

When I visited this district, it was with a friend,
the owner of an estate there, at which we stopped
for a short time and had some refreshment, partaking
of the dinner of the manager and his family. He
was a creole, an intelligent man of an enquiring
mind, and gentlemanly manners, almost self-edu-
cated, having learnt at school only reading, writing,
and the common rules of arithmetic. His wife was
a sensible woman; both were about the age of forty;
they had been married about twenty-four years, and
had a large family. Two or three of the sons had
been brought up like himself, and had already the
management of estates. There were two grown up
daughters, unaffected and lady-like, and one boy,
a fine little fellow only six years of age. There
was an interest, I must confess, felt in conversing
with these sensible unaffected people, in their peculiar
condition and circumstances,—none of them having
ever been out of the island, excepting the father,
who had been at Nevis. I notice them thus particularly, as examples of their class. This estate,

I may mention, was a good instance of the new system. Here I saw three active English horses drawing a cart laden with canes, driven by a negro; and a weeding machine at work, drawn by two horses. Ploughing, too, I was informed, was performed with two horses, guided by reins. One Englishman only was on the property he had been engaged as a labourer, on the condition that, after a certain time, he should be raised to be an overseer, which he then was.

Another day may be pleasantly occupied in a visit to Spooner's Level, about nine or ten miles from Basseterre, high up the mountain range, probably not less than 1,500 feet above the level of the sea. The greater part of the distance, as far as the estate called Philip's, there is a carriage road; from thence there is merely a bridle path. The ascent, (if the weather is at all favorable,) is peculiarly pleasant, passing gradatim and yet rapidly,—the ascent being rapid,—from the cultivated into the wild, from the cane field into the forest, and from the hot or unpleasantly warm air, as that of the lower grounds so often is, into the cool and refreshing atmosphere of the mountain. The kind of wood, where it becomes wild at a certain height, is very like that of the mountain region in Montserrat, as luxuriant or nearly so, and even more abounding in the tree-fern,—indeed, some portions of the forest were formed almost entirely of masses of this beautiful palm-like plant. The Level itself is a cleared portion of land, of a few acres, between seventy and eighty, charmingly situated; its surface irregular, approaching to a plain, (as is indicated by its name,) surrounded by wood, and well supplied with water. Now, it is entirely given up to pasture; a considerable number of cattle, some sheep and horses are kept here, and throughout the year, under the care of a herdsman. It has the reputation of being remarkably healthy; and some years ago, it is said to have been much resorted to for change of climate,—persons for the benefit of its cool and reputed wholesome air, remaining here many weeks together. And in accordance with this, some traces of old cultivation are to be seen amongst the wild shrubs and neglected orange and mango trees, with a water tank overgrown with coarse vegetation, and the ruins of a fallen house. That the spot is deserving of its former reputation, is highly probable. Its coolness is unquestionable; when I was there the thermometer at 1 p. m. was 69°, and an hour after it was 67°. This was on the 21st March; it was raining at the time. In fine weather, the air is described as peculiarly delightful, and the spot altogether charming. The chief objections that I have heard made against it by those questioning its fitness for a sanitary station, are the frequency of rain, its humidity,—the mountains being so often enveloped in mist,—and the difficulty of the approach. The former, I am disposed to think, rather holds good against its agreeableness,—much and frequent rains being almost inseparable from tropical mountain situations,—than its salubrity. The latter, were the island prospering, and were it an object to improve the road, might be overcome without much difficulty, or any very great expense. At present, and for a long period, the only public road connecting the two sides of the island, is and has been over the Level. At one time when the French possessed the two extremities, it was to the English an important line of communication. Since however, that the whole island has become a British possession, it has been considered of little moment, and has been much neglected, indeed I never travelled in any country, not even in the wilds of Ceylon, a road less deserving of the name of road, or ever rode down a descent comparable for difficulty with "the nine turn gut," as the steepest part of the way is called, leading towards the "Old-road," about three miles distant, and to Brimstone Hill.

In going to Spooner's Level, or in returning, the village of Cayon, in the district of the same name, is worth seeing, being a good example of the villages of the country which have sprung up since the abolition of slavery. It stands high, beautifully situated on the side of a deep glen, through which a rivulet finely shaded with trees flows rapidly making many pretty falls in its descent. Large and scattered, it covers a good deal of ground, land in small portions being attached to most of the cottages. Amongst its inhabitants are persons of different trades, all of the colored race, shoe-makers, saddlers, masons, carpenters, wheelwrights, coopers, smiths, sempstresses, but no tailors. I had these particulars from an intelligent man of color who accosted me, when wandering through the village,—he himself, as he said, a shoemaker and saddler, and in addition clerk and sexton. He called the valley the garden of the island, and was evidently proud of it, and of the village, and not least so, of his own little possession, which he took me to see, specially directing my attention to a granadilla plant, and to a mango tree planted by himself, both instances of luxuriant and rapid growth. The former he held to be the finest in the island; it was full of fruit, though only two years old; its stem was about the thickness of a man's arm and its twining vine-like branches supported on a frame, spread over a space of about 20 feet in length and nearly as many in width. The mango tree already of a goodly size, and in bearing had been in the ground about six years. To my enquiries about rent, he rather startled me by what he said. For his own cottage with a portion of land, only 40 feet square, he paid 24 dollars a year. He assured me that land of no greater extent was let here to labourers at a dollar a month. I afterwards heard, and from good authority, of land, an acre, let to 20 labourers, at the rate of £50 a year, and cultivated with profit. It was probably near the rivulet and capable of irrigation,—a use, I was told, to which the stream was applied.

· The eastern portion of the island, the long promontory extending beyond Basseterre already alluded to, may be mentioned as not the least remarkable part of

St. Kitt's,—so peculiar in form and character, and so neglected, consisting of a narrow ridge of low hills, for the most part barren, with—what constitutes its chief attraction, a small salt water lake near its extremity.

Without road through it or path that I ever heard of, it is not easily seen except from the sea. In coasting along I had a partial view of it, in my way to Nevis, which is separated from its extreme point by a narrow channel called appropriately "the Narrows." I started on this little voyage in a four-oared boat engaged for the occasion, and to have the cool and calm of the morning, before break of day, when, as I see noted in my journal, the morning star was shining brightly, its rays reflected from the sea, and occasionally silvering the margin of a passing cloud. As the dawn broke, mention also is made of the beautiful effect of light, of the golden suffusion in the horizon gradually passing by transition through a delicate green into the pure blue of the zenith, and this contrasted, and most impressively, with the near awful cliffs, as they appeared dark, almost black in shade.

As we advanced, Frigate Bay was pointed out, hardly deserving the name of a bay, within which on the low boundary shore were a few cocoa-nut trees, and a little land, said to be in cane cultivation. Next the attention was called to Monk's Bay, so named from a monk, who made the barren side of the hill above, (a fit place for mortification,) his place of abode. Beyond this, we came to a more noticeable inlet, that within which is the salt pan or lake. Here on a beach abound-

ing in thrown up shells and coral, skirted by a cliff
containing selenite imbedded in clay, I landed, sending
the boat round the farthest headland to wait whilst I
took a view of the salt lake.

This lake at the time I saw it, on the 30th of March,
was about two miles in circuit, surrounded by muddy
low ground, occasionally overflowed, and inclosed, except where there is an opening towards the sea, by
hills which would have been naked, but for the great
flowering aloe, and a bushy prickly mimosa scattered
thinly over them. The bottom of the lake being
about six feet below the sea level, there is a tendency
of the sea to flow into it, which is effected in one place
through a sand bank by a process of oozing, giving rise
to a small salt stream, and in another, in tempestuous
weather by the breaking of the waves over the intervening low neck of land, which is paved. It is only
in seasons beyond the average in regard to dryness that
this lake is productive of salt. A strong drying wind
of a month's duration will commonly reduce the water
to the point of crystallization. The salt forms at the
bottom. It is collected by labourers in punts about
eight feet long, and four broad, and by them conveyed
to the shore. Occasionally a crust forms like ice at the
surface impeding the process. The people employed,
have half of what they collect for their labour, or rather
the price of the half, for it is bought on account of the
estate to which the lake belongs:—thus if the price of
the basket is fixed at three pence, three half pence is
fixed for each basket. In some years immense quan-

tities have been obtained, and a labourer has made as much as twelve dollars a week. The salt is sold chiefly to the Americans.

During the operation of making salt in the salines of the Ionian Islands, particularly at Santa Maura, an odour of violets is disengaged ; here the only smell perceived, I was assured, is a disagreeable ammoniacal one, a smell of hartshorn, arising probably from the large quantity of animal matter undergoing a slow decay, shrimps especially abounding in the water in its ordinary state of dilution. This smell is most perceptible, when the salt is in the act of crystallization. Most of this information I had from the manager of the estate, residing in a roomy rude wooden house adjoining. I had some particulars from him also respecting the property, as, how formerly, that is before emancipation, it yielded occasionally 160 hogsheads of sugar, whilst now it seldom produces more than 30, though about the same number of people, about two hundred, are on it, but now, having the privilege of collecting wood to sell in the town, which occupies three or four days in the week, they are in a great measure independent of the hire of 8*d.* a day for working in the cane fields.. On quitting the house, (one as solitary and desolate as a human dwelling well can be, having nothing about it excepting a few cocoa-nut trees, and these in the west, less than in the east, a mark of the presence of social man,) I passed in my way to the boat another small lake or swamp of the same character as the preceding, where salt occasionally forms.

The weather being fine, the passage across was not unpleasant, though even then rather rough. Subject to sudden squalls, caution is required in making it, and a good boat should be used. I heard of more than one accident, in proof of the danger; the most remarkable was that of the loss of a sloop of war which had entered the Narrows under a press of sail, in pursuit of an American privateer, and which suddenly arrested in her career by a squall, went down bodily, and every one on board perished, the privateer making good her escape from the double danger. As the catastrophe was related, it was somewhat dramatic; how when the captain was at dinner on shore, a slip of paper was put into his hands, intimating the appearance of the enemy; how he rose suddenly from the festive board; how, shortly after, his vessel was seen in close pursuit, and how in an instant she disappeared, sinking the moment she was capsized.

# CHAPTER XV.

## NEVIS.

NEVIS, so close to St. Kitt's, is very similar to it,
and not only in its geological structure, but also in the
manner in which it was settled and planted, and now
generally in the condition of its inhabitants, and yet
not without some points and shades of difference.

The island, is commonly described as a single cloud-
capped mountain, rising gradually from the sea level.
It is so, but with the addition in opposite directions of
two hills, processes as it were, of the main mass,
connected by a kind of isthmus, or shoulder. These
are Saddleback hill,—a hill somewhat of the form
implied by the name, at one extremity, and Round-hill
at the other.

In circuit, Nevis is about 24 miles ; in length about
4, in breadth 3, comprising an area of about 24 square
miles. The mountain,—for so it is called there,
having no rival, and requiring no special name to

distinguish it,—is of a conical form, its summit truncated, of an elevation supposed to be not under 2000 feet. Though its declivities are generally gentle and without abrupt cliff-terminations towards the sea, its sides are more or less irregular, presenting great variety of form, with picturesque effect, in abutments, depressions, and gullies. Towards its summit on one side, is a crater-like hollow, a Soufriere, in which sulphur is said to be found, and its common accompaniments.

As regards structure, viewed geologically, the whole island appears to be decidedly volcanic, differing however from St. Kitt's in exhibiting fewer marks of active eruption in beds of ashes and scoriæ, and more of elevation in the compact igneous rocks which are so often met with at the surface, and in various clays, which it may be inferred, are derived from the decomposition of these rocks. Amongst the former, there is one kind having a resemblance to granite, but more in appearance than in reality, being destitute of mica and quartz, and composed of a light colored, fine grained felspar, through which are dispersed minute crystals of augite ; a rock easily quarried and worked and well adapted for building purposes, to which it is beginning to be applied.*

In St. Kitt's, excepting in the Soufriere of Mount

---

* Though I could detect no quartz in this rock, or in any other I examined, it is worthy of remark, that quartz crystals well formed, and of perfect transparency, are found in one part of the island, near to, or on the side of Saddleback hill. From what I could learn, they occur in clay and in the bed of a small stream. Under the name of Nevis diamonds they have been in repute and have been cut and worn as ornaments.

Misery, there are no indications of the proximity of subterranean fire, not a single hot spring or spot the temperature of which is known to exceed the average of the locality. In Nevis there are both. About two miles from Charles Town is a ravine, where in a very limited space, sulphur is pretty abundant, and also alum incrusting variously colored clays ; and where in a fissure or cavity, the temperature is sufficiently high to roast an egg. The baths, about half a mile from the town, are an instance of the warm springs. The temperature of the highest I found to be 108° of Fahrenheit.* They hold in solution in small quantity, potash, lime, magnesia, and silica, with carbonic acid and traces of other ingredients, derived no doubt from the rocky bed from which they rise and through which they percolate.†

* From information I collected, the temperature appears to be variable, lowest in dry weather, highest after rain; I was assured that after a continued rain it had been found as high as 115° Faht.

† I have given an account of the water of these springs or spring—for it is doubtful whether there is more than one—in a letter to professor Jameson, the editor of the *Edinburgh New Philosophical Journal*, to which I beg to refer for particulars. It is to be found in the number for July, 1847.

According to the analysis I made, the waters of the highest temperature collected as it flowed from the pipe, that used for drinking, of specific gravity 10,019, contained about its own volume of carbonic acid. Forty-four cubic inches of it, equal to about 11,120 grains, yielded 1·8 grain of saline matter, readily soluble in water, chiefly bicarbonate of potash with a trace of muriate of magnesia, 1·3 grain of carbonate of lime, ·77 grain of carbonate of magnesia, 1·5 grain of silica, a trace of phosphate of lime, sulphate of lime and of vegetable matter. I stated in the letter alluded to, that I could not detect either iodine or bromine in the water. Afterwards, experimenting on a larger quantity, I satisfied myself of the presence of a minute quantity of the former.

The baths, of which there are two, each sufficiently deep and spacious to

The soils of Nevis are in good repute for fertility. They are commonly spoken of as light and porous, and as containing more clay than those of St. Kitt's. They are probably formed chiefly of volcanic ashes, and of matter derived from the disintegration and decomposition of the varied crystalline igneous rocks of which the island consists, and like the water of the baths are rich in the inorganic elements of plants. In proof of this, may be mentioned a remark of an early writer; that when the ground was first broken up and

swim in, are supplied with water of the same quality as that used for drinking, and probably derived from the same source. One is warm, of the temperature 98°; the other tepid, is 10 degrees lower.

Taken internally, that used as a drink, has been found serviceable in many instances of derangement of the stomach and intestines resisting ordinary treatment, and externally in the form of a bath in cases of obstinate rheumatism.

On exposure of the water to the air, when the carbonic acid it contains in part escapes, some of the carbonate of lime and silica dissolved,—and it may be inferred by the agency of the acid,—is precipitated, forming incrustations,—which are found about the baths and also in the bed of the small stream, chiefly fed by the bath spring.

A building well constructed of stone, of three stories with a spacious verandah or open gallery in front, erected by a philanthropist, a Mr. Huggins, for the use of invalids, adjoins the baths. The middle story alone is now open as an hotel. It is capable of accommodating about 15 persons, has 11 bed rooms, a large common room, and a drawing room. Standing on a rising ground, it commands a pleasant view of part of the island of St. Kitt's and of the intervening sea, and is considered healthy.

The author of the *Natural History of Nevis*, mentions a spot, towards the sea side, in the river flowing from the baths, where about chin-deep, "a man may set one foot upon a spring, so wondrous cold, that it is ready to pierce him to the very heart, and at the same moment place his other foot upon another spring so surprisingly hot that it will quickly force him to take it off again." I made enquiry after these "wondrous cold" and "surprisingly hot springs,"—thinking it probable, making allowance for exaggerated description, that a hot and a cold spring might still exist, the one adjoining the other, but without success.

planted, the canes produced, for two or three years yielded more molasses than sugar: his words are, " When we break up a piece of fresh ground to plant our canes in, the canes for the first two or three years will yield no sugar, so that we find ourselves obliged to distil their juice for rum."*

I am not aware of any exact meteorological obser- vations that have been made here. Its climate, of course, it may be inferred, is very similar to that of St. Kitt's. From such information as I could collect, it would appear that showers are more frequent here than there, and that it is somewhat less subject to drought. Dew is described as being often so abun- dant, as to trickle down the cane leaves on which it forms, and moisten the ground. Hail, occasionally, though rarely seen in St. Kitt's, here, I was assured, has never been witnessed. In common with St. Kitt's, and at the same time, it has suffered from hurricanes and earthquakes; the last earthquake, that of 1843, damaged many buildings.

In point of salubrity, its climate is of the same mixed character as that of St. Kitt's. The windward side, well ventilated by the prevailing winds, and where there is little or no marshy ground or stag- nant water, has the reputation of being remarkably healthy: there, it is said, neither yellow fever nor any other epidemic has ever prevailed. But on the leeward side, especially in the town, Charles Town,

* A Natural History of Nevis, &c. By the Rev. Mr. Smith; 8vo., 1475, p. 238.

situated on that shore, and having in its neighbourhood some low wet grounds and stagnant pools, fevers, which have proved very fatal, and most of all to Europeans, and the newly arrived, are not uncommon.

Considering its extent, the island is pretty well supplied with water in springs and streams; and this, probably, owing to the same cause as the frequency of showers, viz., the mountain acting as a refrigeratory, and condensing the aqueous vapor. The author already referred to, mentions three small streams or rivers besides the bath stream, one of which to windward, he states, is "stocked with the finest mullet and other good fish."

Like Montserrat, and in the same year, Nevis derived its first colony from St. Kitt's. This was in 1628. Its increase and prosperity were rapid and great, mainly owing to the wise administration of Mr. Lake, the immediate successor of Sir Thomas Warner, the leader of the first settlers to St. Christopher's. It is pleasant and refreshing to read the eulogy of this benefactor of Nevis, written by Bryan Edwards,—opportunities for eulogy so rarely occurring in the history of these islands. After mentioning how, under him, (Mr. Lake) "Nevis rose to opulence and importance," the historian of the West Indies continues. 'He was a wise man,' says Du Tetre, 'and feared the Lord.' Making this island the place of his residence, it flourished beyond expectation. It is said that about the year 1640, it possessed four thousand whites: so powerfully are mankind invited by the advantages of a mild and equitable system of government!" He adds; giving way to a generous feeling, and indulging a little in poetical description apt to accompany such feeling,—"Will the reader pardon me if I observe at the same time, that few situations in life could have afforded greater felicity than that of such a governor. Living amidst the beauties of an eternal spring, beneath a sky serene and unclouded, and in a spot inexpressibly beautiful, (for it is enlivened by a variety of the most enchanting prospects in the world, in the numerous islands which surround it,) but above all, happy in the reflection that he conciliated the differences, administered to the necessities, and augmented the comforts of thousands of his fellow creatures, all of whom looked up to him as their common father and protector! If there be joy on earth, it must have existed in the bosom of such a man, while he beheld the tribute of love, gratitude, and approbation towards him in every countenance, and whose heart at the same time told him that he deserved it."* Besides this good administration, circumstances of the island, doubtless contributed to this rapid prosperity; not only its soil and climate, but also the nature of the ground and even its size; the people concentrated and kept together in small space, with a mountain above for refuge from an invading enemy, and the sea at hand for convenience and commerce. Its

* History of West Indies, I. p. 436.

flourishing condition continued for many years, indeed
with little abatement and few reverses up to the
present time, and its white population increasing,—
but for a shorter period. In 1701 the white inhabitants were estimated at about 5,000,*—a number
probably exaggerated; in 1787, at 1,514; in 1805, at
1,300; in 1832, at 500; and now from 165 to 170.
The colored population has fluctuated less, not diminishing during slavery, and this, notwithstanding some
emigration,—tending clearly to show that the climate is at least very favorable to this race.   In
1761 the slaves were estimated in number, at between
10,000 and 12,000; in 1787, at 8,420; in 1805, at
8,000; in 1832, when emancipated, at 8,722; in
1844, (according to the census taken in that year,
including the whites) at 9,671; and at present at
not less then 10,000.† As a whole, the population
of Nevis may be considered, as it is reported officially, a well ordered and peaceable people, and if
not highly prosperous, at least nowise distressed,
and more than commonly united by community of
interests.   No military force is stationed in it, or is
needed, and it is even without a police.

The colored portion, the great majority, constituting
the labouring class, are very favorably spoken of by
the authorities, taking them as a whole, both as to
conduct and condition.   There being little or no crown
land, there is no temptation to squatting, and there is

* Account of European settlements in America.

† So reported in 1848, and as rapidly increasing.  Copies and Extracts,
&c., P. P., p. 239.

none. The old villages on the estates have been nearly abandoned, and dwellings of a better description have supplied their place, either near the old ones, or on detached spots. An inclination has been shewn to purchase small portions of land, but it has not been acted on to the same extent as in St. Kitt's. On working days, the labourers are described as slovenly in their dress and careless of it even to a fault, and in the other extreme, on Sundays and holidays, very much in accordance with the habits of the same class in the West Indies generally. Of a thrifty disposition, many of them belong to benefit societies; in 1845, there were as many as 1,812, when the number of destitute paupers did not exceed 140. They owe much to the exertions amongst them and for many years, of the clergy of the Church of England, and of the Wesleyan missionaries, especially the latter. More than half of the whole population belong to the Wesleyan missionary society; in 1845, the numbers returned as attending its chapel, were 4,000, whilst those attending the parish churches were no more than 1,250. Both clergy and missionaries of late years appear to have been zealous in teaching, and the latter have been tolerably successful in conducting their schools on the self-supporting plan. Whilst in the island, I visited two of these schools. Both of them were kept in the chapel of the society in Charles' Town,—a new building, with a handsome flight of steps ascending to it, and provided with a neat belfry, the whole, so little like a meeting house, that I supposed, at first, it was the parish church. One school was an infant school, consisting of seventy children ; all clean and neatly dressed, and of healthy appearance ; indeed I never saw children of more healthy or happier look. The mistress informed me that they were taught their letters and prepared for the other, the juvenile school. They left the room, which was under the chapel, but not underground,— a spacious, cool, well lighted and ventilated room,— marching to a song. As they passed, most of the boys touched their caps, and not a few shook hands with me, with a "How do you do, Sir?" said in an innocent, cheerful way. The juvenile school was kept in the chapel itself. The boys and girls belonging to it were taught either reading, writing, or arithmetic, and the girls the use of the needle. When I was present, the sewing class were at work. The girls forming it were of an age varying from ten to fourteen. The payment made by the parents differed according to what was taught, from half a bit, to a bit and a half a month, i. e. estimating the bit at five pence, from two pence half-penny to seven pence half-penny. It may be mentioned, as not a little creditable to the people, that this chapel, spacious and well-built, (built of the fine stone of the island) with its well contrived substructure-schoolroom was erected by funds raised by subscription, and chiefly, I believe, from the labouring class.

The agriculture of the island differs more from that of St. Kitt's in the manner in which it is conducted, than in the crops raised. Here, the cultivation of the cane is almost entirely conducted on the metairie or half and half system, and with more success and greater satisfaction than in any other of these colonies in which the same plan has been attempted. The orderly habits of the people, and their being so much under control have probably led to its adoption, as well as, and perhaps even more, the quality of the soil and the nature of the ground, often rocky and intermixed with stones, and likewise the length of time the canes can be continued, viz. as ratoons, from ten to twenty years.* The immediate result, no doubt, is creditable as well as profitable both to the landed proprietors and to the labourers. But whether the system, if persevered in, will be ultimately beneficial in its tendency, is perhaps somewhat doubtful. It seems to be rather a conservative measure than one adapted for improvement,—one of submission to adverse circumstances rather than of enterprising struggle to overcome them. In accordance, the culture of the cane here has been carried on in the old way, the hoe and the hand being the chief implement and power employed. Where canes can be kept ratooning so long, making a

* One specimen of soil which I examined, from an estate remarkable for bearing ratoons for 18 or 20 years, was a light gravelly and stony one, of a greyish brown color, composed of small pieces of glassy felspar or of felspar and hornblende, and of a fine powder or dust, which under the microscope had the character of volcanic dust. It contained no clay. Acted on by an acid, it yielded a little alumine, lime and magnesia. The good quality of this soil is probably owing to a slow decomposition in constant progress,—the fragments yielding a never failing supply of those inorganic elements essential to fertility.

tolerable return, the amount of labour is small, limited in great measure according to the practice here followed to one hoeing after harvest, the laying on of trash, and the giving of some manure before the commencement of vegetation,\* and in case of decay stumping out the old and the replacing it with a young plant.

The land under culture now, probably does not differ much as to extent, from what it was during the period of slavery, and the produce does not appear to be much diminished. Bryan Edwards estimated the produce at a hogshead an acre, the average yearly amount at 4,000 hhds. In 1780, the total was 4,000 hhds of 13 cwt.; in 1799, 3,850 hhds.; in 1805, 2,400 hhds. Recently, for instance, in 1844, the total produce exported, was 2532 hhds. of 15 or 16 cwt.; in 1848, a very seasonable year, 3,734 hhds.; in 1849, 1,814 hhds.

The sugar is made in the boiling houses of the estates, and at the cost of the proprietors. Even in this, the manufacturing branch, I did not hear, nor have I since learnt of any improvements having been introduced, excepting the attempt to substitute steam

---

\* Guano, in some instances has been used; and by one gentleman, a very intelligent planter, sheep manure in a dry state, imported from the Spanish Main. A specimen of it which he sent me, was found to contain, besides vegetable matter, a notable portion of phosphate of lime, with some carbonate of lime and magnesia, potash and ammonia. That it is likely to prove a very useful manure, if it can be obtained at a moderate price, can hardly be doubted. It was rich in soluble matter, seeming to indicate that it had been collected in a dry place and never exposed to rain.

power for cattle, or wind at the mill; and which
was hardly successful, for of the four steam engines
erected, I was assured, one only remained in use,
the other three being out of repair, and no artificer
in the island possessed of the requisite skill to make
them efficient.

The revenue of Nevis, derived chiefly from indirect taxation, is more than equal to its expenditure, and the local treasury is free from debt.
Moreover as denoting welfare, the value of the exports
greatly exceeds that of the imports; the average of
the two, reported in 1851, being as £23,470 to
£16,478.

The government, formed as it is to represent the
three estates, with all the complication of details
belonging to them, though better conducted than
that of Montserrat, appears to be open to nearly
the same objections. One favorable circumstance is
mentioned, that the freeholders (there are only about
158 who pay any direct taxes) are taking an increased
interest in public affairs. There was a time, now
almost forgotten, when the administrative power here
was of more weight and importance,—when the
governor of Nevis, in case of the death or absence
of the governor general, took his place and performed
his functions, and in case of any general meetings,
such as from time to time were held, of the councils
and assemblies of the several islands, for the purpose
of considering and discussing their interests, he took
a prominent part, acting as president: from which

circumstance of precedence, we are told that Nevis was looked up to as a Mother state.*

To invalids or others voyaging to the West Indies, in quest of health or pleasure, Nevis may prove an agreeable resting place. The landing is commonly easy and safe on the sheltered shore close to the town. Better accommodations than are usually to be met with in these islands are to be had either in the town or at the bath establishment. To those who may wish to make trial of the waters, the latter of course will be preferable. To others, whose object is merely to see the island, the former may be more convenient. I remember with pleasure how well I was lodged and treated in a small house in the town, kept by a decent widow something between a lodging house and a village inn, with all the neatness and cleanliness of such 'a public' in the dales of Westmoreland or Cumberland, and with articles of furniture and decoration,—rude colored prints, porcelain or clay figures—very similar.

Small as Nevis is, it affords good scope for pleasant rides and some agreeable excursions. The roads, indicative of an old colony, are little inferior in goodness to those of St. Kitt's, and are most of them, very passable for carriages. They are kept in repair by a labour rate, and by a tax on horses and conveyances. Every labourer has to contribute four days labour yearly, and every one above that

* So says the Rev. Mr. Smith, whose history of Nevis is dedicated to the worthy gentlemen of Nevis, mother of the English leeward Charibee Islands in America.

class double that amount, or a money-equivalent. An inspector has the superintendence of the whole. He was, when I was there, a person of color, with a salary of £120 a year. The plan was said to answer well; it was made as easy as possible to the labourers, by allowing them to perform their task on the roads nearest their dwellings.

In the immediate neighbourhood of the town, the stranger will find pleasant walks, one especially, in the direction of the neat parish church, where the road is well shaded with trees,—a favorite promenade and ride; another, along the beach, where it is planted with cocoa-nut trees, forming an extensive grove, once the ornament of the shore, as well as a source of considerable profit to the proprietors, but at the time of my visit, having a sad appearance from the effects of blight.

If intent on excursions, his first should be round the island. Riding or driving, this may be effected in three or four hours with ease. The road passes, the greater part of the way at a moderate distance from the shore, and is almost level, except where it crosses the ridges north and south connecting the terminal hills with the mountain mass, where in places, especially the southern, it is somewhat steep. In this little journey, an opportunity is afforded of making a general survey of the island, of seeing the five parishes into which it is divided,*

---

* The parishes are St. Paul's, Charles' Town; St. Thomas', Lowland; St. James', Windward; St. George's, Gingerland; St. John's, Figtree. Each sends three members to the House of Assembly.

3 R

and of forming some idea of its cultivation, both as to extent and kind, and of the dwellings and aspect of the people, following their ordinary occupations in their working attire. In making the circuit, I had the advantage of the company of a gentleman, a native, adding much to the pleasure. We stopped only twice, and saw two very different characters, and in very different situations. One was a proprietor in reduced circumstances, dressed in flannel, superintending the making of sugar in his own boiling house,—a young man of gentlemanly manners, who apologized on the ground of dire necessity, for the degrading situation, as he seemed to consider it, in which we had found him. The other was a gentleman past the middle age, also residing on his property, whom we found in a well stored library, somewhat in confusion, as indeed were all things about him,—as if appearances here were of no importance, and might be altogether neglected,—a man of cheerful countenance, free and genial conversation, with the reputation of being a scholar, philosopher, and man of science, and I doubt not, deservedly; as well as a good neighbour; and holding a public office, —an able administrator.

Of the many pleasant rides that may be taken, I shall notice only two. A very agreeable evening one is towards the northern part of the island, and on the higher grounds, from whence there are fine views, both of the mountain and its many ramifications, and of the hilly well cultivated country below its upper wooded region, and also of the near and distant hills of St. Kitt's with the beautiful blue intervening sea, in one direction like an inland lake, realising almost the paradisiacal picture drawn by Bryan Edwards. Another hardly less pleasant, and to some it will be more interesting, is in the opposite direction, to an estate called the Morning Star, so named from a battle fought at dawn on that spot, between a party of militia and an invading French force,—the former conquering. The plantation house of this estate, about three miles from the town, standing high, about 500 feet above the level of the sea, has the advantage of cool air and of an extensive prospect, especially seaward over the boundless ocean. Fine trees shade its grounds, some lofty Cabbage Palms in particular, and a many trunked Banyan, which, though planted only about 60 years, forms a little grove by itself. An estate joins this, similarly situated, on which is the house still standing unaltered, where the hero of the Nile and Trafalgar passed, probably, some of his happiest hours, certainly his most peaceable and domestic ones, after his marriage with Mrs. Nisbet, a native of Nevis. The house belonged to the lady's uncle, and by its situation was well fitted for an invalid, such as Nelson was when he retired here for a while from his active duties, in quest of health. It was pointed out with manifest pride by my companion.

# CHAPTER XVI.

## DOMINICA.

General character.—Geological structure.—Conjectures respecting metalliferous rocks.—Soils.—Climate.—Brief historical notice.—Population.—Condition of the peasantry.—State of education.—Ladies aiding in the cause.—Condition of the scanty white inhabitants.—Produce.—Cultivation.—Fruits and cotton.—Remarks on these.—Suggestions to the passing traveller, &c.

DOMINICA, situated between the two French islands, Martinique and Guadaloupe, about 25 miles from each, is 29 miles in length and 16 in breadth, as commonly estimated, comprising within its area 186,436 acres.

In its general character it more resembles St. Lucia, St. Vincent and Grenada, than any of the leeward group to which it belongs. These even it surpasses in boldness and ruggedness of form, in the height of its mountains,* the depth of its vallies, the abruptness of its shores, and I may add, though with some hesitation, writing chiefly from the report of others, in the wild and picturesque beauty of its scenery.

Though hitherto little explored, there appears to be no doubt, that this island is principally volcanic and

---

* The following are the assigned heights of some of the mountains and hills, Morne Diablotine, 5,314 feet; Laroche, 4,150; Coulisboune, 3,379; Outer Cabrite, (Prince Rupert's,) 542; Inner Cabrite, 430; Morne Bruce, 465; Hospital, (where a rain gauge is kept) 440; Scotshead, 231.

composed, like those others which it resembles, of volcanic products, either elevated or ejected, such as crystalline igneous rocks, the result of fusion, or of tufas
formed of ashes and scoriæ. All the specimens which
I have examined and all the information I have been
able to collect accord with this view of its structure, as
do also its many soufrieres and hot springs ; the former
probably craters of eruption.*

As in St Kitt's there is one instance, that of Brimstone hill, of an associated aqueous and igneous formation, so too there is one in this island. It occurs
where it is most easily observed, viz. above the town of
Roseau, by the side of the road, where a cutting has
been made, ascending to the garrison of Morne Bruce,
about half-way up, and it may be about 300 feet above
the level of the sea. There three several beds are
seen, all conforming and nearly horizontal, the lowest
consisting of coral and shell limestone, like that of
Barbados ; the next of water-worn stones like those of
the sea shore ; the upper of tufa, formed of volcanic
ashes. A like formation I was told occurs in several
places between the town and a promontory to the
southward, the extreme point of the island in that
direction, called Scott's Head.

Of the crystalline rocks, of those at least of which
I procured specimens, the majority were of a light color,

---

* A small lake, probably the crater of a volcano, similar to the Grand
Etang of Grenada, is situate at a considerable elevation amongst the
mountains, about six miles distant from the town ; it is the only one in
the island.

and of a fine grain, composed chiefly of glassy felspar, and containing crystals of augite. Several of them appeared to be well adapted for building stones, and we are told by Attwood in his history of the island, that when in possession of the French, quarries were opened in the Savannah Grande, and that the stone procured was sent in large quantities to Guadaloupe, where some of the principal churches were built of it. He calls it a freestone and describes it as excellent; it only deserves the name of freestone from the ease with which it is worked. Specimens from the Savannah which were sent to me, I found of the same quality as those already mentioned.

It has been asserted that gold and silver mines exist in the island, and that the latter have even been very productive, but on what ground or authority I am ignorant. What I am going to relate may seem in favor of the statement. Amongst the specimens sent me from thence, three were of a very promising kind; one containing silver, another carbonate of copper, malachite; and the third sulphuret of molybdenum. The first was a kind of serpentine, the silver dissemi- nated through it in grains and filaments; the one of copper was in quartz rock; the sulphuret of molyb- denum, a fine crystalline specimen, consisted of little else, there being included in its mass only a very small portion of iron pyrites and of quartz.

Were there certainty that these specimens had been detached from the native rock and not found loose on the surface, the question would be determined as to the presence of metalliferous rocks. But this is not the case; I do not know with any degree of certainty, either where, or how, or by whom, they were found. If found at the surface, lying loose, they may have been foreign, brought from the continent at a distant period by the Caribs. In these islands it is necessary to be always on one's guard in relation to this circumstance. I remember in Barbados having a specimen of orpiment given me as a native mineral, it having been picked up in a wild spot, where, it was thought, it could hardly have been brought, yet where, I have no doubt, it had been dropped or left, judging from its form and still more from the nature of the rocks adjoining. To resolve the question, which the reputed mines and the specimens seemingly somewhat corroborative give rise to, in a satisfactory manner, careful enquiry made by a competent person, seems to be necessary; and almost any honest observant person tolerably instructed might undertake it. The Soufrieres should be the first places examined. At one of them, or in its neighbourhood, it was said that the specimen of copper ore was found; and Pere Labat states that he had heard of a gold mine near the same spot, but could learn nothing precise about it.

Dominica is also reported to abound in mineral springs. Neither respecting them have I been able to obtain any satisfactory information. Two samples were sent me of mineral waters as they were called. Both were slightly saline and in each the principal ingredient was common salt.

Of the soils of Dominica we possess little exact knowledge: they are spoken of as good, and considering their origin, whether formed principally of volcanic ashes, or of ingredients derived from the decomposition of igneous rock, it may safely be admitted that they are not inferior in fertility to those of the other volcanic islands; and the luxuriancy of vegetation, its universal spread, the widely extended forests even to the mountain heights, and from them to the margin of the sea, afford confirmatory proof.

The climate of this island generally appears to differ but little from that of the adjoining islands similarly constituted, such as St. Lucia and St. Vincent. From the greater elevation of its mountains and greater depth of its vallies it probably may be subject to a wider range of temperature, and to local differences in regard to the sensible qualities of its atmosphere. That it would be abundantly supplied with rain, might be expected from its mountainous character; and undoubtedly it is. The many considerable streams, (they are called rivers) which it has,—they are said to be thirty in number besides rivulets,—may be mentioned in proof; there is hardly a valley without one. More frequently the country suffers from excess than deficiency: landslips in connexion with heavy rains are not uncommon. The following table shews the amount of rain that has fallen at Morne Bruce, as measured by the rain gauge for a period of six years.

| | 1846. | 1847. | 1848. | 1849. | 1850. | 1851. | 1852. |
|---|---|---|---|---|---|---|---|
| January ... | | 10·30 | 4·10 | 2·20 | 2·40 | 6·86 | 5·35 |
| February ... | | 2·50 | 5·20 | 2.00 | 1·66 | 6·44 | 4·24 |
| March ...... | | 5·50 | 1·50 | 1·00 | ·91 | 2·00 | 2·23 |
| April ........ | | 3·90 | 3·75 | 3.80 | 2·45 | 0·00 | 1·28 |
| May ........ | | 1·90 | 5·40 | 4·60 | 4·77 | 7·46 | 10·09 |
| June ........ | | 3·50 | 1·80 | 7·30 | 10·62 | 15·23 | 9·95 |
| July ........ | 7·80 | 6.50 | 4·40 | 12·50 | 17·73 | 17·34 | 9·42 |
| August ...... | 14·70 | 14·10 | 11·60 | 7·10 | 11·99 | 11·50 | 13.91 |
| September... | 13·16 | 9·40 | 6·90 | 6·60 | 6·36 | 10·31 | 3·40 |
| October ... | 5·60 | 6.90 | 7·60 | 13·00 | 2·23 | 6·25 | 4·64 |
| November... | 5·70 | 17·00 | 2·50 | 1·90 | 13·84 | 6·74 | 6·08 |
| December... | 11·40 | 9·50 | 11·20 | 7·46 | 10·65 | 6·51 | 18·42 |
| | | 91·00 | 65·95 | 69·46 | 85·61 | 96·64 | 89·01 |

In relation to salubrity, some parts of it, especially those to windward, are in good repute ; others, particularly Prince Rupert's Bay and shores, are of the worst character ; indeed that situation has proved so fatal to Europeans, when the works there were garrisoned by British troops, that for many years, as a measure of necessity, none but black troops have been sent there, on whom the malaria has had no effect. Morne Bruce, now the sole station of our white troops, is commonly healthy, but not exempt from the occasional invasion of yellow fever. At one time dysentery was a prevalent disease in the garrison and a destructive one, owing, it may be inferred, to the impure water used by the men : it has ceased to appear since good water has been supplied.*

* It is worthy of remark, that in the adjoining island of Guadaloupe, the disease has been attributed to the same cause. M. Cornuel, in a paper on the subject, points out that whilst dysentery was prevalent in Basseterre, where turbid water from the hill was used, Grandeterre, separated only by a river, where cistern water was used, was free from it.

A land of mountain, river and forest, this island was long the favorite haunt of the Caribs. The peculiarities which recommended it to them and attached them to it, rather repelled Europeans. When Pere Labat wrote in the beginning of the last century, they, the former, were in entire and undisputed possession. This entertaining author more than once crossed over from Martinique to visit them ; and in his pages are to be found interesting details of their manners and habits illustrating their then condition, and I may add, resemblance to the most warlike inhabitants, the aborigines, of the South Sea Islands. The historian of Nevis, writing about half a century later, makes mention of the French having about that time attempted a settlement there.* These, the first planters, were not allowed to remain long undisturbed ; in 1759, the island was invaded by the English and its conquest, as it was called, effected. Confirmed in possession of it, by the treaty of Paris, concluded in 1763, they retained it till 1778 ; an interval this of much prosperity, numerous allotments of land having then been sold, and estates formed, and sugar plantations commenced,† Rouseau, moreover, having been made a free port in 1766, and having become a great mart of

* Speaking of the Caribs, as then almost extinct, he proceeds, they are "confined to the sorry island of Dominica : nay, I lately heard from a surgeon aboard of a ship of Sir Chaloner Ogle's Squadron, who touched there, that the French had lately made a settlement in Dominica."

† No less than 96,344 acres, about one half of the island, were then sold by the government in allotments from fifty to one hundred acres, yielding the sum of £312,092 : no one was allowed to purchase more than 300 acres. Bryan Edwards, I. p. 408.

commerce, especially in slaves: in that year, [1778] it was retaken by the French and not restored till 1783. Since it has been uninterruptedly a British possession, and in relation to form of government, similar to that of the elder British colonies. But probably from the comparatively few resident English proprietors or settlers from England, its free institutions all along have been conducted with little vigor and have had little effect. In one of the latest reports respecting the island, it is stated that the House of Assembly met only once a week during its session, and that in consequence its proceedings were slow and dilatory.

From the earliest accounts we have of the island to the present time, its population has been very limited. When Pere Labat visited it in 1700, the natives, he states, did not exceed 2000, two-thirds of whom were women and children. In 1788 it would appear from the government returns that the amount of the population of all classes was 16,648, of whom 1,236 were whites, 445 free negroes, and 14,967 slaves. Seventeen years later we have another estimate; then the number of whites were 1594; the free people of color 2,822; the slaves 22,083, making a total of 26,499. Still later, viz. at the time of emancipation, in 1832, the number was reduced to 24,123, of whom 791 only were whites, as many as 4,077 were free negroes; and 19,255 were slaves. The next census was that of 1844, in which the races were not distinguished: the total then returned were 22,469. Of these 714 belonged to the Church of England, 19,046 were Roman Catholics; 2,531 were Wesleyan Methodists; 5 were Moravians; exclusive of 179 unknown as to their religious tenets. Since then it is believed that the colored population has considerably increased, from the births exceeding the deaths.*

Of the two principal classes, the proprietors and labouring, the white and the colored, the condition, from all that we can collect, is very different.

The reports officially made on the latter both by the stipendiary magistrate, and the officer administering the government for several years past, are favorable, hardly less so than in the instances of St. Kitt's and Antigua. Thus, we are informed, that "the condition of the peasantry in general is comfortable and prosperous, as shown in the frequent occurrence of weddings, which are usually conducted in an expensive manner; by their decent appearance, domestic arrangements, habits, and modes of living, and especially in the greater care and attention which they evince towards their children." From another magistrate we learn that " the condition of the peasantry is improving. The profitable result of their labour places at command, to an extensive degree, the comforts of life; little effort is requisite to obtain necessaries; the unlimited occupation of land, a most generous soil, and the favorable

* The distribution of the population according to Districts was as follows; District A, comprising the parishes,— St. David's, St. Patrick's, St. Mark's, St. Luke's, St. George's, St. Paul's, containing 14,964 souls;—District B, comprising the parishes,—St. Joseph's, St. Peter's, 3,166;—District C, comprising St. John's, St. Andrew's, 4,339.

season should make them happy, and generally they appear to be so; any complaint of poverty may be received as a fiction." By a third magistrate we are assured that "the condition of the peasantry is in every respect comfortable; they have more of ground-provisions and fruit than they can consume, and have ready for sale the remainder, and from their well known habits of economy, it is supposed they save all their money wages." The Lieutenant Governor states,—"In reference to the principles of religion, conduct, and peaceful demeanour of the peasantry, I feel happy in expressing myself favorably." These statements relate to them in 1845, when tradesmen were earning from 1s. 2d. to 1s. 9d. a day, and field labourers from $7\frac{1}{4}d.$ to 9d. a day,—the latter about the present average, and they might make by task and job work, which was mutually preferred, from $10\frac{3}{4}d.$ to 1s. 6d. a day, independent of house and ground allowed them free of rent; when there were only 254 paupers out of the whole population, and only 183 convicted during the year for any offences, most of them of a trivial kind. The later reports mostly accord with the preceding: the Lieutenant Governor writing in 1849, concludes his dispatch—noticing "the quiet and very orderly conduct of the people throughout the island;" remarking further, that, "although like most of the West India Colonies, the island has partaken of the general depression consequent on the low price realized on the sale of the staple production of the soil, yet I may add, that industry has never been carried to a greater extent than at present, nor has the population been more contented and happy." This state of contentment and well doing is the more remarkable as there is no labour contract, and as from the nature of the country and the extent of crown land unoccupied, there is a temptation to squatting; a temptation indeed not always resisted, yet not given way to, it would appear, to any great extent. A partial outbreak which occurred amongst the labourers, accompanied by some acts of violence against property in 1844, requiring the calling out of the militia, as it was supposed, for its suppression, may seem to indicate a kind of feeling contrary to that for which credit has been given: but duly considered as to its origin, (a misconception of the census act,) and how easily order was restored and maintained, and goodwill revived, the event may be adduced rather in confirmation.

As regards education, the peasantry here seem to be pretty much on a par with their fellows in the other islands; not neglected and yet but little advanced. In 1845 there were 1,224 children attending day schools, and 764 attending Sunday schools. At this time and till 1851, the grant by the local government in aid of education, was only £300 annually. In the latter year, it was raised to £800 sterling, to be equally apportioned to the several schools, whether Church of England, Roman Catholic, or Wesleyan missionary. It is pleasing to see recorded that ladies take an active part in advancing the education of the people: thus in one report it is stated that "within the last year, (1848) a few ladies and Sisters of Charity arrived in the island, and have under the vicar-general, established a school for female children, which promises to be of great utility;" and in another, earlier, for 1845, that, "instruction is given in four localities to thirty or forty children by females of the Roman Catholic persuasion." Would that examples of the same kind were more common in these colonies and afforded also by protestant ladies! A rivalship such as this would be doubly beneficial, and not least so to those making the exertion, in a country and state of society where there is so little out of their own families to engage worthily the attention and excite the interest of women; and where in consequence, they, especially the unmarried, too often become the prey of ennui, and fall into an idle and frivolous life,—that low life, but little removed from mere animal or brute existence.

In this island, as in Nevis, and from the same cause, the metairie system of cultivation being a good deal followed, no great disposition has been shown to purchase land and establish freeholds and free villages. In 1845 the number of persons livings in villages built since emancipation was 342; the number paying direct taxes only 133. Particular instances however, even then, are given denoting some desire on the part of the peasantry to possess land and houses in their own right; one is mentioned of twenty lots of land having been sold at £10 3s. 4d. an acre; another of a respectable rural constable purchasing five lots.

Friendly societies here appear to have had little encouragement, I can find mention made of two only, and these on a small scale, no more than one hundred and twenty-three persons being connected with them.

Of the other class,—the white inhabitants and the planters, I can give but little information, having, though twice in the island, but only for a few hours, had no opportunity to become acquainted with them. Of the proprietors of estates, most of the English, are I believe, absentees,* trusting the management of their property and its cultivation to attorneys and persons acting under them, and most of the French if resident, in straitened and reduced circumstances. Neither in the official reports on the state of the island, or in conversation with those connected with it, have I been able to find any proof of advancing intelligent exertion similar to that witnessed in Barbados, Antigua, and St. Kitt's. No attempt that I have heard of has been made to form an agricultural society, or to establish a library, or, in brief, to accomplish anything tending to promote the advancement of knowledge, the acquisition of science, without which, how is it possible at the present time that any people can be successful, even as regards the lowest, their worldly and material interests! I may perhaps, in making these statements, appear to those acquainted with the island, to express myself too strongly, inasmuch as an effort has been made within

---

* In a recent communication from thence, it is stated that the number of resident proprietors, (worthy of the name) does not exceed two.

the last few years to institute in the town of Rouseau, a higher school than the ordinary ones, an academy or grammar school, intended for the education of the children of the upper class; but even this, though aided by a proportionally large annual grant of money from the island treasury, does not appear to have been successful, and has been considered by those who ought to be competent judges, the Lieut. Governor and Governor General as ill-timed and injudicious.

The staples of the island were coffee and sugar. The former chiefly cultivated by the French, by whom it had been first introduced from Martinique, the latter more by the English. At one time, when the island was most flourishing, when the sugar estates were not more than fifty, and their produce one year with another might average 3,000 hogsheads, the number of coffee plantations was about two hundred, and the quantity of coffee exported annually to Great Britain, was from four to five millions of pounds.*

Now, from various circumstances combined, such as have had effect in British Guiana, already mentioned, and others peculiar to Dominica, as the hurricane of 1834, and since then the devastations committed by the blighting white fly, the decline of these beautiful plantations has been rapid and great. Very many are

---

* The *History of the Island of Dominica*, by Thomas Attwood; London, 1791. About this time coffee sold in England for from £4 15s. to £5 5s. per cwt. The rapid spread of this plant in the West Indies promoted by high prices, is almost as marvellous as that of the cotton plant in America, we are told by Pere Labat, a good authority on the subject, that it was first cultivated in Martinique, in 1724, and by means of plants brought from Paris from the Royal Garden.

abandoned, the labourers, we are told becoming virtually the proprietors ; some are let at a very low rent, more on the metairie system ; one only in 1845 is reported as cultivated by hired labour. In that year the crop had fallen off to 53,610 pounds, and before it had even been less, as it has been since. In a late report a prospect has been held out of some improvement, which it is to be hoped, will be confirmed, as both the soil, climate, and nature of the ground seem peculiarly fitted to this plant. The time of course must come, if the sound system of free trade be continued and carried to its utmost pitch, when the crops grown in every country will be those most congenial, those which can be grown with least risk of failure, with largest profit, on the natural, not on the forcing plan.

The sugar estates have suffered less, more attention and more labour having been bestowed on them ; and consequently though the quantity of sugar has diminished it has been in a less degree ; indeed it would appear that some of the estates are as productive now as they ever were in the period of slavery. From the manner in which the reports are made in this island, and the returns accompanying them being half-yearly, the whole of unusual brevity and paucity of information, it is difficult to arrive at exact results, and to compare in a satisfactory manner the amount of produce of one year or one period with that of another.

As to the manner in which the agriculture of the

island is conducted there is little that has come to my knowledge requiring remark. It has already been mentioned that the metairie plan is most followed ; it is conducted much in the same way as in Nevis, and with similar results, —the labourers contented and well remunerated, the proprietors deriving little profit and for the most part discontented. And here, likewise, as in Nevis, few efforts appear to have been made to introduce a superior culture, and implemental, in place of manual husbandry. Guano, that facile and expensive manure has, it is said, been pretty largely used; but this, unless skilfully applied and on principle, can hardly be considered an exception in relation to agriculture; no more than some reputed improvements in the course of being made in the machinery of the mills, of foreign invention, in relation to the manufacturing processes. On most of the estates, the planters have the very great advantage of water power, and also that of water carriage, the greater number of them being near, or on the coast; I believe there is not a single steam engine employed, and only one windmill.

Besides sugar and coffee, other articles in smaller quantities are produced and exported from hence. Of these, cocoa, ground-provisions, arrowroot, and fruits, are the principal. For these minor and mixed objects of culture, the sheltered and well watered vallies are peculiarly favorable, especially for tropical fruits. Even now, thanks to the early French settlers, Dominica is more productive of fruit, than any other of these islands, with the exception of Grenada. Were this

cultivation extended and attention paid to the propagation of the more valuable kinds and their improvement,
such as the pineapple, the mango, the orange, especially the shaddock, that most wholesome of fruits, as
well as most grateful, great might be the profit ; as
these are fruits which will all bear exportation, and by
means of steam-packet communication, might with
ease and little cost be brought to England, where, (if
gathered at a proper time and well preserved, (they
could hardly fail to be in request, and to sell at a
price amply remunerative.

Cotton, we are told by Pere Labat, was in his time
abundant amongst the Caribs, leading to the inference,
that it was a native plant of the island. An attempt
to resume the culture has lately been made, but under
what circumstances is not mentioned. Should the trial be
continued, as it is to be hoped it will be, and should it
ultimately succeed, that is prove a profitable crop, capable of being raised on land least fitted for the cane or
the coffee shrub, it may be an important acquisition ;
and such result may encourage the renewal of its
culture, with like selection of ground in the other
islands, from which it has been driven by the hitherto
more lucrative sugar cane.

From my very limited knowledge of this island,
I cannot presume to point out the parts most interesting, or to attempt any description of those best worthy
of being visited by the passing traveller. I may
briefly remark, that even in passing, should it be
by day, as the steamer commonly runs along shore,

he will have a tolerable opportunity of seeing the general character of the country,—its lofty mountains rising amongst clouds,—its charming cultivated vallies descending and opening out towards the sea,—its steep upland slopes partially cultivated, more frequently wooded, forming a succession of landscapes, of more than common impressiveness and beauty.

The detention of the packet here is short, commonly hardly allowing of time to land and to ascend to Morne Bruce. If the stay permitted of this, he, the traveller, would be amply repaid by the ascent ; for I do not know any spot in the West Indies from which there are finer views of mountain scenery, this fortified hill being backed by others of great boldness, and skirting a valley of peculiar wildness and beauty as well as depth, with the additional charm of a river, often a torrent—as after rain,—coursing through it. Should it be his intention to remain some time in the island to explore it—a task much needed, if undertaken by a competent person,—it would be prudent to be provided with one or more letters of introduction to influential residents, from whom information might be obtained as to the best routes and best modes of proceeding, and aid in the finding of quarters, a great difficulty, one almost insurmountable, without his throwing himself on the hospitality of the planters.

# CHAPTER XVII.

## WEST INDIAN TOWNS

Their sites, how commonly objectionable.—Some of their defects and peculiarities.—Character of adjoining country.—Population.—Its character.—Occupations in town and country.—Public buildings.—Churches and burying
grounds.—Prisons.—Hospitals.—Codrington college.—Town habits.—Modes of
conducting business.—Brief notice of military stations.

RELATIVE to the towns in the West Indies, I shall
offer but a few remarks; and, I regret to think that
these can seldom be laudatory,—no just principle
having been acted on either in the selection of their
sites, or in their construction,—expediency, or immediate profit in the way of business, having been
in all respects more considered than the sanitary
circumstances, every where so important, and more
especially so in a tropical climate.

Most of the towns, with the objects just mentioned
in view, are situated on the leeward coast, close to
the sea, and mostly in low situations equally unfavorably for ventilation and drainage, for coolness consequently, and the absence of malaria or noxious
effluvia. Not one of them that I am acquainted
with, is provided with sewers, or is efficiently drained,

or is well supplied with water,*—great and fatal omissions in regard to the health, comfort, and welfare of their inhabitants.

In their general appearance and aspect, with one or two exceptions, there is little that is attractive; —even the best of them have few pretensions to beauty; they are mostly inferior in this respect to second rate country towns in England. Being commonly built of wood, and the houses often provided with galleries and verandahs, they are somewhat oriental in appearance, and the worst of them and the worst parts of them, often reminded me of a Turkish town or village, exhibiting not only a similarity in the external forms of the buildings, but also in the neglect of neatness, order, and cleanliness in the streets and purlieus.

Though in the selection of their sites, there is no reason to suppose that beauty of country in the immediate neighbourhood, has had any consideration, yet many of them are not without this advantage, and some possess it in an eminent degree, especially Kingstown, St. Vincent; George Town, Grenada; Castries, St. Lucia; Rouseau, Dominica; James Town, Nevis; and most of all Port of Spain, Trinidad. Even those towns which are not so formed, such as St. John's, Antigua; Basseterre, St. Kitt's; Bridgetown, Barbados; are not altogether without the charm

* Probably by this time Port of Spain is an exception. Preparations for some years have been making for bringing water into the town; when I was last there,—that was in 1848,—the iron pipes were provided.

of surrounding scenery, that which depends on gentle outlines and high cultivation,—beauty, which by many, whose tastes are not formed for the picturesque, may be preferred to the bolder and wilder features of landscape which belong to those first mentioned. This remark applies in some measure even to George Town, Demerara; and to New Amsterdam, Berbice; towns which are peculiar; they may not inaptly be designated garden-towns, almost every house being detached in a garden, "with flowering shrubs and fruit and ornamental trees," forming wide streets regularly laid out and planted on the banks of canals, and bounded in one or more directions by a great expanse of cane fields, with the character and charm of garden cultivation.*

None of them, it is worthy of remark, and it may help to explain some of their peculiarities and deficiencies, are, though burghs, corporate towns or have any special privileges.

The following table, showing the population of the several towns, will give a pretty accurate idea of their respective sizes. It is formed from the census of 1844,—but is applicable, it is believed, with few exceptions to the present time, the exceptions being chiefly those which have owed an increase of their

---

* This description does not apply to the trading portion of George Town, its water street, where the houses are closely compacted and with the same neglect of sanitary measures as is witnessed in our sea port towns at home,—and with like bad consequences. The inhabitants of this street are those of the town population who have suffered first and most from yellow fever.

inhabitants to immigration, such as Port of Spain, Trinidad, and George Town, British Guiana. The total population of each colony, from the same census, is given for the purpose of comparison.

| | MALES | FEMALES | TOTAL | TOTAL OF COLONY. |
|---|---|---|---|---|
| Barbados, Bridgetown* | 7,846 ... | 11,516 ... | 19,362 ... | 122,198 |
| Grenada, Town of St. George .. | 1,921 ... | 2,476 ... | 4,397 ... | 28,923 |
| St. Vincent, Kingstown | 1,903 ... | 2,866 ... | 4,769 ... | 27,248 |
| St. Lucia, Castries† | ... | ... | 4,000 ... | 21,C01 |
| Tobago, Scarborough | 605 ... | 869 ... | 1,474 ... | 13,208 |
| Trinidad, Port of Spain | 6,656 ... | 8,953 ... | 15,609 ... | 59,815 |
| British Guiana, George Town ... | 8,483 ... | 10,103 ... | 18,586 ⎱ | 98,133 |
| ——————— New Amsterdam | 1,610 ... | 1,850 ... | 3,460 ⎰ | |
| Antigua, St. John's | 3,744 ... | 5,277 ... | 9,021 ... | 36,178 |
| Dominica, Rouseau | 1,545 ... | 2,336 ... | 3,881 ... | 22,469 |
| Montserrat, Plymouth | 447 ... | 682 ... | 1,129 ... | 7,365 |
| Nevis, Charlestown | 755 ... | 1,051 ... | 1,806 ... | 9,571 |
| St. Kitt's, Basseterre | 1,908 ... | 2,785 ... | 4,693 ... | 23,177 |
| | 37,423 | 50,764 | 92,187 | 469,286 |

From this table we see how large is the population of the towns compared with that of the country in the several colonies, constituting 19·6 per cent. of the whole; and, also how large is the female portion compared with the male, the former being in excess to the amount of 23,341;—results little to have been expected in countries such as these, most of them under-peopled, mainly agricultural, and suffering, it is imagined, from want of labourers. They seem to denote, and I fear too truly, that the condition of the people—of the town resident popula-

* According to the census of 1851, the population of Bridgetown was 20,026; of the Town of St. George, Grenada, 4,567; of Kingstown, St. Vincent, 4,983; of Port of Spain, Trinidad, 17,563; of George Town, Demerara, 25,508.

† From Brees' *History of the Colony*; sexes not given.

3 U

tion—is unsound, idle, and licentious in no small degree. Details might be given in confirmation. We are informed by the governor of Trinidad, that in Port of Spain, 8,000 of the inhabitants are without occupation, without apparent means of earning a livelihood.* The like fact is reported of Barbados; according to the last census, of the total population of Bridgetown (20,026) 11,871 are without fixed employment.†

The town population is of course very miscellaneous, and in its composition very like that of

---

* Lord Harris in a despatch of the 18 May, 1852, states, "In the return attached to the Blue Book, page 152 a, the large number of 26,987 persons are placed under the head of 'no employment.' This is, I am sorry to say, the fact, and is a strong evidence of the habits of a large portion of the population, for if all the children under ten years of age be deducted from the amount, which would not be strictly correct as some of them are employed, there remains a surplus of more than 10,000 persons, out of a population of less than 70,000, having no employment, and of these 8,000 turn out to be inhabitants of Port of Spain." He adds, "I think it necessary, particularly, to call attention to this fact, because it must be remembered that in a community such as this, there are no idlers among the better class, so that a seventh of the whole population of the colony, nearly a fourth of the adult population and more than half of the total population of the chief town, are composed chiefly of persons in the lower ranks of life, and having no visible means of gaining an existence." Reports P. P., 1852, p. 164.

† Of the total population, 73,098 are returned as having "no fixed employment." Mr. Milner, Superintendent General of the Colonial Bank, by excluding certain classes reduces the number to 22,829, and expresses his opinion that this number need not excite our surprise considering the facility of living, and that though without "fixed employment;" they may have some occasional employment. To some extent this may be true, so as to diminish the feeling of surprise, but hardly to do away with it altogether and neutralise the effect of Lord Harris's remarks, especially as the sick and infirm—3,559,—are excluded, of whom 709 belong to the town. Reports before quoted, p. 74.

provincial towns in England, with this difference, that the number of occupations is much more limited, denoting fewer wants or a less advanced stage of society. The two following examples are given in illustration;—one from the census returns of Grenada, for 1851, shewing the distribution of the population in the town and country, including that of Cariacou, classed according to the occupations followed;—the other, shewing the same distribution of the whole population of Trinidad from the census of the same year. This selection is made, more attention having been paid to the subject of occupations in the drawing up of the returns for these islands than in any of the others.

I.

| OCCUPATION. | TOWN. | COUNTRY. |
| --- | --- | --- |
| Clergy.. .. .. .. .. .. .. | 4 .... | 10 |
| Law . .. .. .. .. .. .. | 1 .... | 1 |
| Physic .. .... .. .. .. .. | 5 .... | 8 |
| Government officers .. .. .. .. | 25 .... | 30 |
| Planters .. .. .. .. .. .. | 21 .... | 371 |
| Field labourers .. .. .. .. .. | 36 .... | 13,074 |
| Merchants and shop keepers .. .. | 72 .... | 80 |
| Hucksters and petty traders .. .. | 180 .... | 62 |
| School teachers .. .. .. .. .. | 23 .... | 26 |
| Writing clerks .. .. .. .. .. | 59 .... | 24 |
| Artificers .. .. .. .. .. .. | 433 .... | 1,421 |
| Mariners and fishermen .. .. .. | 130 .... | 337 |
| Sempstresses .. .. .. .. .. | 545 .... | 1,250 |
| House servants .. .. .. .. .. | 391 .... | 614 |
| Washerwomen and laundresses .. | 410 .... | 224 |
| Porters and jobbers .. .. .. .. | 170 .... | 19 |
| Others, variously employed .. .. | 396 .... | 1,186 |

| | | |
|---|---|---|
| Infants and others unemployed .. | 1,366 .... | 8,108 |
| Sick .. .. .. .. .. .. .. | 65 .... | 336 |
| Infirm .. .. .. .. .. .. | 107 .... | 798 |
| Soldiers (British) .. .. .. .. | 128 .... | 125 |
| | 4,567 | 28,104 |

## II.

| | |
|---|---|
| Public officers .. .. .. .. .. .. .. | 32 |
| Clergymen .. .. .. .. .. .. .. .. | 34 |
| Tradesmen .. .. .. .. .. .. .. .. | 2,191 |
| Servants .. .. .. .. .. .. .. .. | 1,403 |
| Sempstresses .. .. .. .. .. .. .. | 1,329 |
| Carters .. .. .. .. .. .. .. .. | 86 |
| Shopkeepers .. .. .. .. .. .. .. | 348 |
| Writing clerks .. .. .. .. .. .. .. | 204 |
| Police constables .. .. .. .. .. .. .. | 49 |
| Schoolmasters .. .. .. .. .. .. .. | 90 |
| Hucksters .. .. .. .. .. .. .. .. | 284 |
| Engineers .. .. .. .. .. .. .. .. | 22 |
| Solicitors at law .. .. .. .. .. .. .. | 9 |
| Barristers at law .. .. .. .. : .. .. | 13 |
| Planters and proprietors .. .. .. .. .. | 2,001 |
| Mariners .. .. .. .. .. .. .. .. | 72 |
| Medical men, (14 in the county of St. George, including Port of Spain; 2 in county Victoria; 1 in county St. Patrick).. .. .. .. .. .. | 17 |
| Merchants .. .. .. .. .. .. .. .. | 103 |
| Midwives .. .. .. .. .. .. .. .. | 14 |
| Druggists .. .. .. .. .. .. .. .. | 12 |
| Military men .. .. .. .. .. .. .. | 6 |
| Naturalists .. .. .. .. .. .. .. .. | 5 |
| Agents .. .. .. .. .. .. .. .. | 13 |
| Hotel-keepers .. .. .. .. .. .. .. | 2 |
| Butchers .. .. .. .. .. .. .. .. | 8 |
| Cigar makers .. .. .. .. .. .. .. | 16 |
| Fishermen .. .. .. .. .. .. .. .. | 62 |

| | | |
|---|---|---:|
| Auctioneers. | .. .. .. .. .. .. .. | 74 |
| Mahomedan Priests | .. .. .. .. .. .. | 2 |
| Watchmakers | .. .. .. .. .. .. .. | 2 |
| Music master | .. .. .. .. .. .. .. | 1 |
| Land surveyors .. | .. .. .. .. .. .. | 6 |
| Contractors.. .. | .. .. .. .. .. .. | 21 |
| Labourers .. .. | .. .. .. .. .. .. .. | 28,884 |
| No employment .. | .. .. .. .. .. .. .. | 26,989 |

The distribution of the inhabitants, distinguishing between town and country, according to trades and professions, we see by the above to be very irregular; and it is, as might be expected, most so in the colonies of greatest extent and most recently settled, such as Trinidad and British Guiana. In these, certain trades and professions are in a great measure limited to the towns, the condition of the country population, thin, scattered and poor, not affording encouragement to their introduction into the distant rural districts. Thus in the instance of medical men in Trinidad, of the total returned (17) no less than 13 are resident in Port of Spain, 4 only residing and practising in the country, and in consequence many of the counties are destitute entirely of medical aid : even in the town a large portion of the population is in the same predicament, accompanied by a great proportional mortality, especially in infancy, and in all ages when epidemics prevail.*  The effects of such a partial

* See the "Report on the measures necessary to be taken in the colony of Trinidad for the better preservation of the public health," &c.  The author, Dr. Gavin states, "that not more than eight persons who die (in the town) receive medical aid, and that the great mass of the poor in their illness never receive medical aid at all." p. 13.

distribution, it need hardly be remarked, concern the interests of society generally, and is one of the great drawbacks to settling in a new country, especially where slave-labour is excluded. In the chapter on Trinidad, some facts and circumstances are mentioned which may be referred to in illustration. In slave colonies, be it remembered,—and great is the advantage materially—with capital at command, an industrial community for the mere purpose of gain, may be formed, as it were, instanter ; skill of all kinds as well as labour may be purchased and located.

Of the public buildings belonging to towns in the West Indies, it may generally be said, that in accordance with the occupations of their inhabitants, they are all of that kind which the commonest wants of society require, such as places of worship, courts of law, prisons and hospitals ; and of these no more than is absolutely needed ; no other wants, such as concern either the health, amusement or instruction (excepting elementary) being cared for : there are no baths, no theatres, no museums, no public libraries,* and I may add, no grounds set apart for exercise, such as planted public walks, distinct from the high roads, or, with two or three exceptions, any open spaces where such

* The exception is that already noticed, p. 103, in Barbados. There are collections of books in several of the colonies, but they belong to societies and are supported by subscription by their members; two exist in Bridgetown, one in St. Kitt's,—the best, and I believe the oldest in the West Indies; one in Antigua and one in New Amsterdam. The works they contain are mostly of the class of belles-lettres, history and travels, and are very deficient in works of science, and the useful arts,—not excepting agriculture.

exercise can be taken; and the exceptions are spaces reserved, it may be inferred, rather as parade grounds or for military purposes, than with any regard to the health or pleasure of the inhabitants.

As to the style of the public buildings, the churches are almost the only ones that have any pretensions, I will not say to architectural beauty, but rather to architectural neatness*. The largest and most respectable, as regards construction, are the old cathedral church of St. Michael's, Barbados, the new cathedral in St. John's, Antigua, and the Protestant and Roman Catholic churches in Port of Spain, Trinidad.

Of their churchyards also, or burying grounds—for they are commonly identical, the few exceptions being those belonging to garrisons and exclusively for the military,† little can be said in commendation. None of them are neatly kept; too frequently they are overrun with shrubs and rank weeds; in their general aspect affording no tokens of that lasting family respect and affection for the departed, which is commonly witnessed in well regulated communities, and whenever witnessed, forcibly impresses the mind, that if there be such attention to the dead, such marks of humanity bestowed on them, order and propriety will not be wanting amongst the living;—and yet this neglect,—not to pass too hastily a censure, may in a great measure

---

* The public offices in Port of Spain, Trinidad, may be an exception; they promised to be so when I was last in that island, viz., in 1848 when they were only in progress.

† Before emancipation the slaves were not buried in consecrated ground: it is hardly known now where they were buried.

arise not from any want of proper feeling, but rather
from the circumstance of so many of the principal
proprietors being absentees, their families at least non-
resident and from the rapidity of vegetation, and the
difficulty of keeping it under in a tropical climate:
of this I well remember a forcible example when seeking the grave of a young officer, in the burying ground
attached to Fort Canje, Berbice.  Though assisted
by two men of the company of the Royal Artillery to which he belonged, who were present at
the interment and helped to convey the body to the
place, we had difficulty in discovering it, the ground
was so overgrown with rank vegetation, the growth of
a few months.  Such neglect, however, is to be regretted, and the more so considering, that were due
attention paid and taste exercised with such a choice of
shrubs and trees, especially of the palm-kind, how easy it
would be to shade and dress these grounds in a manner
fitting the feelings which ought to be associated with
them.

The prisons and hospitals, the buildings next in note
to the churches, vary much in the different colonies.
The former commonly are superior to the latter, excepting in Barbados, where the civil hospital, when I
was there, was admirably conducted, a model in most
respects of neatness, cleanliness and order.*  In another

* Yet the ground round this hospital was not drained; it was without
a sewer, and in consequence it was becoming saturated with what flowed
from the water-closets;—an occurrence this so common, that when I
pointed it out to the medical officers of the institution, it seemed to excite
little attention and no apprehension.  The same apathy, I may remark,
is too often witnessed at home.

island, St. Kitt's, the Civil Hospital was as much open to censure for opposite qualities, which, even now, I fear, are only partially corrected. When I visited it, it was in a very discreditable state; dirty and disorderly and ill provided, more likely to be productive of, than to promote the cure of disease, to increase than to alleviate suffering; and this owing, not so much to want of funds, as to neglect and bad management. Under any circumstances lamentable, in this instance it was the more so, as the building new and neat and not ill fitted for the purpose for which it had been constructed, was a boon to the island from a deceased Lieutenant Governor, the late Mr. Cunningham, who provided the funds expended in its erection. In Antigua, I witnessed a better example: in a building of a very humble and unexpensive kind, I had the pleasure to see the sick well attended and well cared for, under an official medical officer, and the superintendence of the benevolent clergyman who was most instrumental in establishing it; this was the infirmary at St. John's.* Of the

* The following is from my note book, written immediately after visiting the Hospital in St. John's, or the Infirmary as it is called, in June, 1848. Found the infirmary in good order in part, the best part of it, that in which sailors are placed; some of the small rooms were rather crowded and not so clean as they should be: they were occupied chiefly byPortuguese labourers with sore legs, and especially required cleanliness.

The infirmary consists of small rooms, very like the houses of the native labourers, under a common roof, each separated by a wooden partition, each capable of holding three beds; and of a larger building of two stories, partly of stone, partly of wood, in which there are two or three wards of a larger size in the upper story. The buildings, especially the first mentioned, are of a very economical kind, and well fitted for the purpose to which

other hospitals, and of the several prisons, and of the three lunatic asylums,—one in Antigua, another in Barbados, and a third in Demerara,—I shall not attempt any description, inasmuch as the information I should have to give respecting them, could be interesting to few if any readers, or instructive except perhaps in the notice—a thankless office—of glaring defects.*

applied. It was originally intended to have had the infirmary of stone altogether, and according to the ordinary expression ' a respectable building.' It was so begun ; but was thrown down by the last destructive earthquake, giving a lesson which has not been forgotten. I am not quite sure of the number of patients under treatment; I think they were thirty. The cost of the establishment I was informed by the Archdeacon, —who was so good as to come to me when he heard I was going round,— the last year was about £90 a month,—nearly the same as the expenses of the infirmary in St. Kitt's, which, I was told by Mr. D., were for the year £900 sterling,—a grant from the local government. The infirmary in St. John's is supported chiefly by subscription. It is well conducted, because well looked after, especially by the Archdeacon ; who, I believe, visits it almost daily, and probably often more than once a day. He is a spiritually minded man, but not careless of secular affairs. According to common report, the Archdeacon of St. Kitt's is devoted mainly to church matters, having a morning service daily, that is little attended ;—what a pity he cannot give his attention also to the latter : he is on the committee of management. The Antigua infirmary had a very humble beginning, having sprung from a soup-charity, in a year of drought and scarcity. Its founder, the Archdeacon pointed out with natural and manifest interest, the boiler,—an old ship-boiler, that was first used, and was still in use. The establishment, such as it is, has grown up with the wants of the people, and the expanding benevolence of its supporters. It was much needed after the introduction of immigrants.

    * The Civil Hospital, in George Town, Demerara, the largest in these colonies, liberally supported and solely by the local government, is, I believe, as a whole deserving of commendation : in no civil hospital either at home or abroad with which I am acquainted, are the medical records so carefully kept, as in this, under the superintendence of the Surgeon General to the colony, Dr. Blair.

Under the head of Barbados, mention is made of Codrington College; and its situation is briefly described. The building is worthy of the site; for architectural effect, plain as it is, it is the most pleasing in the colony. A good drawing of it is to be found in Sir Robert Schomburgk's *History of Barbados*. In the same work both the building and the collegiate establishment are fully described. Those who wish for information respecting it, I would refer to this source and to an account of it in a separate form, in minute detail drawn up by the Bishop of Barbados.* Originally intended by the philanthropical individual who so munificently endowed it, for the education of missionaries for the propagation of the gospel in foreign parts, (to the society bearing this name, the property was bequeathed in trust) it is to be hoped that the intentions of the founder, too long neglected, will at length be acted on, and that the college will really become what it was designed to be, a missionary one. Were men educated here expressly as missionaries and especially Africans, and instructed in medicine† as well as divinity, how efficient they might prove in promoting civilization in Africa and in introducing

---

* *Codrington College in the island of Barbados.* London, 1847. Printed for the Society for the propagation of the Gospel, and sold by Rivington.

† According to General Codrington's will the professors and scholars maintained in the college, should "be obliged to study and practise physick and chirurgery, as well as divinity, that by the apparent usefulness of the former to all men, they might both endear themselves to the people, and have the better opportunity of doing good to men's souls, whilst taking care of their bodies."

christianity there. Further, it may not be amiss to
remark, that this college from its salubrious situation
and the qualities of the mild tropical climate it enjoys,
is deserving the attention of students, at our univer-
sities, such as may be threatened with pulmonary
consumption, or are labouring under any chronic
ailment of the air passages, so common in the northern
regions. The voyage itself is likely to be beneficial,
especially in the cooler months,—our winter ones;
and the course of reading need not be interrupted;
in moderation it might be carried on in connexion
with due attention to health. The suggestion is
offered, I should add, ignorant entirely whether it
would accord with the arrangements of the institution
to receive students, under such circumstances of
health, and merely for a limited time. Medically
viewed, should it be practicable, I have no doubt of
its being beneficial.

Compared with towns in this country, there are some
peculiarities belonging to those of the West Indies,
which may be briefly mentioned;—I more particularly
allude to the habits of the people and the modes of
conducting business. Business in them, is not as with
us, protracted into night; but, as in the east, at least
in Turkey, is finished not later than, or before sunset.
Every shop is then closed; every workshop is deserted:
the merchant and the superior tradesman retires to his
country villa, the small trader and artificer betakes
himself to his family and evening meal, to his ease and
recreation. Another marked difference, is the absence

of the seductive gin-palaces and public-houses, not less tempting to excess and dissipation.  In these towns, it is true there are licensed dealers in spirits, and also in the villages, where a dram of rum may be had, but where there are no conveniences for convivial meetings and deep potations.  A third is, the absence of pawn-brokers' shops, affording facilities to the thoughtless, extravagant, and improvident, to exceed their means, for the sake of present gratification, gliding from diffi-culties into want and pauperism.  Further, the good system commonly prevails in trade, too much neglected at home, of ready-money dealings ; the labourer here is not supplied with what he wants on credit, laying him open to imposition and tempting him to exceed his means ; he must go to market or shop with money in his hand, and with it, knowing well its value, makes the best bargain he can, and on this 'vantage ground of " money in hand," he is not often imposed on.  Another peculiarity in accordance with the state of society and marking it, is that several trades and branches of busi-ness are often conducted by one person, or as one concern.  A watch-maker, may be a land-surveyor, a printer, engraver, a civil engineer,—in brief when a man of ingenuity, he may be a kind of factotum : and in the same shop or " store, " ( the name commonly applied to a shop when on a large scale,) almost every article is to be had that is wanted, whether it be for the family of the planter or labourer, whether for food or raiment ; articles almost invariably imported, and in the instance of made-up clothes, from their cheapness, interfering with native industry. As the towns are without corporations, so generally the trades exercised in them are without guilds.

Of the towns individually, whether improving as are some, chiefly George Town, Demerara, and Port of Spain, or decaying or stationary as are most of the others, I shall not task the patience of the reader by entering into any minute details: even the best of them are hardly of sufficient importance to need such description. Nor shall I attempt to give any account of the military posts or stations, most of which, like the towns they adjoin and for the protection of which they were designed, are in unhealthy situations, selected with little or no regard to sanitary requirements, and in consequence have from time to time been the scenes of destructive epidemics decimating and more than decimating the susceptible white troops by whom they have been occupied, making the Command a dread and opprobrium second only to the Western Coast of Africa, "the white man's grave."*

* It is interesting and important to find how, even on this coast, the health of Europeans may be preserved by proper precautions and by the preventive use of bark, or what is better, of the sulphate of quinine;—see a paper on the subject, especially deserving of the attention of merchants sending ships to that coast, by Dr. Bryson, published in *Medical Times and Gazette*, for January 7th, of the current year.

# CHAPTER XVIII.

## CONCLUDING.

In the introductory chapter, the rise and progress of these colonies have been briefly sketched; and, in the subsequent ones, their general condition has been more fully described. Of their future, hopeful views, I cannot but think, may be taken of improvement and advancement, without being too sanguine,—speculating on what is probable and practicable.

What are their advantages and disadvantages, and what are the measures likely to promote the one, and to remove or diminish the other? These are subjects well deserving of consideration and discussion; and if discussed temperately can hardly fail of being useful.

Sufficient, I apprehend, has already been stated, both generally and particularly of their advantages in relation to soil and climate, affording proof that they are second to none in their agricultural capabilities, not even to the most productive of the foreign colonies, which of late, with slave labour, have become their great rivals.

Their disadvantages or defects, those which are least doubtful have also been more or less adverted to : on account of their importance, even at the risk of some repetition, it seems desirable to recur to and consider them somewhat more in detail and in order.

They may be classed under a few heads :—

1. As to the great body of the people, the labourers ; their position, can hardly be maintained to be, such as it ought to be, either as regards their own interests or the interests of the landed proprietors. The evils of their present position, that in which they were immediately placed on emancipation, is strongly exemplified in the recent history of many of the colonies, and not only in the large ones, as British Guiana, Trinidad, and Jamaica, but also in some of the smaller ones, as Tobago and Montserrat,—in brief, wherever the circumstances have been similar.

So long as the labourers hold their cottages as tenants-at-will, liable to be expelled at a day's notice ; so long as the planters are insecure of their labour from day to day ; so long as land is apportioned to the labourer in lieu in part, or altogether of money wages, neither the planter nor labourer is likely to be contented, nor fair and honest labour attainable. Why should there not be ordinances or laws regulating these matters ? Lord Grey, in his recent work, *The Colonial Policy of Lord John Russell's Administration*, has alluded to the unguarded and precipitate manner in which emancipation was effected, and the evils resulting from the neglect of precautions to insure just dealings between the two classes.

2. As regards education, and instruction, it must, I fear, be admitted, that the measures taken have been far from adequate, and that the consequences have been since the time of slavery either a very slow improvement or none at all, and a great danger of relapse into barbarism as brutal as that which prevailed before emancipation, when the labourers were treated as brutes. This is a danger which is most of all to be apprehended in the larger colonies, such as British Guiana, Trinidad, and Jamaica. Is not this a subject also for legislation? Why should not a system of instruction be attempted similar to that which has been so useful in Ireland,—that which, "in secular matters was intended to be combined, in religious matters separate," to be paid for (either in part, that which relates to the foundation of the schools, or altogether) out of funds raised by a general rate? The source of dissension which has existed in Ireland, and has occasioned so much opposition there, notwithstanding that in the old way of educating the people by "Hedge schools," no attention whatever was paid to religious or moral training,* our colonies would in a

---

* See *Traits and Stories of the Irish Peasantry*; by William Carleton. Alluding to the moral and religious deficiencies of these schools, (vol. I. p. 312) he remarks;—"Now, when we consider the total absence of all moral and religious principles in these establishments, and the positive presence of all that was wicked, cruel, and immoral, need we be surprised that occa-

great measure be exempt from, though not altogether.
Of this unfortunately we have proof in what has recently occurred in British Guiana, where an improved
system of education proposed, applicable to the children
of all sects and of all creeds, similar to that most
approved and followed in the United States, has been
arrested owing to the opposition made to it by influential parties on the ground,—that which was intended to
neutralize opposition,—of the system being secular,
religious instruction being to be left to the ministers of
religion.*    Where, opposition of this kind is made
and has effect might not another plan be proposed,
not less national than the Irish and as suitable to a
large colony, viz. that instituted by the committee of
council of education and applicable to all denominations,

sional crimes of a dark and cruel character should be perpetrated? The truth
is, that it is difficult to determine, whether unlettered ignorance itself were
not preferable to the kind of education which the people then received."
    * The prosperity of the United States—well worthy the attention of such
objectors—is strongly illustrative of the influence of education in advancing a people, and that of a system, which may be called secular, but with
the object "not merely to teach the pupil to read, to learn the news of
the day, to write and cypher, to keep his accounts, but to receive that
thoroughly mental discipline, which may prepare him for any sphere in
which he may be called to move ; that development of the mind which will
elevate and ennoble his aspirations ; that cultivation of the faculties which
will awaken a quenchless thirst for knowledge ; that influence on the
mental powers which will incline them to the truth, as delicately as the
needle seeks the pole.    Its object is to make strong minds, courageous
hearts, prompt, active, energetic men."   " In relation to obedience, diligence, stillness, decorum, manliness of manners, respect to superiors, the
pupil should be disciplined most thoroughly."   From the *Report of the
Winchendown School Committee, Massachusetts.*  See *Edinburgh Review,*
July, 1853, article, *Popular Education in the United States,*—States, be it
remembered, which are taxed for education as much as most of our
colonies for all the purposes of local government.

at least, of christians? Reflecting on the alacrity to learn and the quickness of attainment displayed by colored children when under a good system of instruction, can it be doubted that the happiest results would be obtained both morally and industrially, had the rising generation the benefit of a training under such a system.

The deficiency of education, at present, is not limited to the labouring classes; it extends throughout society generally. One college only deserving the name exists in the West Indies, Codrington College, and that now restricted in its use, viz. to the preparing of young men for the church. How great might be the advantage of a college having schools connected with it, for instruction in the higher branches of knowledge, and in the sciences administering to the useful arts! Why might not law and medicine be taught as well as divinity? As regards the teaching of medicine, no serious obstacle could be apprehended in either of the large colonies, where there is a civil hospital affording the means for practical clinical instruction. Such a proposal to some may appear utopian and chimerical; but I cannot think it will be viewed in that manner by those who have witnessed, as I have, the result even of one trial, such as the teaching of the elements of chemistry to the boys—most of them of the colored class, in Harrison's school in Barbados, made when Sir Wm. Reid was governor, and I believe by his desire; when they were examined, and that was after a short course only, the intelligence shown by many of them, and their progress were remarkable. For a considerable time, amongst the most enlightened of the planters of Barbados, a want has been felt of a model-farm. Is not a college with which such a farm might be associated, a greater desideratum, in which not only the elements of the sciences connected with agriculture might be taught, but also those connected with the other useful arts, especially mechanics?* Examples and illustrations we are not wanting of an encouraging kind, shewing how in small communities, as in some of the second and third-rate German towns, such as Giessen, with a population of little more than 7,000 inhabitants, schools of science may be formed and flourish and become centres of attraction, when under the direction of competent teachers.

3. As to sanitary measures ; have not these hitherto been neglected even more than the educational? Have we not proof of this, in the condition of the towns, without adequate drainage, sewerage, or supply of wholesome water, and in the low increase of the population? Those who have any doubts on the subject will do well to refer to the reports of the Civil Inspectors sent to these colonies, on the occasion of the out-

* As already mentioned, the only periodical published in the West Indies, is the *Barbados Agricultural Reporter.* Of this monthly journal, the subscription to which is only six shillings per annum, (it had been as low as one dollar) the impression, I am informed, is only 300 copies,—a number exceeding the demand, no more than 235 being disposed of; viz., 183 in Barbados, 43 to subscribers in the other islands, and 9 to individuals in England, interested in West Indian agriculture. So small a demand for this the only periodical that there is, and it, on the staple produce and industry of these colonies, needs no comment to the reflecting mind.

break of cholera in Jamaica. During the time of slavery, every planter intent on his own interest engaged a medical practitioner to attend his labourers, and there was an infirmary on every estate. Since emancipation, no provision having been made by the government to afford medical aid, the labourers seem more and more to have become indifferent to it, and the consequence has been, especially in the instance of children, a large proportional mortality. Is not this another matter needing administration and legislative interference? Ought there not to be a competent medical officer appointed to each parish or district, and a dispensary established in it, at the public cost, defrayed by means of a parochial rate? Were there a medical school, or schools instituted, there would be no difficulty in affording the aid required. The study and practice of the profession could hardly fail of having an elevating effect on the minds of those thus engaged—the colored Creoles, those above the rank of labourers, who at present have no liberal profession open to them.

4. Is not reform needed in the manner in which the cultivation of the larger properties is so generally conducted, viz. by being entrusted to paid managers resident on the estates, under the direction of attornies living at a distance, and acting commonly for absentee proprietors? A spirited writer in a letter addressed to Governor Barkly, published in the British Guiana *Royal Gazette*, says, "the day which sees the disappearance of the last attorney and of the last manager in

Guiana will cast the first ray of returning prosperity upon her rich and fertile cane fields." The reflection is applicable to these colonies generally.* How great would be the advantage could these estates be rented, let for a term of years, after the manner followed in England, and more especially in Scotland; and as there corn-rents have been found useful, so in these colonies sugar-rents might be found to answer, i.e. a rate of rent determined by the average market price of the produce.†

* Surinam affords now a striking example of the great evils of the old attorney system of conducting the business of estates. There the attorneys are called Administrators, and such is their administration, that, notwithstanding a soil of great fertility, and most of the natural advantages of British Guiana, and some which British Guiana does not possess, the condition of that colony—still a slave-colony,—is at its lowest ebb, so much so, that property in estates has become in a manner valueless and unsaleable. These administrators acting for absentee proprietors in Holland, are allowed ten per cent. on the gross proceeds of the estates, and are said to increase it at least 30 per cent. in a variety of ways: few, if any, of the estates, it is stated, yield any income to the proprietors. I have before me the journal of a gentleman, a planter of Demerara, who visited Dutch Guiana in 1850, giving details of a very instructive kind, shewing the evils of a bad system of management, and of total failure of agricultural success even with slave labour;—illustrating well, what our colonies might have become had slavery been continued in them, and had matters been conducted without improvement after the manner in use in the olden time. He describes labour as scarce, and daily becoming more so, from a wasting slave population—the deaths exceeding the births,—and from an increasing free colored population residing chiefly in the towns, leading a life of idleness; labour, by them, being held in contempt: the properties are mostly mortgaged, and the administrators only, and at the cost of the proprietors, flourishing.

† One property in Barbados, the Codrington College Estate, which under the old system of management, yielded only about £ 800 a year, is now let on a long lease for £ 1,700 a year, and the lessee, I have it on good authority, is well content, and making a profit by its culture.

5. Is there not a neglect of interest also in the manner in which the agriculture of these colonies is now so restricted,—limited in many of them solely to cane-cultivation, and in very few of them comprising any other crop of importance, regardless, as it were, of the varieties of soil, of situation and of peculiarities of climate, and equally so of the great capacity of a tropical climate for varied culture? Were the natural advantages made available, how very many valuable productions might be grown, respecting which, trials of them having been already made, there could be no question as to success? It would be tedious to enumerate them, they are so many, whether as manufacturing materials, such as cotton, the cocoa-nut palm, the ground nut,* &c., or as articles of food, such as besides cocoa and coffee, rice,† and various kinds of starch

---

* Ground nut (*arachis hypogæa*); this curious plant, which flowers in the air and matures its fruit (an oily nut in a thin capsule) under ground, has hitherto been cultivated in the West Indies only for the table. On the west coast of Africa, where it is indigenous, it has become, on account of the oil it yields, an important article of export, and may prove the means of conducing to the civilization of that country. The Governor of Gambia, in a despatch of the 1st May, 1851, states, " The demand for ground nuts has led to the cultivation of large tracts of land ; and if the trade were to extend every ten years as rapidly as it has done since its commencement, viz., 43 tons exported in 1837 to 8,636 tons exported in 1847, whole tracts of country, at present covered with jungle or traversed by rude and savage tribes, would then be cultivated and reclaimed, whilst all the blessings which spring from agriculture and bringing with them peace and plenty, industry, civilization, and improvement, would necessarily follow." Reports P. P. for 1850, p. 201.

The cultivation of this plant requires more care than labour; the old, young, and infirm might be employed in weeding and taking care of it.

† That the colony is peculiarly fit for the cultivation of rice there can be no doubt, irrigation in all seasons being, to a great extent, practicable.

and meal, and many kinds of fruit, especially of the orange kind, fresh and preserved, &c. It is lamentable to reflect that even in these colonies themselves, with few exceptions, fruits which might be had in abundance, and of the best kinds, are scarce, and the best of them expensive; and even more lamentable, the fact, that the majority of them, not excluding Trinidad and British Guiana, are dependent, in a great measure, on foreign countries for the greater part of the necessaries of life,—for corn, for meat, for fish, salted,—to say nothing of the luxuries and comforts.

6. Is not the immigration system for procuring labour, open to serious objection? A system, so costly, so ephemeral, and so little connected with the true and best interests of society! Might not the loans advanced by the Imperial government to introduce this forced and temporary kind of labour more applicable to a tem-

In the catalogue of articles contributed to the Great Exhibiton in 1851, under the head of grain, it is stated that "three crops of rice can be obtained *annually* from *one sowing*, the new crop ratooning or springing from the old roots after each reaping;" and, in the introductory remarks to the catalogue, it is mentioned, that a gentleman who formerly turned his attention to this cultivation in Essequibo actually succeeded in rearing three crops in the year. It is with pleasure I add, that there is an attempt now in progress to introduce this crop as a staple. In a letter of the 17th of August, 1853, from my friend Dr. Blair, he states, " In the month of April last I joined a small company whose object is the establishment of rice as a staple of the colony;" adding, "The labour required for it is light, and many, we believe, who would be unable or unwilling to labour in the cane cultivation might be induced to industry in the new staple." He continues; "We have had time to plant 75 acres, only before the rainy season set in. When I left the colony (on the 1st of August) the rice was looking beautiful, and harvest was expected to commence in about a fortnight after,—four months after sowing.

porary work, such as a railway, than to continuous and regular cultivation, have been better applied to the improving and cheapening the means of cultivation, by the making of roads and bridges, the draining of lands, rendering them fit for implemental husbandry and other measures of the like kind ?*

7. The defence of these colonies till a late period, was mainly intrusted to the natives, in the manner mentioned in page 66, in the instance of Barbados. Since emancipation the militia force has not been called out, except on one or two occasions, in any of the islands. Its place has been supplied by white troops, regiments of the line, and a detachment of the royal artillery and by one or more West India regiments, composed of Africans, and recruited on the coast of Africa. West India service is held in abhorrence both by officers and men of the regular army, even now that the period of service is reduced to three years ; and not without reason, on account of the climate often so fatal to life,—the mortality amongst our white troops ranging on an average from 5 to 25

---

* For remarks confirmatory of the above and for details, I would refer those interested in the subject to two works, *Demerara after fifteen years of Freedom ;* by a Landowner; London, 1853, and *Eight years in British Guiana*, (a Journal); by Barton Premium, a Planter of the Province; London, 1850; and to the *Barbados Agricultural Reporter*, the number for June, 1853 : The author of *A Journal of a tour in Dutch Guiana*, an extract from which is published in that number, remarks, "I believe that £50,000 laid out in supplying steam engines of sufficient magnitude to drain any coast-district in Demerara would do more to ensure the success of the free system in Guiana, than millions of the public money expended in bringing immigrants." He assigns his reasons for this opinion and they are many and cogent.

per cent. per annum,—and on account of the circumstances in which they are commonly placed, so unfavorable to discipline and efficiency, without any compensating advantages, such as are enjoyed in the East Indies, or any opportunity to earn distinction, or even to be employed except in the merest routine of garrison duties.

Such being the risks run by these troops, it has often been suggested and urged that they should be withdrawn entirely, and that, as on our settlements on the Western Coast of Africa, Africans, by means of additional West India regiments, should supply their place, either alone, or in conjunction with a native militia, composed of men of color ;—both of them fitted for the climate, the mortality in the West India regiments not exceeding 2 per cent., being about the same as that of British troops on home-service, and both of them, judging from their conduct on many trying occasions, being trustworthy and efficient.

At present the men of these regiments are enlisted for a limited period, viz. ten years. At the expiration of this period, were a portion of land allotted to them, many might be induced to become settlers, and this in the colonies in which they would be most useful, as in Trinidad and British Guiana, where there is abundance of crown land that might be thus appropriated,* and where

* The experiment has been tried and has succeeded in India. " There cannot be (writes Sir John Malcolm) a more interesting spectacle than the great district stretching a hundred miles along the Ganges, which is inhabited by the discharged soldiers (to whom small portions of land have been granted) of the company's army. This district, a few years ago, had

military service might still be required of them, similar to that of our home pensioner-corps. To render them as useful as possible, both as soldiers and afterwards as settlers, more attention than is at present given, should be paid to their education. It would be well were they taught besides the use of arms, the English language and the use of implements of husbandry, not neglecting moral and religious instruction.* Further, might not these regiments be organized after the manner of the native regiments in India, and have a class of officers of their own color? Were this boon granted and another,—the granting of commissions to serve in these regiments to the sons of planters, their fidelity might still more be secured. Justice and confidence seldom fail to beget fidelity and gratitude : "What was the cause (remarks Tacitus) of the fall of the Lacedæmonians and Athenians, but powerful as they were in arms they spurned their subjects as aliens."

8.—These colonies are remarkable for the little connexion existing between them; civilly, as well as

---

been a mere jungle, abandoned for ages to tigers and robbers : it is now covered with cultivated fields and villages ; the latter of which are filled with old soldiers and their families, in a manner which shewed their deep gratitude and attachment for the comfort and happiness they enjoyed." *Malcolm's India*, p. 526.

* To be made even efficient as soldiers, the black troops need instruction. The majority of the Africans composing the West India regiments are ignorant of the English language ; or, at most are acquainted with it only in the most imperfect manner. There is a school at the head quarters of each regiment, or a school room, but not always provided with a teacher : I can speak from my own knowledge of negligence of this kind, which is likely to be concealed from the higher authorities.

territorially, they are for the most part insulated. Even the islands constituting the several governments are in a great measure independent of each other and have but little intercourse,—for instance Barbados, with St. Vincent, or St. Vincent, with St. Lucia, or the last mentioned, with either Tobago or Grenada,—and so of the rest, to the disadvantage, it is believed, of all, and that in very many ways. Even the Governors General appear to interfere but little with the proceedings of the Lieut. Governors, and when, as has occasionally been the case, they have interfered, they have not always had the support of the colonial office.

Situated as these colonies are with interests common to all of them and with some peculiarities belonging to each, it can hardly be doubted that a closer union would prove advantageous to all. One general officer suffices for the whole military command ; might not one governor general also suffice for the civil command ? *

There was a time, when general assemblies were held in the leeward islands, formed of delegates from the several islands, which met periodically with the intent of considering and discussing their common interests. Might not such assemblies be again constituted with marked benefit,—assemblies composed of representatives from all the colonies ?

* Of the evils of want of union, examples might easily be given, especially in civil affairs : I shall mention only one, and that the manner in which the last census was made, (when made, for in some instances it was neglected) as if to prevent comparison, the form of return for each colony being different : other instances will readily occur to those acquainted with the laws and statistics of these islands.

Steam navigation, so admirably adapted for these seas, by means of which, communication even against the trade winds is rendered easy, has hitherto been little used. The only steam vessels hitherto available, have been the packets of the West India Company, engaged for a special service, the conveyance of mails. Is it right that the government should be without a steamer? With such an aid, how greatly would union be promoted, and how much more easily and efficiently might duties be performed! It is only those who have experienced the delays and difficulties belonging to the present mode of conveyance, who can have any tolerably accurate idea of the good that would result from the change. How greatly would the transport of troops be expedited! What facility would be afforded to circuit courts of justice!* How greatly diminished would be the difficulties of establishing a general penal settlement, or a union lunatic asylum, both of them very much needed! † At present the

---

* Till very lately there was a Chief Justice in St. Lucia, Nevis, and Montserrat: now the duties of this officer in the first mentioned island are performed by the Chief Justice of Barbados, and of the second by that of St. Kitt's, and of the third by the Chief Justice of Antigua; and it is understood, in a manner giving great satisfaction. In most of the islands the legal duties can occupy but a small portion of time, so that an exclusive appointment and confinement to the colony must be irksome to a zealous man. I know one or two so situated, who, had they more ample occupation as Circuit Chief Justices would be better satisfied, and probably even without any augmentation of salary: I do not allude to the late and last Chief Justice of St. Lucia; though, even he, had he had duties to perform that would have sufficiently occupied his time, and engaged his restless energies, would, perhaps, have led a more innocent life and have been the cause of less mischief.

† At present there is only one Penal Settlement, and that for British

fares for intercolonial passages by the company's steamers are so high, as to the majority, to amount almost to a prohibition; for instance, twenty dollars, rather more than four pounds, from Barbados to St. Vincent, the nearest island, and forty-five dollars, more than nine pounds to St. Thomas.*

Lastly, is the connexion with the mother country on a right and safe footing? Have the interests of these colonies been duly attended to by the imperial government and not unduly interfered with? The colonists are of opinion that their true interests have not been duly attended to, and that they have been unduly interfered with; and have become, it is believed, generally discontented in consequence. They have, I apprehend, truth on their side. Not to go back to a period anterior to emancipation,—confining attention to the present times only and to the latest measures affecting the interests of these colonies, can it fairly be admitted that the great change in the sugar duties (their equalization to

Guiana alone. It has been proposed to form another for the other colonies on one of the Grenadines: but whether approved or not, I have not learnt. All the smaller islands, and more than one of the more populous, are without Lunatic Asylums. Fortunately, insanity is comparatively rare in these colonies; unquestionably much more so, than in the United Kingdom, at least in the working classes, especially in our large towns, where, in consequence of various causes co-operating, it appears to be alarmingly increasing.

* Many other advantages might result;—to specify only one, it would be no longer necessary to have a large supply of stores at the several stations for the public service, more than are immediately required. In the instance of medical stores and of ammunition, great is the waste and loss from the effects of climate, and want of care, where more than ordinary care and precautions are required to counteract the deteriorating influence of moisture and high temperature.

take place next year, 1854,—according to act of parliament,) without distinction as to the kind of labour by which the staple produce is raised, whether slave or free —is conducive to their welfare? Can it be fairly admitted that the taxing of skill in our colonies, is for their benefit, as in the instance of the *ad valorem* duties on sugars, according to which, the better the quality of the article, the more pains, the more science, the more skill employed in its manufacture, the higher the duty imposed on it, when brought to this country? That the equalization of the duties has checked cane cultivation in our colonies,—Barbados I believe being the only exception,—and encouraged it in the slave colonies can hardly be doubted,* no more than that the

* Ought not the history of cotton cultivation to be a warning to our statesmen, as to the risks incurred by placing slave-grown and free-grown produce on an equality? In 1803, when the greater portion of the cotton used in our manufactories was supplied from the West Indies, these colonies, including British Guiana, imported into Great Britain, taking the average of three years, 10,224,085 lbs; whilst now, that the consumption of the material has so amazingly increased, that it weekly exceeds this quantity,* the yearly supply from those colonies had dwindled to below a million of pounds—in 1850, it was 228,913 lbs, the greater proportion,—the very many millions consumed being furnished by the United States, raised by slave-labour. Now, as cotton-cultivation has supported slavery in the southern states of the union, encouraged as that cultivation has been by our government, is it not likely that sugar cultivation similarly encouraged, will have a like effect in preventing emancipation in the Spanish colonies and in the Brazils? And further, is there not a danger of the extension of cane cultivation in these colonies even to its extinction in some of our colonies?

As the cultivation of cotton for the English market and the markets of Europe has been the main support of slavery in the United States, the disuse of manufactured cottons has been proposed with a view to the

* In 1851 the consumption of cotton wool in the United Kingdom was upwards of 760 millions of pounds. See a lecture upon cotton, &c.,—by T. Bazley, Esq.

*ad valorem* duties, however much they may have encouraged skill and the application of science at home to the refining processes, have had the contrary effect in the boiling houses of the planters.*

abolition of slavery;—an impracticable measure;—one, perhaps less so would be the imposing of a duty on slave-grown cotton and on that alone, —a measure that would at the same time promote the revival of its culture in our colonies and its extension in India and in foreign colonies with free labour; and were France, Germany, and Switzerland to act on the same principle, the greater, of course, and the more certain would be the effect.

The effect of the abolition of the differential sugar duties in promoting the slave trade is strikingly exemplified by the facts adduced by Lord Clarendon and by Lord Carlisle in the House of Lords, on the occasion of petitions being presented to put a stop to that traffic. The former on the 24th of May, (1853) stated, "that within the present year, no less than six slavers returning to Cuba with their cargoes, had been captured by the naval force at the Havannah; and that in the case mentioned by Lord Brougham, who had presented one of the petitions, the slave ship started with 1,300 slaves on board, of whom 200 had been killed in a revolt on the passage, and 300, and only this number, had been rescued by the British Consul out of the 1,100 landed! Lord Carslile stated (May 30th) that between November and February last, 5,000 slaves were known to have been landed in Cuba; equivalent, it may be remarked, in money worth to about half a million sterling,—and he expressed his repentance for his complicity in that political movement—the abolition of the differential duties as remarked on by the Bishop of Oxford. Lord Carlisle's words, as given in the *Evening Mail,* are well worthy of being quoted. " I, my lords, of course, cannot deny my complicity in the adoption of that policy ; and at this time of day, I still feel great doubt whether in so doing I was right or wrong. This I know, that in the whole course of my parliamentary experience I never acquiesced in any measure with so much hesitation, and there is none with regard to which at this day, I should be so glad as to be quit of all responsibility."

* The evils of the *ad valorem* duties have long been a subject of complaint and remonstrance, as tending to diminish the profits of the planters, and to increase the cost of sugar to the consumers. The matter has been well considered by Bryan Edwards in his *History of the British West Indies.* According to an estimate carefully made by him, it would appear that the average loss " arising chiefly from unavoidable waste at sea by drainage," is in the instance of raw or common Muscovado sugar one eighth part : " in

Lord Grey, in his concluding observations " on the Colonial Policy of Lord John Russell's Administration,"

other words, a hogshead of sugar weighing nett 16 cwt. when shipped in Jamaica, shall, when sold in London, be found to weigh 14 cwt. only. The difference therefore is a dead loss both to the public and the planter. The former lose the use of two hundred pounds weight, and the latter is deprived of its value, which at 40s. per cwt., may be stated at £3 6s. per hogshead, the merchants charges deducted." He points out other sources of loss connected with this class measure—(vol. ii. p. 462, and seq.) When he wrote, the difference of duties of the several kinds of sugars was greater than at present; but the reasoning and weight of argument are much the same.

Very recently the subject has again been brought forward, and is still under agitation, and it is to be hoped the consideration of Government, chiefly owing to the strong representations made, shewing the impolicy of the measure and the evils connected with it, by two of the most experienced of the West India governors, Sir Charles Grey and Sir Henry Barkly. " The bad effect of the high duty on refined sugars (says the *Spectator* of the 1st of October, 1853, from which I quote) is forcibly pointed out by the late governor of Demerara." "The discouragement which the existing arrangement of duties offers to our improved system of manufacture will (he states) be best conceived from the following facts : first, the process of spoiling sugar,—when it seems better than would be likely to pass the lowest standard—is not of unfrequent occurrence on estates where the vacuum pan is used ; second, that a gentleman in charge of an estate, on which vast expense had been incurred from steam clarifiers, bag and charcoal filters, vacuum pans and pneumatic pumps, assured me, "that for a further trifling outlay of about £100 he could, were it not for the quasi prohibition duty, ship the whole of his crop (1,000 tons) of a quality equal to refined sugar, though made, *bonâ fide* by a single process from the raw material." The reply to Sir Henry Barkly's despatch to the Colonial Minister, the Duke of Newcastle, has been considered as favorable, inasmuch as the admission is made that the effect of the existing duties is to discourage improvement. The words are (authorized by Lords of the Treasury, and involving a principle) "To impose a discriminating duty upon distinct kinds of a given produce, such as the produce of vineyards varying in richness, different qualities of tea or tobacco, would appear to be a legitimate application of *ad valorem* duties; but to strike with a superior duty one pound of sugar which by a better mode of manufacture contains more saccharine matter than another pound obtained from the same raw material, is to inflict direct discouragement upon improvement." Unfortunately hitherto,

4 A

alluding to the period and events, remarks, "a commercial revolution deeply affecting the interests of many of our most important colonies has been safely passed through, not it is true without much distress and loss to individuals which I deeply lament, but with great advantage to the permanent welfare of these colonies and of the mother country."*

This is a conclusion, which he evidently applies to the sugar producing colonies, and he appears intent on supporting it by shewing that the production of sugar in our colonies, has, with the exception of Jamaica, a tendency to increase rather than to diminish; and this tendency, he infers, denotes that it cannot be without profit; an inference also made by the late Chancellor of the Exchequer on the memorable occasion of his announcing himself a renegade in withdrawing his support as a protectionist from the West India interests. Lord Grey compares the amount of produce exported from the several colonies in 1850 and 1851, shewing a great increase. The results are given in the subjoined note.*

all that is thus censured, has been practised, the duties on wines, teas and tobacco, chiefly the produce of foreign countries, having been levied irrespective of quality,—encouraging consequently amongst foreigners science and skill;—but on sugars, chiefly the produce of our own colonies, the opposite, with the effect insisted on that of checking improvement. Surely it is not the intention of government in calling attention to the evil to persist in enforcing it!

* Op. cit. ii. p. 303.

| † | 1850. | 1851. |
|---|---|---|
| Antigua ... ... ... ... ... | 123,485 ... ... ... | 200,235 cwts. |
| Barbados ... ... ... ... | 524,651 ... ... ... | 583,840 „ |
| Dominica ... ... ... ... | 51,816 ... ... ... | 60,239 „ |

This demonstrated increase, I believe, has been
owing to three causes; mainly, to a more favorable
season in one year than in the other; to greater and
increasing exertion and economy practised by the
planters in consequence of their necessities; and to
increase of available labour from increase of population
from immigration. The inference that there has been
some profit to the planters may be true or not: we know,
and know too well, that the extension of a concern,—
take the instances of railways and manufactories,—
does not necessarily imply profit,— both, in too many
instances having been conducted with diminishing
profits, and in some, with actual loss. But, granted
that there has been some profit, it does not follow that
it has been an adequate one, sufficient with a favorable season to compensate for losses resulting from
a bad season and various drawbacks and contingencies.
Take the most flourishing example, that of Barbados,
where the increase of late years has been so great, and

| | 1850. | | | | 1851. | |
|---|---|---|---|---|---|---|
| Grenada ... | ... | ... | ... | ... | 92,803 | ... ... ... | 121,381 | cwts. |
| Montserrat | ... | ... | ... | ... | 1,607 | ... ... ... | 7,675 | ,, |
| Nevis | ... | ... | ... | ... | 15,508 | ... ... ... | 33,309 | ,, |
| St. Christopher's | ... | ... | ... | 70,717 | .. ... ... | 122,029 | ,, |
| St. Lucia | ... | ... | ... | ... | 53,903 | ... ... ... | 69,930 | ,, |
| St. Vincent | ... | ... | ... | ... | 139,567 | ... ... ... | 163,409 | ,, |
| Tobago | ... | ... | ... | ... | 44,297 | ... ... ... | 45,130 | ,, |
| Trinidad ... | ... | ... | ... | ... | 366,214 | ... ... ... | 441,772 | ,, |
| British Guiana | ... | ... | ... | 525,297 | ... ... ... | 595,200 | ,, |
| Tortola | ... | ... | ... | ... | 1,406 | ... ... ... | 3,070 | ,, |
| Mauritius | ... | ... | ... | ... | 1,003,296 | ... ... ... | 1,000,269 | ,, |
| British India ... | ... | ... | ... | 1,359,690 | ... ... ... | 1,574,173 | ,, |
| Jamaica ... | ... | ... | ... | ... | 574,796 | ... ... ... | 627,823 | ,, |
| | | | | | 4,949,053 | | 5,649,784 | |

where, if anywhere, the planters are competing with a chance of success against foreign planters, with slave labour; even there, there are doubts regarding the results,—doubts whether the present revival of some prosperity will continue and not be cut short by drought or the more ruinous hurricane. In one of the last letters received from that island, my friend, a very intelligent planter, and not despondingly disposed, alluding to their state and treatment, finishes with the words, " *Perfer et obdura* must be our motto," without on his part, the completion of the sentence, which, no doubt, Lord Grey would make, regarding the circumstances, the hardships and grievances that gave rise to the use of the words,—*dolor hic tibi proderit olim.*

Such doubts, it is believed, would be entirely removed, were the total abolition of the differential duties arrested, as by reducing in August, 1854, the duties on our colonial sugars to 7*s* or 8*s* the cwt., without any reduction of duties on the slave-grown sugars, and the abolition on the former altogether, and on them only of the *ad valorem* duties; thus, whilst encouraging and aiding our planters and insuring to the home population cheap sugar, affording somewhat less encouragement, or, it may be, a check to the extension of cane cultivation by slave-labour.*

---

* For facts in confirmation of the accuracy of the above conclusion see an interesting and valuable *Narrative of a recent visit to Brazil ;* by John Candler and Wilson Burgess, to present an address on the slave-trade and slavery, issued by the Religious Society of Friends, London, 1853. Friends' Book and Tract Depository 84 Houndsditch.

By means of slave-labour, they ascertained that sugar can be pro-

That the connexion between the mother country
and the colonies should not be broken, seems hardly
to admit of question, provided,—and how necessary
is the proviso, that the two are beneficial to each
other; that the one is not enacting the part of a harsh
step-mother, but of a kind and considerate parent,—nor
the other the part of burdensome spendthrifts, but of
industrious, dutiful and loving children; and it may
be added, out of leading strings, self-controlling, and
self-supporting, and not in perpetual tutelage. That
the mother country and these colonies should be in
this relation, also can hardly be questioned, considering how well the resources of the one are adapted
to supply the wants of the other; and how even with
the abrogation of the old tyrannical monopoly system
under which our colonies were too long enchained,
they, the two, are by nature fitted to be aids to each

duced in the Brazils at a cost of 10*s.* per cwt. and under with profit,
even in some situations and circumstances at so low a rate as 7*s.* 6*d.*

The admission of slave-grown sugar into our markets, they show, greatly
encouraged the slave-trade : before that (I quote from their pamphlet)
"the number of Africans landed in Brazil, so far as can be ascertained,
amounted in the four years from 1842 to 1845, both inclusive, to 78,830;"
since, "in the four years 1846 to 1849, it rose to the awful number of
220,496."

Though from their statements there appears to be a sincere intention on
the part of the Brazilian Government to abolish the foreign trade in slaves,
there is less inclination, if any, to abolish slavery,—on account of the interfering interests of the body of planters and of the inhabitants generally,
with some worthy exceptions,—interests which, if the duties on sugars be
equalized, as proposed, next year (1854) will be strengthened, in fact, a
motive will be afforded by our government of the strongest kind to perpe-
tuate slavery, and *vice versa*, were the duties on sugars the growth of free
labour to be reduced, those on slave-grown sugars remaining the same,
with a continuance, in the instance of the latter, of the *ad valorem* duties.

other, and in the free commerce of the world to be each others best customers.

That the attachment of these colonies, especially of those primarily settled, is strong, as little admits of doubt, or that it will be durable—perdurable under treatment such as they have a right to expect, such as an enlightened and just government must feel it right to bestow,—inasmuch as it is founded on interest,—the advantages of imperial protection; and not only on common interests, but also on feeling, the pride and pleasure of being British subjects, the love of country and home; the old country being so considered and so named even by those who have never left their own particular island and who never expect to set foot on British ground. *Civis Romanus sum*, was never a prouder boast, and probably rarely connected with the same affectionate and home feeling,—approaching in its quality to the religious.

It is sad to think that circumstances should ever have occurred warranting even a suspicion that there exists a party at home, liable to the accusation of having at heart the destruction of these colonies and of our colonies generally.* May the future prove

* The party above alluded to is that which has been called " the Agricultural Confederacy." The author of *Eight years in British Guiana,* —a ruined man, and not from his own imprudence, thus speaks, morbidly excited by his distresses, of the intention of this party : " In short every thing shows that the destruction of our colonies is the object at which they aim, because this being effected, they imagine they will have unrestricted intercourse with foreign countries, and an immensely increasing demand for manufactured goods. Policy short-sighted and wicked. It is grasping at the shadow and dropping the substance."

that such a suspicion is without foundation, or at least that such a design is without the approval or support of the British public, or of the government, by whatsoever party in the state administered.

PRINTED BY T. KNAPP, FARINGDON.